D1000139

Prentice Hall Advanced Reference Series

Physical and Life Sciences

PRENTICE HALL
Polymer Science and Engineering Series
James E. Mark, Series Editor

MARK, ALLCOCK, AND WEST *Inorganic Polymers*
MARK AND ERMAN, EDS. *Elastomeric Polymer Networks*
RIANDE AND SAIZ *Dipole Moments and Birefringence of Polymers*
ROE, ED. *Computer Simulation of Polymers*
VERGNAUD *Liquid Transport Processes in Polymeric Materials: Modeling and Industrial Applications*

FORTHCOMING BOOKS IN THIS SERIES (tentative titles)

CLARSON AND SEMLYEN, EDS. *Siloxane Polymers*
FRIED *Polymer Science and Technology*
GALIATSATOS, BOUE, AND SIESLER *Molecular Characterization of Networks*
VILGIS AND BOUE *Random Fractals*

INORGANIC POLYMERS

James E. Mark
University of Cincinnati

Harry R. Allcock
The Pennsylvania State University

Robert West
University of Wisconsin

Prentice Hall
Englewood Cliffs, New Jersey 07632

Library of Congress Cataloging-in-Publication Data

MARK, JAMES E. (date)
 Inorganic polymers / James E. Mark, Harry R. Allcock, Robert West.
  p. cm.—(Prentice Hall advanced reference series. Physical and life sciences) (Polymer science and engineering series)
  Includes bibliographical references and index.
  ISBN 0-13-465881-7 :
  1. Inorganic polymers. I. Allcock, H. R. II. West, Robert (date). III. Title. IV. Series. V. Series: Prentice Hall polymer science and engineering.
 QD196.M37 1992
 546—dc20
              91-9936
              CIP

Editorial/production supervision
and interior design: Kathryn Gollin Marshak
Cover design: Lundgren Graphics
Prepress buyers: Kelly Behr and Mary Elizabeth McCartney
Manufacturing buyer: Dave Dickey
Acquisitions editor: Michael Hays

© 1992 by Prentice-Hall, Inc.
A Division of Simon & Schuster
Englewood Cliffs, New Jersey 07632

The publisher offers discounts on this book when
ordered in bulk quantities. For more information, write:
  Special Sales/Professional Marketing
  Prentice-Hall, Inc.
  Professional and Technical Reference Division
  Englewood Cliffs, New Jersey 07632

All rights reserved. No part of this book may be
reproduced, in any form or by any means,
without permission in writing from the publisher.

Printed in the United States of America
10 9 8 7 6 5 4 3 2

ISBN 0-13-465881-7

Prentice-Hall International (UK) Limited, *London*
Prentice-Hall of Australia Pty. Limited, *Sydney*
Prentice-Hall Canada Inc., *Toronto*
Prentice-Hall Hispanoamericana, S.A., *Mexico*
Prentice-Hall of India Private Limited, *New Delhi*
Prentice-Hall of Japan, Inc., *Tokyo*
Simon & Schuster Asia Pte. Ltd., *Singapore*
Editora Prentice-Hall do Brasil, Ltda., *Rio de Janeiro*

Contents

Foreword

POLYMER SCIENCE AND ENGINEERING SERIES

One of the most exciting areas in chemistry, chemical engineering, and materials science is the preparation, characterization, and utilization of polymers. The growing importance of polymers has been truly astounding, to the point that it is difficult to imagine our lives without them. They are under development in virtually every industrialized country in the world with activities accelerating rather than abating.

Not surprisingly, the amount of information relevant to polymer science and engineering is increasing correspondingly, making it more and more difficult to enter new fields in this area or even to remain abreast of developments in one's current field. There is thus a real need for authoritatively written, easily accessible books on polymer science and engineering, both for the relatively uninitiated and for the better-informed professional.

The present series of books was inaugurated to help meet this need. It will cover the organic chemistry of polymers, the relevant physical chemistry and chemical physics, polymer processing and other engineering aspects, and the applications of polymers as materials. The level will range from highly introductory treatments that are tutorial and therefore particularly useful for self-study, to rather advanced treatments of more specialized subjects. Many of these books will be exceedingly useful as textbooks in formal courses at colleges and universities.

Considerable attention will be paid to polymers as "high-tech" materials. This is in response to the fact that the most exciting applications of polymers no longer involve huge tonnage amounts of materials. Rather, they involve situations in which polymers generally are not present in large amounts, but are absolutely critical for the functioning of the system. Examples are polymer matrices and encapsulants for the controlled release of drugs and agricultural chemicals, biopolymers and synthetic polymers for biomedical applications, conducting polymers for batteries and other electrical devices, polymers having non-linear optical properties for optoelectronic applications, high-temperature polymers for use in outer space and in other hostile environments, ultra-oriented polymers for high-strength materials, new types of polymer-based composites, photosensitive polymers for microlithography, and inorganic and organometallic polymers for use as ceramic precursors. These are all rapidly developing fields, and there is a particularly great need for authoritative, comprehensive treatments of these subject areas.

It is hoped that this series of volumes will meet these needs, and be of lasting value to the polymer community.

J. E. Mark
University of Cincinnati
Series Editor

Preface

This book has been written to provide an introduction to the field of inorganic polymers. It is not a comprehensive survey; instead it aims to present an overview and perspective of selected areas to develop general principles. For a long time, there has been a need for an introductory book on this subject to provide a starting point for those who wish to understand the underlying principles of inorganic polymer chemistry before studying one of the more specialized volumes. A knowledge of the basic concepts presented in a typical undergraduate course in chemistry will be sufficient to understand the material in this book. Some familiarity with the fundamentals of polymer science would be helpful but not essential, since many of these fundamentals are covered in the first two chapters.

The organization of the book has been designed so that it may be used either as a textbook for a one-term or one-semester course, or for an individual wishing to study the subject on his or her own. For this reason, a very brief introductory chapter is followed immediately by a section on the characteristics and characterization of polymers, with many examples taken from the field of inorganic polymer chemistry. The remaining four chapters focus on the synthesis, reaction chemistry, molecular structure, and uses of three well-developed areas of inorganic polymers. Chapter 3 deals with phosphazene polymers which, from a synthetic and structural viewpoint, constitute the broadest and most versatile of the known inorganic macromolecular systems. Chapter 4 covers the chemistry and technology of organosiloxane (silicone) polymers. These compounds have been studied since the mid-1940s and, in spite of their somewhat restricted molecular diversity, they have a rich chemistry and technology associated with them. For this reason, Chapter 4 includes more technological material than will be found in the other chapters. Chapter 5 is devoted to a survey of polysilanes. This is one of the newer areas of inorganic polymer research, and is a field that has developed rapidly during recent years, partly because of its relationship to ceramic science and other areas of solid-state science. The final chapter (Chapter 6) deals with a variety of other inorganic polymer systems, particularly species that contain the elements boron, sulfur, selenium, tin, aluminum, and transition metals. These polymers are, for the most part, less well developed than their counterparts that contain phosphorus or silicon, but they are the starting points for the long-range development of this field.

We recommend that the book be read in the order of the chapters. Individuals who already have a detailed knowledge of macromolecular characterization may wish to scan the first two chapters briefly and then read Chapters 3–6 in detail. In any case, we hope that the reader will enjoy delving into this fascinating and important field and that the material presented will serve as a stimulus for the generation of many new ideas.

J. E. Mark
H. R. Allcock
R. West

INORGANIC POLYMERS

1

Introduction

1.1 WHAT IS A POLYMER?

A polymer is a very-long-chain macromolecule in which hundreds or thousands of atoms are linked together to form a one-dimensional array. The skeletal atoms usually bear *side groups*, often two in number, which can be as small as chlorine or fluorine atoms or as large as aryl or long-chain alkyl units. Polymers are different from other molecules because the long-chain character allows the chains to become entangled in solution or in the solid state or, for specific macromolecular structures, to become lined up in regular arrays in the solid state. These molecular characteristics give rise to solid-state materials properties, such as strength, elasticity, fiber-forming qualities, or film-forming properties, that are not found for small molecule systems. The molecular weights of polymers are normally so high that, for all practical purposes, they are nonvolatile. These characteristics underlie the widespread use of polymers in all aspects of modern technology. Attempts to understand the relationship between the macromolecular structure and the unusual properties characterize much of the fundamental science in this field.

1.2 HOW POLYMERS ARE DEPICTED

Polymers are among the most complicated molecules known. They may contain thousands of atoms in the main chain, plus complex clusters of atoms that form the side groups attached to the skeletal units. How, then, can we depict such molecules in a manner that is easy to comprehend?

First, an enormous simplification can be achieved if we remember that most synthetic polymers contain a fairly simple structure that repeats over and over down the chain. This simplest repetitive structure is known as the *repeating unit*, and it provides the basis for an uncomplicated representation of the structure of the whole polymer.

For example, suppose that a polymer consists of a long chain of atoms of type A, to which are attached side groups, R. The polymer chain can be represented by the formula shown in Structure 1.1. The two horizontal lines represent the bonds of the main chain. The brackets (or parentheses) indicate that the structure repeats many times. The actual number of repeating units present is normally not specified, but is represented by the subscript, n. If only a few repeating units (for example, 5–20) are present, n is usually replaced by x. Note that this formula says nothing about end groups that may be present. If the polymer chain is very long, the end groups represent only a small component of the molecule, and are ignored in the formula. The formula shown in Structure 1.1 can also represent a cyclic (or macrocyclic) structure in which, of course, no end groups are present.

When the repeating unit contains two or more different skeletal elements, the formula can be expanded as shown in Structure 1.2. If different repeating units bear different side groups (R and X), a formula such as Structure 1.3 may be used. However, beyond a certain point, the complexity of the molecule cannot be represented by a simple formula. For example, Structure 1.4 tells us nothing about whether the R groups on adjacent repeating units are *cis-* or *trans-* to each other. Such information is usually best described by supporting information in the text rather than by adding to the complexity of the formula.

$$\left[\begin{matrix} R \\ | \\ -A- \\ | \\ R \end{matrix}\right]_n \qquad \left[\begin{matrix} R \\ | \\ -A-B- \\ | \\ R \end{matrix}\right]_n$$

1.1   1.2

$$\left[\left(\begin{matrix} R \\ | \\ A \\ | \\ R \end{matrix}\right)_x \left(\begin{matrix} X \\ | \\ A \\ | \\ X \end{matrix}\right)_y\right]_n \qquad \left[\begin{matrix} R \\ | \\ -A-B- \\ | \\ X \end{matrix}\right]_n$$

1.3   1.4

$$\left[\begin{array}{c} Cl \\ | \\ +P=N+ \\ | \\ Cl \end{array}\right]_n \qquad \left[\begin{array}{c} CH_3 \\ | \\ +Si-O+ \\ | \\ CH_3 \end{array}\right]_n \qquad \left[\begin{array}{c} CH_3 \\ | \\ +Si- \\ | \\ \bigcirc \end{array}\right]_n$$

1.5   1.6   1.7

$$\left[S \right]_n$$

1.8

The *naming* of polymers in this book follows an accepted practice used by the vast majority of polymer chemists (though not by specialists in nomenclature). In the system used here, the name of the repeating unit is preceded by the word "poly." If parentheses or brackets are needed to avoid ambiguity, they are used. If not needed, they are left out. For example, Polymer 1.5 is named poly(dichlorophosphazene), Species 1.6 is called poly(dimethylsiloxane), and Polymer 1.7 is poly(methylphenylsilane). Species 1.8 is polysulfur.

1.3 REASONS FOR INTEREST IN INORGANIC POLYMERS

Polymer chemistry and technology form one of the major areas of molecular and materials science. This field impinges on nearly every aspect of modern life, from electronics technology, to medicine, to the wide range of fibers, films, elastomers, and structural materials on which everyone depends.

Most of these polymers are organic materials. By this we mean that their long-chain backbones consist mainly of carbon atoms linked together or separated by heteroatoms such as oxygen or nitrogen. Organic polymers are derived either from petroleum or (less frequently) from plants, animals, or microorganisms. Hence, they are generally accessible in large quantities and at moderate cost. It is difficult to imagine life without them.

In spite of the widespread importance of organic polymers, attention is being focused increasingly toward polymers that contain inorganic elements as well as organic components. At the present time, most of this effort is concentrated on the development of new *chemistry*, as research workers probe the possibilities and the limits to the synthesis of these new macromolecules and materials. But in certain fields, particularly for polysiloxanes, both the science and the technology are already well established, and technological developments now account for a major part of the siloxane literature. For other systems to be discussed in this book, technological developments are emerging from the chemistry at an accelerating rate.

Why, with the hundreds of organic polymers already available, should scientists be interested in the synthesis of even more macromolecules? The reasons fall into two categories. First, most of the known organic polymers represent a compromise in properties compared with the "ideal" materials sought by engineers and medical researchers. For example, many organic backbone polymers react with oxygen or ozone over a long period of time and lose their advantageous properties. Most organic polymers burn, often with the release of toxic smoke. Many polymers degrade when exposed to ultraviolet or gamma radiation. Organic polymers sometimes soften at unacceptably low temperatures, or they swell or dissolve in organic solvents, oils, or hydraulic fluids. At the environmental level, few organic polymers degrade at an acceptable rate in the biosphere. Finally, the suspicion exists that the availability of many organic polymers may one day be limited by the anticipated scarcities of petroleum. It is generally accepted that polymers that contain inorganic elements in the molecular structure may avoid some or all of these problems.

The second set of reasons for the burgeoning interest in inorganic-based macromolecules is connected with their known or anticipated *differences* from their totally organic counterparts. Inorganic elements generate different combinations of properties in polymers than do carbon atoms. For one thing, the bonds formed between inorganic elements are often longer, stronger, and more resistant to free radical cleavage reactions than are bonds formed by carbon. Thus, the incorporation of inorganic elements into the backbone of a polymer can change the bond angles and bond torsional mobility, and this in turn can change the materials properties to a remarkable degree. Inorganic elements can have different valencies than carbon, and this means that the number of side groups attached to a skeletal atom may be different from the situation in an organic polymer. This will affect the flexibility of the macromolecule, its ability to react with chemical reagents, its stability at high temperatures, and its interactions with solvents and with other polymer molecules. Moreover, the use of noncarbon elements in the backbone provides opportunities for tailoring the chemistry in ways that are not possible in totally organic macromolecules. Many examples of this feature are given in the later chapters of this book. Thus, the future development of polymer chemistry and polymer engineering may well depend on the inorganic aspects of the field for the introduction of new molecular structures, new combinations of properties, and new insights into the behavior of macromolecules in solution and in the solid state.

Thus, inorganic polymers provide an opportunity for an expansion of fundamental knowledge and, at the same time, for the development of new materials that will assist in the advancement of technology. Throughout this book an attempt has been made to connect these two aspects in a way that will provide a perspective of this field. For example, the superb thermal stability of several poly(organosiloxanes) can be understood in terms of their fundamental chemistry. The controlled hydrolytic degradability of certain polyphosphazenes, which depends on molecular design to favor specific hydrolysis mechanisms, is the basis for their prospective use as pharmaceutical drug delivery systems. The unusual energy ab-

sorption characteristics of polysilanes is indicative of surprising electronic structures, and this underlies the interest in some of these materials for use in integrated
circuit fabrication.

1.4 TYPES OF INORGANIC POLYMERS

A glance at the Periodic Table or at an inorganic chemistry textbook will convince
the reader that, of the 100 or so stable elements in the table, at least half have a
chemistry that could allow their incorporation into macromolecular structures. This
will undoubtedly come to pass in the years ahead. However, at the present time,
most of the known inorganic polymer systems are based on relatively few elements
that fall within the region of the Periodic Table known as the "Main Group"
series. These elements occupy groups III (13 in the IUPAC nomenclature), IV
(14), V (15), and VI (16) and include elements such as silicon, germanium, tin,
phosphorus, and sulfur. Of these, polymers based on the elements silicon and
phosphorus have received by far the most attention. This is the reason why silicon-
and phosphorus-containing polymers are considered in the greatest detail in this
book.

Specifically, the greatest emphasis in the following chapters is placed on
polyphosphazenes (Structure 1.9), polysiloxanes (Structure 1.10), and polysilanes
(Structure 1.11). The last chapter introduces a wide variety of other polymers that

$$\left[\begin{array}{c} R \\ | \\ P = N \\ | \\ R \end{array}\right]_n \qquad \left[\begin{array}{c} R \\ | \\ Si - O \\ | \\ R \end{array}\right]_n \qquad \left[\begin{array}{c} R \\ | \\ Si \\ | \\ R \end{array}\right]_n$$

1.9   1.10   1.11

contain elements such as germanium, sulfur, boron, aluminum, and tin and a variety of transition metals. These polymers are expected to provide the basis for
many of the new advances of the future.

1.5 SPECIAL CHARACTERISTICS OF POLYMERS

Polymer molecules have many special characteristics that may be unfamiliar to
some readers of this book. For this reason, the following chapter has been devoted
to a summary of the special techniques used for the characterization and study of
macromolecules. The remaining chapters deal with the synthesis, reaction chemistry, molecular structural, and applied aspects of selected inorganic polymer systems.

2

Characterization of Inorganic Polymers

2.1 MOLECULAR WEIGHTS

2.1.1 Introduction

2.1.1.1 Importance. One of the most important properties[1-3] of a polymer molecule is its molecular weight. This is the characteristic that underlies all the properties that distinguish a polymer from its low molecular weight analogs. Thus, one of the most important goals in the preparation of a polymer is to control its molecular weight by a suitable choice of polymerization conditions.

Many properties of a polymeric material are improved when the polymer chains are sufficiently long. For example, properties such as the tensile strength of a fiber, the tear strength of a film, or the hardness of a molded object may increase asymptotically with increases in molecular weight, as is shown schematically in Figure 2.1. If the molecular weight is too low, say, below a lower limit M_l, then the physical property could be unacceptably low. It might also be unacceptable to let the molecular weight become too high. Above an upper limit M_u, the viscosity of the bulk (undiluted) polymer might be too high for it to be processed easily. Thus, a goal in polymer synthesis is to prepare a polymer so that its molecular weight falls within the "window" demarcated by M_l and M_u. This is frequently accomplished by a choice of reaction time, temperature, nature and

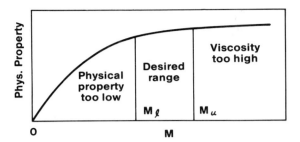

Figure 2.1 Typical dependence of a physical property on the molecular weight of a polymer. Reprinted with permission from J. E. Mark, *Physical Chemistry of Polymers*, ACS Audio Course C-89, American Chemical Society, Washington, DC, 1986. Copyright 1986, American Chemical Society.

amount of catalyst, the nature and amount of solvent, the addition of reactants that can terminate the growth of the polymer chains sooner than would otherwise be the case, addition of complexing agents such as crown ethers, or the presence of an external physical field, such as ultrasound.

2.1.1.2 Statistical aspects.   The termination of the growth of a particular chain molecule is a statistical process. If termination happens soon after the chain starts to grow, then, obviously, the chain will be short. If the chain evades termination for a while, it will be longer. Because of this, polymers are usually characterized by a *distribution* of molecular weights. This distinguishes them from their low molecular weight analogs and causes any experimentally determined molecular weight to be an average. Such an average molecular weight is generally determined by dissolving the polymer in a solvent, followed by measurement of some physical property of the resultant solution. An additional complication arises from the fact that different properties can depend on molecular weight in very different ways. Because of this, it is necessary to define the different types of averages, since these emphasize different ranges of the molecular weight distribution.

2.1.2 Types of Molecular Weights

The first type of molecular weight is the *number average*, and is defined by

$$M_n = \frac{\sum N_i M_i}{\sum N_i} \qquad \text{(Eq. 2.1)}$$

where N_i is the number of molecules that have a molecular weight M_i. This type of average is also called the "mean" value of a distribution. (The same types of summations are carried out in calculating the average grade in an examination, wherein 5 students who score 90 points each would contribute 5×90 points to the numerator and 5 events to the denominator, etc.) This type of average is obtained by use of any technique that "counts" particles, for example,

- end-group analysis
- vapor pressure lowering
- boiling point elevation
- melting point depression
- osmotic pressure

These are the so-called "colligative" properties, where the adjective signifies a "tying together" of all properties that have a particular characteristic. Specifically, they depend only on the *number* (molar) concentration of particles present.

The *weight-average* molecular weight is defined by

$$M_w = \frac{\sum N_i M_i^2}{\sum N_i M_i} \qquad \text{(Eq. 2.2)}$$

in which the N_i weighting factor that appears in eq. (2.1) has been replaced by $N_i M_i$, which is proportional to the *weight* of polymer that has the specified M_i. A measurement of the intensity I of light scattered from a polymer chain in solution is the most common way to obtain this type of average, and it depends on the fact that I is proportional to the *square* of the molecular weight of the polymer chain that is the origin of the scattering.

These two types of molecular-weight averages are representative of the type obtained in "absolute" methods, in that well-established thermodynamic equations can be used to convert the experimental data directly into a value of the molecular weight. However, some other methods require calibration. The most important of these "indirect" methods involves a measurement of the intrinsic viscosity. This quantity is a measure of the extent to which a polymer molecule increases the viscosity of the solvent in which it is dissolved. The viscosity method can be calibrated to yield a viscosity-average molecular weight, defined by

$$M_v = \left(\frac{\sum N_i M_i^{1+a}}{\sum N_i M_i} \right)^{1/a} \qquad \text{(Eq. 2.3)}$$

A solution viscosity measurement is a hydrodynamic-thermodynamic technique, and the extent to which a polymer molecule increases the viscosity of a solvent depends on the nature of its interactions with that solvent (as well as on its own molecular weight). These interactions are characterized by the quantity a that appears in eq. (2.3). The calibration of the method, using samples of the same polymer having known molecular weights, in essence determines its value. The disadvantage of this calibration requirement is offset by the simplicity of the experimental measurements.

2.1.3 Experimental Techniques

2.1.3.1 Colligative properties. A *chemical* method for determining number-average molecular weights involves the analysis of end groups. If the polymer was prepared in such a way that each chain has either one or two labeled ends, then analysis for these ends is equivalent to counting the chains. For example, the ends could be hydroxyl groups or radioactive initiator fragments, and the analysis could involve titration, spectroscopy, or measurements of radioactivity. The number of chains gives the number of moles of chains, and the weight of the sample divided by the number of moles gives the desired molecular weight. The method works best for relatively low molecular weights (below about 25,000) where the number density of chain ends is not too small.

Vapor pressure lowering, boiling point elevation, and freezing point depression are very similar thermodynamically. For example, the increase in boiling point ΔT_b is interpreted thermodynamically by using the boiling point elevation constant K_b to obtain the molality of the solution, as stated in the equation

$$\Delta T_b = K_b m \qquad \text{(Eq. 2.4)}$$

where m is the molality. The weight concentration c of the same solution is, of course, known from the weights of its components. The factor that converts the molar concentration to the weight concentration is simply the desired molecular weight. This is shown in the equation

$$m(\text{moles}/\text{kg solvent}) \times M(\text{g}/\text{mole}) = c(\text{g}/\text{kg solvent}) \qquad \text{(Eq. 2.5)}$$

These methods also become less reliable as the molecular weight increases because fewer solute particles are now present in the system at fixed weight concentration.

The final colligative property, osmotic pressure, is different from the others, as is illustrated in Figure 2.2. In the case of vapor pressure lowering and boiling point elevation, a natural boundary separates the liquid and gas phases that are in equilibrium. A similar boundary exists between the solid and liquid phases in equilibrium with each other in melting point depression measurements. However, to

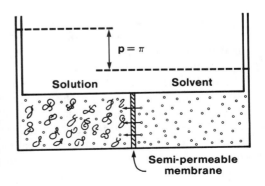

Figure 2.2 Schematic diagram of a membrane osmometer. Reprinted with permission from J. E. Mark, *Physical Chemistry of Polymers*, ACS Audio Course C-89, American Chemical Society, Washington, DC, 1986. Copyright 1986, American Chemical Society.

establish a similar equilibrium between a solution and the pure solvent requires their separation by a semipermeable membrane, as illustrated in the figure. Such membranes, typically cellulosic, permit transport of solvent but not solute. Furthermore, the flow of solvent is from the solvent compartment into the solution compartment. The simplest explanation of this is the increased entropy or disorder that accompanies the mixing of the transported solvent molecules with the polymer on the solution side of the membrane. Flow of liquid up the capillary on the left causes the solution to be at a hydrostatic pressure that is higher than that on the solvent side. The back pressure that is just sufficient to prevent further transport of solvent through the membrane is called the osmotic pressure π of that solution.

For an ideal solution (one that obeys Raoult's law)

$$\pi = RTm'$$ (Eq. 2.6)

where m' is the molarity (moles/cc of soln). Also, in parallel with eq. (2.5),

$$m'M = c$$ (Eq. 2.7)

where c is now a weight per unit volume. Thus, $\pi = cRT/M$, which is analogous to the ideal gas law, $p = nRT/V = cRT/M$.

Since polymer solutions are markedly nonideal, osmotic pressure data are taken at low concentrations and are extrapolated to infinite dilution ($c \rightarrow 0$). In the case of membrane osmometry, the relevant equation is

$$\frac{\pi}{cRT} = \frac{1}{M} + A_2 c + A_3 c^2 + \cdots$$

$$\cong \frac{1}{M} + A_2 c$$ (Eq. 2.8)

where the second virial coefficient A_2 provides a measure of the polymer-solvent interactions. The larger its value, the stronger are these interactions, and the more nonideal is the solution. Some typical results are shown schematically in Figure 2.3.

In membrane osmometry, molecular weights above a million are essentially impossible to measure because there are too few particles in a given weight of polymer. On the other hand, polymers with molecular weights less than 25,000 can cause problems by their diffusing through the membrane.

A related technique, vapor phase osmometry, is based on the idea of *isothermal distillation*. Such an osmometer is shown schematically in Figure 2.4. In effect, the vapor phase replaces the membrane in membrane osmometry. Thus, it permits passage of the solvent but not the polymer. A droplet of solvent is placed on one thermistor and a droplet of solution on another thermistor. The solvent molecules can vaporize and do so preferentially from the pure solvent droplet. The different rates of evaporation generate different temperatures, and the difference in temperature ΔT between the two droplets is measured.

A problem with the technique is that, because of thermal losses, the tem-

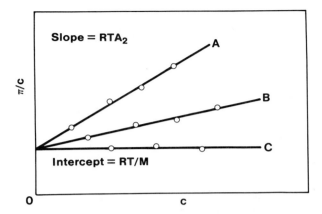

Figure 2.3 Typical osmotic pressure data, where solvent A is a very good solvent (strong polymer-solvent interactions), B is a moderately good solvent, and C is a Θ solvent (in which polymer-solvent interactions have been adjusted to nullify the excluded volume effect). Reprinted with permission from J. E. Mark, *Physical Chemistry of Polymers*, ACS Audio Course C-89, American Chemical Society, Washington, DC, 1986. Copyright 1986, American Chemical Society.

perature never really stabilizes and a thermodynamic analysis cannot be applied. It is customary, therefore, simply to take the relatively constant value of ΔT that appears near the maximum in a plot of ΔT against time.

This is not an absolute method, and it has to be calibrated. A plot is made of temperature difference against molar concentration, using a relatively low molecular weight solute of high purity and known molecular weight. Once the calibration curve has been established, any measured value of ΔT can be converted to a molar concentration, and thus a molecular weight can be obtained, using an equation such as (2.5) or (2.7).

The molecular weight of the solute should be small, but not so small as to allow it to be volatile.

2.1.3.2 Light scattering.

A typical photometer used in light-scattering measurements is shown in Figure 2.5. The symbol S locates the source of the light (often a laser), F represents filters which are sometimes needed for making the radiation more nearly monochromatic or decreasing its intensity, L is a collimating

Figure 2.4 Schematic diagram of a vapor phase osmometer. Reprinted with permission from J. E. Mark, *Physical Chemistry of Polymers*, ACS Audio Course C-89, American Chemical Society, Washington, DC, 1986. Copyright 1986, American Chemical Society.

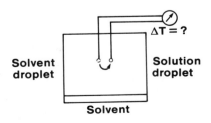

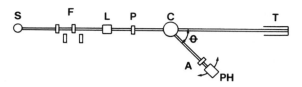

Figure 2.5 Sketch of the major components of a light-scattering photometer. Reprinted with permission from J. E. Mark, *Physical Chemistry of Polymers*, ACS Audio Course C-89, American Chemical Society, Washington, DC, 1986. Copyright 1986, American Chemical Society.

lens, C is the sample cell, T is a trap for the nonscattered incident beam, and PH is the photocell for measuring the intensity of the scattered radiation. The photocell is mounted on a turntable so that intensities can be measured as a function of scattering angle θ as well as concentration. P and A are a polarizer and analyzer, respectively, and can be used to measure the optical anisotropy of the polymer.

The quantity of primary interest is the *Rayleigh ratio*, which is defined by

$$R_\theta'' = \frac{i_\theta r^2}{I_o} \qquad \text{(Eq. 2.9)}$$

where i_θ is the intensity of scattered beam, I_o the intensity of incident beam, and r the distance from detector to scattering cell. The ratio i_θ / I_o is in the vicinity of 10^{-5}, which can complicate designing a photometer of this type. The Rayleigh ratio also contains r^2, the square of the distance between the detector and the scattering cell. Since the intensity of light decreases inversely with the square of distance, this makes the Rayleigh ratio independent of this arbitrary design feature.

Two corrections are required to the Rayleigh ratio as defined by eq. (2.9). Because the solvent, as well as the polymer, scatters the radiation, and only the polymer contribution is of interest, each value of the Rayleigh ratio is corrected as shown by the equation

$$R_\theta' = R_\theta''(\text{soln}) - R_\theta''(\text{solvent}) \qquad \text{(Eq. 2.10)}$$

An additional factor, $\sin \theta$, is required to correct for the increase in scattering volume that occurs with changes in scattering angle away from the reference value of 90°. A final factor $1/(1 + \cos^2 \theta)$ is required to take into account the fact the horizontal and vertical components of the light are scattered to different extents. These final two corrections are incorporated in the equation

$$R_\theta = R_\theta' \left[\frac{\sin \theta}{(1 + \cos^2 \theta)} \right] \qquad \text{(Eq. 2.11)}$$

The scattering function Kc/R_θ is the quantity directly related to the physical properties of interest. It consists of an optical constant K times the concentration, divided by the Rayleigh ratio. The constant K is defined by

$$K = \left(\frac{2\pi^2 n_o^2}{N_{avo}\lambda^4} \right) \left(\frac{dn}{dc} \right)^2 \qquad \text{(Eq. 2.12)}$$

and consists of the numerical constant π, the index of refraction of the pure solvent n_o, Avogadro's number, the wavelength λ of the light, and the dependence of the index of refraction of the solution on concentration.

There are two uses of the scattering function. If its value at zero scattering angle is plotted against concentration, then the intercept is the reciprocal of the molecular weight of the polymer, and the slope of the line is twice the second virial coefficient. The relevant equation is

$$\left(\frac{Kc}{R_\theta}\right)_{\theta \to 0} = \frac{1}{M} + 2A_2c + \cdots \qquad \text{(Eq. 2.13)}$$

Also, if the scattering function at zero concentration is plotted against $\sin^2(\theta/2)$, then the intercept gives a check on the molecular weight, and the slope gives the radius of gyration. The detailed mathematical relationship is given by

$$\left(\frac{Kc}{R_\theta}\right)_{c \to 0} = \left(\frac{1}{M}\right)\left\{1 + \left[\frac{16\pi^2}{3(\lambda/n_o)^2}\right]\langle s^2 \rangle \sin^2\left(\frac{\theta}{2}\right) + \cdots\right\} \qquad \text{(Eq. 2.14)}$$

In interpreting the light scattering from a polymer solution, the required extrapolations to zero scattering angle and zero concentration are usually carried out simultaneously. The result is called a Zimm plot and is described in Figure 2.6. The scattering function is plotted as the ordinate, and the abscissa is a composite quantity: $\sin^2(\theta/2)$ + a constant (100) times the concentration. The constant is arbitrary and merely determines how spread out the data will be.

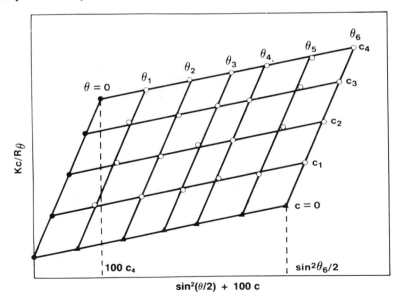

Figure 2.6 A Zimm plot, in which light-scattering data are simultaneously extrapolated to zero scattering angle and zero concentration. Reprinted with permission from J. E. Mark, *Physical Chemistry of Polymers*, ACS Audio Course C-89, American Chemical Society, Washington, DC, 1986. Copyright 1986, American Chemical Society.

The open circles in the uppermost line in Figure 2.6 locate data taken at the highest concentration. These data are extrapolated over to give the point that corresponds to zero scattering angle at this concentration. Since at $\theta = 0°$, $\sin^2(\theta/2) = 0$, the abscissa of the desired point, located by the filled circle, is simply 100 times the constant value of the concentration. This is repeated for the other (approximately parallel) constant-concentration lines to give the other points that correspond to zero angle. These points are then extrapolated to give an intercept which yields the molecular weight. The slope gives the second virial coefficient.

The more nearly vertical lines in the grid correspond to fixed scattering angle, with the line farthest to the right corresponding to the widest angle. Each such line is extrapolated downward to give the point that corresponds to zero concentration at this fixed value of θ. At $c = 0$ the abscissa of the desired point is simply sin-squared of half that particular angle. In this way, the location of each of the filled triangular points is determined. They are themselves then extrapolated to give the intercept and to yield a check of the molecular weight. The slope gives the mean-square radius of gyration.

A Zimm plot does simultaneously what is described in eqs. (2.13) and (2.14). An advantage is in being able to find the intercept that is most consistent with both the concentration dependence and the angular dependence.

Advantages of the light-scattering technique include the wide range of molecular weights that can be determined, and the ability to measure dimensions of the polymer coil in solution, and the ability to study the interactions between the polymer and solvent in the solution. The main disadvantages are the complexity of the equipment and the absolute need to remove all dust particles, which scatter tremendously because their effective molecular weights are extraordinarily high.

2.1.3.3 Viscometry. Viscometry is by far the easiest technique for the characterization of polymers in solution. This can be seen from the simplicity of the typical (glass) viscometer shown in Figure 2.7. It is used to obtain the viscosity of a liquid by the use of Poiseuille's equation, which is

$$\eta = \frac{\pi p r^4 t}{8LV} \qquad \text{(Eq. 2.15)}$$

The experiment involves a measurement of the amount of time t required for a volume V of liquid to flow through a capillary of radius r and length L when the

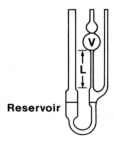

Reservoir

Figure 2.7 Schematic diagram of a viscometer. Reprinted with permission from J. E. Mark, *Physical Chemistry of Polymers*, ACS Audio Course C-89, American Chemical Society, Washington, DC, 1986. Copyright 1986, American Chemical Society.

pressure difference is p. The viscosity η may then simply be calculated from this equation, in which the π is the numerical constant.

In polymer solution viscometry, it is not necessary to determine absolute values of the viscosity; relative values are sufficient. These quantities are called "viscosities," but the terms are misnomers because they are generally unitless ratios and therefore do not have the units of viscosity. In any case, the relative viscosity is simply the ratio of the viscosity of the polymer solution to the viscosity of the pure solvent at the same temperature. Since the time t is the only quantity on the right-hand side of eq. (2.14) that varies significantly, the relative viscosity is simply the ratio of the two efflux times, and is given by

$$\eta_{rel} = \frac{\eta_{soln}}{\eta_{solvent}} = \frac{t_{soln}}{t_{solvent}} > 1.0 \qquad \text{(Eq. 2.16)}$$

The relative viscosity is larger than unity since the dissolved polymer molecules impede the flow of solvent. The specific viscosity, defined by

$$\eta_{sp} = \eta_{rel} - 1.0 \qquad \text{(Eq. 2.17)}$$

quantifies the viscosity increase itself by subtracting unity from the relative viscosity. The higher the concentration of polymer in the solution, the higher will be the viscosity, and the specific viscosity therefore has to be divided by the concentration, to give the quantity η_{sp}/c. The units of concentration chosen are typically g/dl, where a deciliter is a tenth of a liter. This is done so that the intrinsic viscosity obtained from these quantities generally has a value that is conveniently between 0.1 and 10.

The two extrapolations used to obtain the intrinsic viscosity are described by eqs. (2.18) and (2.19),

$$\frac{\eta_{sp}}{c} = [\eta] + k'[\eta]^2 c \qquad \text{(Eq. 2.18)}$$

$$\frac{\ln(\eta_{rel})}{c} = [\eta] - k''[\eta]^2 c \qquad \text{(Eq. 2.19)}$$

and by Figure 2.8. The two Huggins constants k' and k'', which determine the slopes of the two extrapolation lines, are frequently reported in viscometric studies. Unfortunately, they do not yield a great deal of useful molecular information. It is known, however, that k' and k'' should add algebraically to $1/2$, with k'' usually being much smaller in magnitude than k' and negative.

Typical viscometric plots are shown schematically in Figure 2.8, where the two extrapolations yield the intrinsic viscosity. For many polymer-solvent systems a value of 2 dl g^{-1} would correspond to a molecular weight in the region of a million.

To obtain molecular weight values from intrinsic viscosities, it is necessary to use the calibration illustrated in Figure 2.9. It is based on the Mark-Houwink

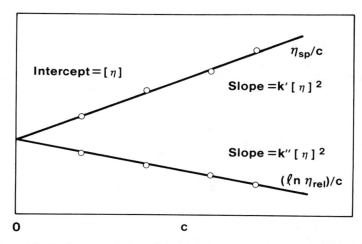

Figure 2.8 Double extrapolations of viscosity data to obtain the intrinsic viscosity $[\eta]$. Reprinted with permission from J. E. Mark, *Physical Chemistry of Polymers*, ACS Audio Course C-89, American Chemical Society, Washington, DC, 1986. Copyright 1986, American Chemical Society.

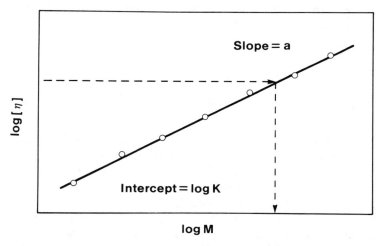

Figure 2.9 Calibration of intrinsic viscosity results. Reprinted with permission from J. E. Mark, *Physical Chemistry of Polymers*, ACS Audio Course C-89, American Chemical Society, Washington, DC, 1986. Copyright 1986, American Chemical Society.

relationship, which is given by

$$[\eta] = KM^a \qquad \text{(Eq. 2.20)}$$

where K and a are constants that depend on polymer, solvent, and temperature. Typical values are K the order of 10^{-4} and $a \cong 0.7$. A typical calibration curve is shown schematically in Figure 2.9. The intercept gives a value for the constant K, and the slope yields a value for the constant a. These constants have been tabulated for many polymer-solvent-temperature combinations.

The main advantages of the viscometric technique are (1) the simplicity of the method and (2) the range of molecular weights that can be measured. The main disadvantage is the fact that the method must be calibrated.

2.1.4 Uses for Molecular Weights

There are several important reasons for wanting to know molecular weights in polymer science. From the viewpoint of inorganic polymers, the main uses are for the interpretation of molecular weight–dependent properties and for the elucidation of polymerization mechanisms. The latter involves characterization of the molecular weight distribution, which is the subject of the following section.

2.2 MOLECULAR WEIGHT DISTRIBUTIONS

2.2.1 Importance

The subject of molecular weight distributions is of great practical as well as fundamental importance. For example, a small amount of material of either very low molecular weight or very high molecular weight can greatly change solid- and liquid-state properties, and thus affect the processing characteristics of a polymer. Therefore it is necessary to develop quantitative ways for characterizing molecular-weight distributions.

2.2.2 Representation of Molecular Weight Distributions

One way to characterize a distribution is shown in Figure 2.10, where W_M is the weight fraction of polymer that has a molecular weight in an infinitesimal interval about the specified value of M. This is the so-called differential representation of the distribution, and the two curves shown are for a relatively narrow distribution (A) and a relatively broad one (B). Such curves provide all the quantitative details of the distribution. For example, they indicate what weight % of the polymer has a molecular weight above or below a specified value of M, or between two specified values (as shown by the hatched segment under one of the curves). Also, the maximum in the curve locates the most probable value of M. It

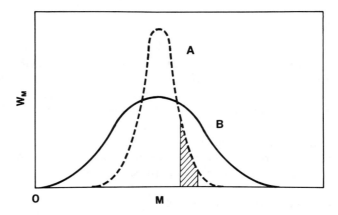

Figure 2.10 Typical differential distribution curves for the molecular weight. Reprinted with permission from J. E. Mark, *Physical Chemistry of Polymers*, ACS Audio Course C-89, American Chemical Society, Washington, DC, 1986. Copyright 1986, American Chemical Society.

is also immediately obvious whether the distribution is unimodal (one maximum) or multimodal (often bimodal). Such information can help to elucidate a polymerization mechanism, particularly if it is available for polymers having different extents of polymerization (usually obtained by using different polymerization times). For example, a bimodal distribution could indicate that some of the polymer was formed at a catalyst surface and some in solution. Bimodal distributions have, in fact, been observed in both polyphosphazenes[4] and polysilanes.[5] Illustrative results for a phosphazene polymer are shown in Figure 2.11.[4]

The simplest measure of the breadth of a distribution is the ratio of two different types of average molecular weight. Specifically the ratio of M_w to M_n is by far the most widely used for this purpose, and is called the *polydispersity index*. It has a minimum value of unity (for a monodisperse material in which all the chains have exactly the same length). The extent to which it exceeds unity is a measure of the breadth of the distribution. Typical values are in the range 1.5–2.0, but many polymerizations yield considerably larger values.

2.2.3 Experimental Determination

The most direct way to obtain a molecular weight distribution such as the ones illustrated in Figures 2.10 and 2.11 is to separate the polymer into a series of fractions each of which has a considerably narrower distribution. An advantage of this approach is the fact that these fractions can also be used to determine structure-property relationships that are not complicated by excessive polydispersity. This is shown by the family of curves making up the outer curve in Figure 2.12. The simplest technique for achieving such a separation is based on the decrease in polymer solubility which occurs as the molecular weight is increased, as is illus-

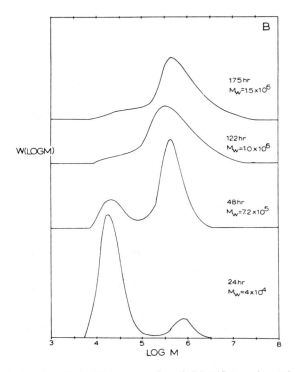

Figure 2.11 Differential distribution curves for poly[bis(trifluoro-ethoxy)phosphazene] shown as a function of extent of reaction (polymerization time). Reprinted from G. L. Hagnauer, *J. Macro. Sci.-Chem.* **1981**, *A16*, 385, p. 390, by courtesy of Marcel Dekker, Inc.

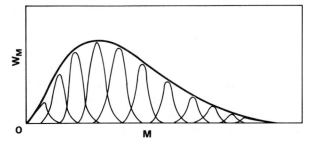

Figure 2.12 Resolution of a broad distribution curve into contributions from its constituent fractions. Reprinted with permission from J. E. Mark, *Physical Chemistry of Polymers*, ACS Audio Course C-89, American Chemical Society, Washington, DC, 1986. Copyright 1986, American Chemical Society.

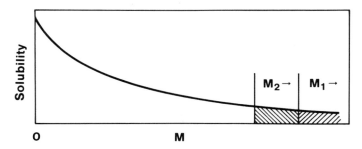

Figure 2.13 The decrease in solubility with increase in molecular weight that serves as the basis for the fractional precipitation technique. Reprinted with permission from J. E. Mark, *Physical Chemistry of Polymers*, ACS Audio Course C-89, American Chemical Society, Washington, DC, 1986. Copyright 1986, American Chemical Society.

trated in Figure 2.13. Thus, lowering the temperature of a polymer solution can decrease the solubility of the polymer sufficiently to cause precipitation of the highest molecular weight material. This fraction might have a molecular weight of M_1 or higher, as is illustrated by the corresponding region in the figure. Removal of this fraction, followed by subsequent cooling of the remaining solution, could then be used to produce a fraction in which the molecular weights are primarily in the range M_2 to M_1, and so on. It is also possible to decrease the solubility of the polymer by adding a nonsolvent to the solution. This method is used more frequently than the method that uses cooling.

Another technique for separating a polymer into fractions is called *solvent gradient elution*. In it, the unfractionated polymer is deposited as a very thin coating on high-surface-area beads in a chromatographic column. A very poor solvent for the polymer is then allowed to flow over the particulate material, thereby extracting only the lowest molecular weight (most soluble) material. A controlled amount of a good solvent for the polymer is then added to the first solvent, and this mixture is used to extract polymer of somewhat higher molecular weight. Repetition of this process *n* times, up to complete dissolution of the polymer, then yields *n* fractions of the material. The main disadvantage of this technique results from the fact that it works best only if an equilibrium can be established between dissolved and insoluble polymer. This is inherently difficult since shorter chains may be prevented from escaping into solution because they are entangled with longer molecules in the sample. On the other hand, the fractional precipitation technique is generally carried out in dilute solution, where each chain can participate in the partitioning process between solution and precipitated phase completely independently of the other chains. The main advantage of the gradient elution technique is the ease with which it can be automated.

The most modern technique for fractionating polymers is called *gel permeation chromatography (GPC)*. The packing in the chromatographic column consists

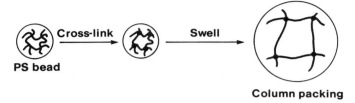

Figure 2.14   Preparation of swollen, cross-linked beads for gel permeation chromatography. Reprinted with permission from J. E. Mark, *Physical Chemistry of Polymers*, ACS Audio Course C-89, American Chemical Society, Washington, DC, 1986. Copyright 1986, American Chemical Society.

of cross-linked polystyrene beads swollen with a convenient solvent, as illustrated in Figure 2.14. The polymer to be fractionated is dissolved in the same solvent, and the solution is then forced through the packing. The smaller the polymer molecule, the more likely it will be able to enter the various "caves" or interstices between the chains that make up the swollen network structure. This slows the passage of the smaller polymer chains through the packing. Thus, each subsequent fraction collected has a molecular weight smaller than its predecessor. This technique is also relatively easy to automate.

The determination of molecular weights and their distributions is almost always the first technique used to characterize polymers, and this is especially true for inorganic polymers. Additional techniques used to characterize other features of polymeric materials are described in the following sections.

2.3 OTHER STRUCTURAL FEATURES

2.3.1 Backbone Bonding

A knowledge of the molecular weight of a polymer provides information about the number of backbone or skeletal bonds per molecule.[6,7] But the skeletal bonds require additional characterization. The most important item of information that needs to be determined is the length of each bond. In this regard, inorganic polymers are quite different from their organic counterparts. Specifically, virtually all covalent bonds between pairs of atoms in inorganic polymers (Si, P, N, etc.) are longer than the $C-C$ bond found in organic polymers. Thus, inorganic polymers are much less "congested" and, as a consequence, much more flexible. They are more flexible both in the equilibrium sense, which means that they can form more compact random coils, and in the dynamic sense, which means they can readily switch between different spatial arrangements. The first of these factors has a powerful influence on the melting point of a polymer. The second influences the temperature below which the polymer becomes a glass.

The melting point of any crystalline material is given by

$$T_m = \frac{\Delta H_m}{\Delta S_m}$$ (Eq. 2.21)

where ΔH_m is the heat of fusion and ΔS_m is the entropy of fusion. Since inorganic polymers can adopt very compact random-coil arrangements of high entropy, their entropies of fusion are frequently very high and their melting points relatively low. Of course exceptions can occur, particularly when unusually strong intermolecular attractions give an atypically large heat of fusion.

The glass transition temperature T_g is the temperature below which the polymer is glasslike because long-range motions of the polymer chains are no longer possible. The more flexible the polymer, in the dynamic sense, the lower is the temperature to which it can be cooled before this flexibility is "frozen out." Thus, the high dynamic flexibility generally enjoyed by inorganic chains frequently generates relatively low glass transition temperatures as well. The glass transition temperature is of considerable practical, as well as fundamental importance. For molded objects, it closely approximates the empirically defined "brittle point" of a polymer. This has little influence on the properties of films and fibers, however. After all, glass wool is far below its glass transition temperature at room temperature, yet it is clearly not brittle. The same is true for polystyrene film, which is used extensively as a packaging material.

2.3.2 Branching and Cross-linking

Under some conditions, branches can grow from the chain backbone, as shown in Figure 2.15. This could occur, for example, during a polymerization process, as in the case of the formation of some phosphazene polymers. It could also occur subsequently, through processes such as high-energy irradiations and the generation of free radicals. Because branch points represent irregularities in the chain structure, they can greatly suppress the tendency of a polymer to crystallize. Branching can occur to the extent that a *network* is formed, as in the preparation of thermosetting epoxy resins, or in the curing of elastomers. This is illustrated in Figure 2.16. In some cases, such cross-linking can be highly disadvantageous, since it can make the material impossible to manipulate or fabricate.

Figure 2.15 Branching in the homopolymer poly(A).

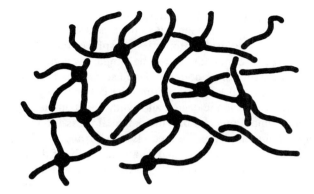

Figure 2.16 Network formation in a polymer, with the dots representing cross-links. Reprinted with permission from J. E. Mark, *Physical Chemistry of Polymers*, ACS Audio Course C-89, American Chemical Society, Washington, DC, 1986. Copyright 1986, American Chemical Society.

2.3.3 Chemical and Stereochemical Variability

Homopolymers possess only one type of repeat unit in the chain structure. Thus, except for the chain ends, the composition is constant throughout the chain. It is also possible to prepare chemical *copolymers*, in which two chemically different repeat units are distributed along the chain backbone. In terpolymers, there are three different types of repeat units. A chemically copolymeric polyphosphazene is shown in Figure 2.17. Thus, copolymers involve a composition variable since, for example, an AB copolymer could contain from 0 to 100 mol % of repeat unit A. In this respect an analogy can be made with mixtures or solutions of small molecules. Chemical copolymers have an additional variable that is unique to systems that consist of covalently bonded chains. This is the order, or sequencing, in which the different units are organized along the chain. For example, even at 50-50 equimolar composition, a copolymer could have any of the following sequential structures:

1. ABABABABABABABABABABAB   Alternating
2. ABBAAABABBBBAABBBAAAAB   Random
3. AAAAAAAAAAABBBBBBBBBBB   Diblock
4. AAAAAABBBBBBBBBBBAAAAA   Triblock

Figure 2.17 Sketch of a copolymeric phosphazene.

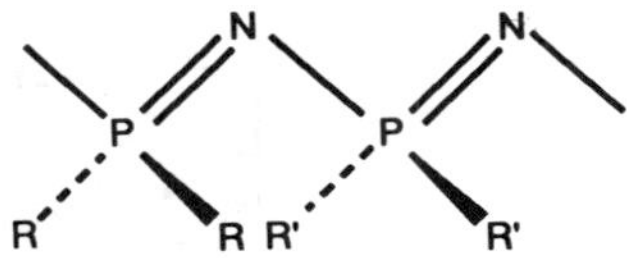

The alternating structure (1) does not generate unusual properties since it is equivalent to that of a homopolymer having AB as the repeat unit. A random copolymer structure (2) can give rise to useful properties, in spite of the fact that many of the properties of this type of copolymer are some simple average of the properties of the all-A and all-B parent homopolymers. One use for this type of structure is to introduce sufficiently irregular sequences into the polymer to prevent crystallization. This approach is used in the design and synthesis of ethylene-propylene elastomers. Other copolymer units can provide potential cross-linking sites, as do the vinyl side groups in dimethylsiloxane-methylvinylsiloxane copolymers.

The two blocklike structures (3 and 4) are generally of the greatest interest. If the blocks are relatively long, typically 20 or more repeat units, the two types of sequences will be incompatible and the system will undergo microscopic phase separation. Domains that contain repeat units of one type only, and having dimensions the order of 200 Å, will be formed within a continuous phase of the other type. Such multiphase materials can have very important mechanical properties. For example, in the triblock copolymer, chains that have crystalline or glassy end blocks and elastomeric center blocks can form thermoplastic elastomers. The crystalline or glassy domains at the ends of the elastomeric sequences serve the same role as the cross-links in a conventially cross-linked material. However, these elastomers are reprocessible in the sense that heating them above the melting point or glass transition temperature of the end blocks temporarily breaks up the network structure. Two important examples are the styrene-butadiene-styrene copolymers (the Kratons®),[7] illustrated in Figure 2.18, and diphenylsiloxane-dimethylsiloxane-diphenylsiloxane copolymers.[8]

If the chain backbone contains atoms that are unsymmetrically substituted, then stereochemical variability becomes possible. Stereochemical copolymers have repeat units that are chemically identical but stereochemically different, and in a sense these parallel the chemical copolymers already described. If substituted skeletal atoms in adjacent repeat units have the same atomic configuration, then the placement is said to be meso; placements with opposite configurations are called racemic. Polymers having long meso sequences are called isotactic and those with long racemic sequences, syndiotactic. Because of their stereoregular structure, these polymers can generally crystallize. If meso and racemic placements occur at random along the chain, the polymer is described as being atactic and is generally incapable of crystallization. Some examples of stereochemically variable chains

PS   **PBD**   **PS**

Figure 2.18 Example of a reprocessable elastomer, with the cross-linking polystyrene domains shown by the circles. Reprinted with permission from J. E. Mark, *Physical Chemistry of Polymers*, ACS Audio Course C-89, American Chemical Society, Washington, DC, 1986. Copyright 1986, American Chemical Society.

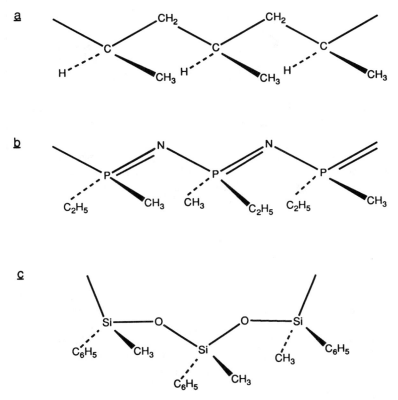

Figure 2.19 Sketches of some stereochemically variable chains, of polypropylene (two meso placements), poly(methyethylphosphazene) (two racemic placements), and poly(methylphenylsiloxane) (meso placement followed by racemic placement).

are shown schematically in Figure 2.19. The importance of crystallinity in polymers is discussed later, in Section 2.6.

2.4 CHAIN STATISTICS

2.4.1 Basic Goals

In this research area, the goal is to characterize the spatial arrangements or configurations of a chain molecule.[9] This is most commonly accomplished in terms of the overall size of the molecule, typically in a random-coil state. More specifically, the quantities of interest are the mean-square end-to-end distance or dimension $\langle r^2 \rangle$ of the chain, as illustrated in Figure 2.20, or its mean-square radius of gyration. These measures of the size of the polymer domain can be obtained di-

Figure 2.20 Representation of the end-to-end distance of a random-coil polymer. Reprinted with permission from J. E. Mark, *Physical Chemistry of Polymers*, ACS Audio Course C-89, American Chemical Society, Washington, DC, 1986. Copyright 1986, American Chemical Society.

rectly from light scattering or neutron scattering or, somewhat indirectly, from solution viscosities. The experimental results are then compared with theoretical predictions obtained by the use of realistic models of the chain molecule.

2.4.2 Illustrative Models

The simplest model of this type is called the freely jointed chain, and is illustrated in Figure 2.21. In it the skeletal bonds are joined end to end, but are completely unrestricted in direction. This is clearly a situation not found in a real polymer (bond angles in real polymers are relatively fixed). It is also assumed that the chains have zero cross-sectional area, that is, that the chains are unperturbed by excluded-volume effects. These effects arise because each atom of a chain excludes from the space it takes up, all other atoms from all chains present in the system. They are related to excluded-volume effects occurring even in systems as simple as real gases, but now have an intramolecular component. The expression for the mean-square end-to-end distance of such an idealized chain is particularly simple,

$$\langle r^2 \rangle_o = nl^2 \tag{Eq. 2.22}$$

where $\langle r^2 \rangle_o$ is called the unperturbed dimensions and n is the number of skeletal bonds, of length l. Values of $\langle r^2 \rangle_o$ calculated for more realistic models, or obtained for real chains by experimental measurements, are then compared to nl^2, in what is called the *characteristic ratio*, $\langle r^2 \rangle_o / nl^2$. The extent to which this ratio exceeds unity is a measure of the extent to which the unperturbed dimensions of a revised chain model, or of a real polymer chain, differ from those of the freely jointed model. In the limit of a very large value of n, it is frequently given the symbol C_∞.

Figure 2.21 Sketch of a freely jointed chain. The bonds, represented by the straight lines, are uncorrelated in direction. Reprinted with permission from J. E. Mark, *Physical Chemistry of Polymers*, ACS Audio Course C-89, American Chemical Society, Washington, DC, 1986. Copyright 1986, American Chemical Society.

The first of the more sophisticated models takes into account the fact that bond angles in molecules are very nearly constant. It also assumes that rotations about the skeletal bonds are unhindered (no energy barriers). This model has become known as the freely rotating chain. For tetrahedral bonding, this modification approximately doubles the characteristic ratio, to 2.2. The ratio is generally increased further if it is assumed that different rotational angles have different energies. The final refinement involves taking into account the cooperativity of neighboring rotational isomeric states: in other words it assumes that the conformation of one skeletal bond influences the conformation of the neighboring skeletal bonds. The extent to which the final two modifications affect the characteristic ratio depends on the detailed geometry of the chain and its conformational preferences. Most experimental values of the characteristic ratio are in the range 4–10. Melting points generally increase with increase in the value of the ratio, since it is an approximate measure of the equilibrium stiffness of a polymer chain.

2.4.3 Other Configuration-Dependent Properties

Although the unperturbed dimensions have been used most commonly to characterize spatial configurations, other properties are also being used for this purpose. Among these are

- mean-square dipole moments
- strain birefringence and dichroism
- optical rotation
- NMR chemical shifts
- cyclization equilibrium constants
- stereochemical equilibrium constants

In a typical analysis of a polymer chain, the *experimental* values of configuration-dependent properties and their temperature coefficients are compared with the results of rotational isomeric state *calculations*. These comparisons yield values of the energies for the various rotational states about the backbone bonds, and these conformational preferences can then be used to predict other configuration-dependent properties of the chains. It is also possible to obtain such conformational information from potential energy calculations, using the methods of molecular mechanics.[9-11]

2.4.4 Minimum Energy Conformations

Completing the conformational analysis of a polymer chain in either way can yield an important bonus. Such an analysis immediately identifies the conformation having the lowest intramolecular energy. This is important, because the lowest energy conformation is almost always the one adopted by the chain when it crys-

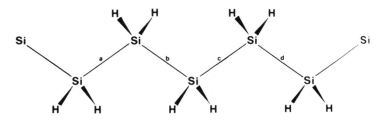

Figure 2.22 Sketch of a bond sequence in polysilane itself. Reprinted with permission of the American Chemical Society.

tallizes. Thus, the conformation assumed by a crystallized sequence in a polymer chain depends almost entirely on intramolecular interactions, even though crystallization is a highly cooperative process, in which *inter*molecular interactions must occur in profusion. Apparently, only after a regular conformation is chosen by a single chain does the question arise about how it is going to pack with its neighbors in a crystalline lattice.

Very early examples in this area are the predictions that polyethylene should crystallize in a planar zigzag conformation, poly(oxymethylene) in a helix having nine repeat units per five turns, some isotactic poly(α-olefins) in helices having three repeat units per single turn, and a number of polypeptides and proteins in the now-famous α-helices. All these predictions, and many others, have been confirmed experimentally.[9]

Another example that is much more recent and more relevant to inorganic materials pertains to conformational energy calculations carried out on the polysilane sequence sketched in Figure 2.22.[12, 13] The results, as is customary, are presented as a map, as shown in Figure 2.23.[13] In this representation, contours of equal energy are plotted as a function of the two rotational angles about adjacent skeletal bonds in the central portion of the chain sequence. As shown in the figure, the minimum energy conformations should occur at rotational angles of approximately ±120°, to generate a helix very similar to that formed by poly(oxymethylene). Thus, it is predicted to be very different from the all-*trans*, planar zigzag conformation preferred by its hydrocarbon analog, polyethylene.

2.5 SOLUBILITY CONSIDERATIONS

2.5.1 Basic Thermodynamics

It is generally more difficult to find a solvent that will dissolve a polymer than it is to find a solvent for a small molecule such as, for example, the monomer from which the polymer was prepared. This is understandable by a consideration of the free energy change for the dissolution process:

$$\Delta G_{\text{dis}} = \Delta H_{\text{dis}} - T\Delta S_{\text{dis}} \qquad \text{(Eq. 2.23)}$$

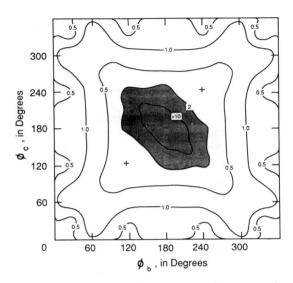

Figure 2.23 Conformational energy map for the polysilane sequence shown in Figure 2.22, where ϕ_b and ϕ_c are the rotational angles about skeletal bonds b and c in this figure. The contour line labeled 2 encloses regions corresponding to very high energy, greater than 2 kcal mol^{-1}. The two plus signs locate the conformations of minimum energy. Reprinted with permission from W. J. Welsh, L. DeBolt, and J. E. Mark, *Macromolecules* **1986,** *19*, 2978, p. 2980. Copyright 1986, American Chemical Society.

where ΔH_{dis} and ΔS_{dis} are the enthalpy and entropy of dissolution, respectively. Usually, the interactions between unlike species are unfavorable, and the enthalpy change is positive (the energy increases). Since the free energy change must be negative for the process to occur (at constant temperature and pressure), the second term must be negative and predominate over the first. That is, dissolution processes are generally entropically driven.

2.5.2 Problems Specific to Polymers

The foregoing conclusion is unfortunate for the case of polymeric solutes, because their entropies of dissolution are unusually small. The repeat units cannot become as disordered as can the corresponding monomer molecules since they are constrained to be part of a chainlike structure. Such disordering is particularly difficult if the chain is stiff. Thus, in this situation dissolution is even less likely. Crystalline polymers are also more difficult to dissolve than are their amorphous counterparts since the enthalpy of dissolution also contains a large, positive contribution from the latent heat of fusion.

The foregoing equation also explains why two different *polymers* are seldom miscible. Both solute and solvent are now polymeric, and thus both suffer from the entropy decreases described. It also explains why it is necessary to heat a mixture in which the solute does not dissolve at room temperature. This increase in T increases the magnitude of the last term in Eq. (2.23), and this is the term that generally makes the free energy change negative.

2.5.3 Solubility Parameters

2.5.3.1 Definitions.   As already mentioned, interactions between unlike species are generally unfavorable. This observation is, in fact, the basis for the rule of thumb that "like dissolves like." This rule is obviously very qualitative, but it has now been successfully extended to provide a quantitative basis for finding potential solvents for a polymer. In the theory of Hildebrand, it is acknowledged that ΔH_{dis} is almost certainly going to be positive. The goal is then to find a means to make ΔH_{dis} as small as possible. Specifically, it is given by the equation

$$\Delta H_{dis} = v_1 v_2 (\delta_1 - \delta_2)^2 \qquad \text{(Eq. 2.24)}$$

where it is written as the product of the volume fraction of solvent, the volume fraction of polymer, and the square of the difference between the values of a molecular characteristic called the *solubility parameter*, δ, for the solvent and polymer, respectively. The quantity δ is defined by

$$\delta \equiv \left(\frac{\Delta E}{V} \right)^{1/2} \qquad \text{(Eq. 2.25)}$$

that is, as the square root of the cohesive energy density, which is the ratio of the molar energy of vaporization to the molar volume of the liquid. Thus, the quantity δ is a measure of the strength of the interactions between the molecules of a substance, since vaporization involves greatly increasing the average distance of separation between them. Two molecules are "like" one another if the strengths of their interactions are similar. The two values of the solubility parameter will then be similar, and this will minimize the positive value of ΔH_{dis}, as shown in eq. (2.24). The utility of the approach is based on the fact that values of δ have been determined for many solvents and polymers by both experiment and approximate calculations. Solubility is likely to occur when a polymer and solvent have solubility parameters within one and a half units of each other, when the units are in Joules and cubic centimeters.

2.5.3.2 Experimental determination, solvents.   It is relatively easy to determine the solubility parameter of a solvent. The molar volume can be obtained from pycnometry, or a value can possibly be found in the literature. Also, since most solvents of interest have significant volatility, their heats of vaporization can be determined calorimetrically. The experimentally determined heat of vaporization can be converted into the desired energy of vaporization through a conversion term that is simply the change in pressure-volume product for the process. To an adequate approximation, this is simply RT, where R is the usual gas constant.

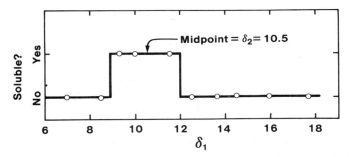

Figure 2.24 Solubility bar graph for determining the solubility parameter δ_2 for a polymer. Reprinted with permission from J. E. Mark, *Physical Chemistry of Polymers*, ACS Audio Course C-89, American Chemical Society, Washington, DC, 1986. Copyright 1986, American Chemical Society.

2.5.3.3 Experimental determination, polymers.

Polymers degrade before they vaporize, so indirect methods must be used to determine their solubility parameters. The simplest of these methods is a two-valued bar graph, as is illustrated in Figure 2.24. A series of solvents of known solubility parameter δ_1 are used. Each is tested to determine if it is a solvent for the polymer. The "yes" or "no" answers are then plotted at different heights along the δ_1 abscissa. The value of δ_1 corresponding to the midpoint of the "yes" bar is taken to be the solubility parameter δ_2 of the polymer. This is a reverse application of eq. (2.24). The argument is now that solvents that are found to dissolve the polymer must have solubility parameters close to the unknown solubility parameter of the polymer.

An alternative approach is to measure the minimum temperature required to bring about dissolution of the polymer, again in a series of solvents of known solubility parameters. This is illustrated in Figure 2.25. The solubility parameter

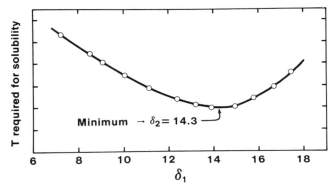

Figure 2.25 The dissolution temperature method for determining polymer solubility parameters. Reprinted with permission from J. E. Mark, *Physical Chemistry of Polymers*, ACS Audio Course C-89, American Chemical Society, Washington, DC, 1986. Copyright 1986, American Chemical Society.

of the polymer is taken to be the value of δ_1 that corresponds to the smallest required increase in temperature.

If a cross-linked sample of the polymer of interest is available, then it is possible to determine δ_2 from swelling equilibrium measurements. Portions of the sample are weighed and are then placed into each of a series of solvents of known solubility parameters. The polymers are allowed to swell by imbibing the solvent. After the swelling is complete, each sample is reweighed, and the weights and specific volumes of polymer and solvent are used to calculate values of the ratio of the swollen to dry volume of the sample. These values are then plotted against the solubility parameters of the solvents, as shown in Figure 2.26. The largest value of the ratio, or the highest degree of swelling, must be that obtained by using the best solvent for that polymer. Therefore, the solubility parameter of the polymer must be at least approximately equal to δ_1 for this particular solvent. A similar approach would be to search for the maximum intrinsic viscosity when the *uncross-linked polymer* is dissolved in the same series of solvents.

If short chains, or oligomers, of the polymer of interest are available, it is possible to use extrapolation techniques to obtain δ_2. The oligomers are chosen to have low enough molecular weights that the required energies of vaporization can be determined experimentally. The resultant values of δ_2 are then plotted against the reciprocal of the molecular weight, and the curve is extrapolated to infinite molecular weight, as shown in Figure 2.27. The intercept corresponds to the solubility parameter for the high-molecular-weight polymer.

A final method for obtaining solubility parameters, for solvents as well as for polymers, involves a theoretical approach called a group additivity scheme.[14] First it is necessary to determine how much each atom or chemical group contributes to the solubility product of a molecule. (For example, the contribution from a methylene group is simply the change in solubility parameter observed in changing from ethane to *n*-propane.) Such contributions have been tabulated extensively,

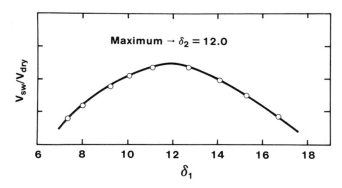

Figure 2.26 The network swelling method for determining polymer solubility parameters. Reprinted with permission from J. E. Mark, *Physical Chemistry of Polymers*, ACS Audio Course C-89, American Chemical Society, Washington, DC, 1986. Copyright 1986, American Chemical Society.

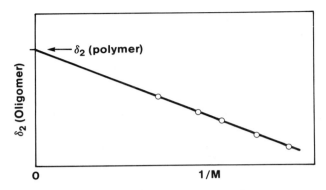

Figure 2.27 Extrapolating solubility parameters for oligomers to obtain polymer solubility parameters. Reprinted with permission from J. E. Mark, *Physical Chemistry of Polymers*, ACS Audio Course C-89, American Chemical Society, Washington, DC, 1986. Copyright 1986, American Chemical Society.

in the literature. The appropriate group contributions are then added together to predict the solubility parameter of the entire solvent molecule. In the case of a polymer, the contributions from the groups in the *repeat unit* are added. Since the solubility parameter is a ratio of molar quantities, the value for the repeat unit is equal to that of the polymer.

2.5.3.4 Trends.   Actual values of solubility parameters show the same trends for both solvents and polymers. Nonpolar molecules and repeat units have weak intermolecular forces, small energies of vaporization, and therefore small solubility parameters. As might be expected, increased polarity increases the solubility parameter, and hydrogen bonding gives the largest values of all.

2.5.3.5 Applications.   The fundamental idea is to determine the solubility parameter of the polymer, and then use tabulated results to identify a number of solvents that have solubility parameters close to this value. The list of potential solvents is then narrowed to two or three candidates. Solvents that are too volatile, too toxic, too flammable, too expensive, and so on can be removed from the list. Other criteria would depend on the nature of the studies to be pursued. If the objective is to carry out light-scattering measurements, the need for maximizing the contrast factor would make the index of refraction of the solvent an additional important consideration.

It is important to note that these predictions are for the possible miscibility of two *amorphous* materials. They do not apply to crystalline polymers because of the neglect of the already-mentioned positive heat of fusion. However, the predictions would be useful for temperatures above which the polymer would no longer be crystalline.

It should also be mentioned that polymer-solvent interactions can be characterized by the second virial coefficients that appear in eqs. (2.8) and (2.13) and by the free energy of interaction parameter χ_1 that appears in the Flory-Huggins theory of polymer solution thermodynamics.[1-3]

2.5.4 Useful Modifications of Structure

A final comment concerns two chemical modifications which could be used to increase the solubility (and processability) of rigid polymer chains. The first involves inserting some relatively flexible "swivel joints" into the chains. This increases chain flexibility, configurational entropy, and the entropy of mixing. The entropy of mixing can also be increased, for constant backbone rigidity, by employing very flexible side chains to mix with the solvent. Both changes decrease the free energy of dissolution, and thus increase the chances for miscibility.

2.6 CRYSTALLINITY

2.6.1 Importance

As mentioned earlier, crystallinity[15, 16] in a polymer is important in many applications. It is a way by which the chemist introduces solidity or hardness, but it can also have a beneficial effect on a variety of other mechanical properties. The most important of these is probably the impact resistance of the material. Because polymer chains are so long, different segments of the same chain become incorporated into different microcrystallites. Figure 2.28 illustrates the situation. As a result, much of the polymer that connects crystallites is badly entangled and poorly positioned for crystallization. This is the reason why "crystalline" polymers typically contain 20% to 50% of amorphous material, and are therefore better termed

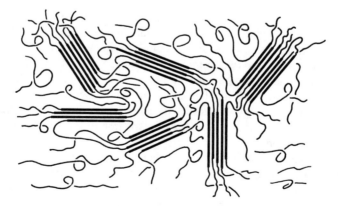

Figure 2.28 Sketch of the simplest model of a partially crystalline polymer, where the parallel straight lines represent the crystallites. Reprinted with permission from J. E. Mark, *Physical Chemistry of Polymers*, ACS Audio Course C-89, American Chemical Society, Washington, DC, 1986. Copyright 1986, American Chemical Society.

"partially crystalline." The crystallites are of great importance because they give the polymer the solidity or rigidity required in many applications, for example, when they are used in molded objects. The amorphous regions are also important. The chains in these regions have enough mobility that they can absorb impact energy through their skeletal motions, using frictional effects to convert the energy harmlessly into heat. A partially crystalline polymer is therefore much tougher than the same polymer would be if it were 100% crystalline.

2.6.2 Requirements

Because crystallinity is frequently highly desirable, it is important to establish the structural features that are conducive to achieving it. Most important, considerable regularity in chain structure is generally required. Specifically, crystallization is usually impossible for chain sequences that contain defects such as branch points, cross-links, chemical irregularities (comonomeric units), stereochemical irregularities (atactic placements), head-to-head placements instead of head-to-tail placements, and so on. An interesting exception occurs when two different side groups are similar in size and can replace one another in the crystalline lattice. In this case, even chains that are irregular in this sense can undergo crystallization.

Strong intermolecular attractions are also conducive to crystallization. They increase the heat of fusion, since fusion generally involves increasing the distance of separation between chains. The stronger the attractions, the higher the heat of fusion. Because the melting point is directly proportional to the heat of fusion, as shown in eq. (2.21), it too is increased. The higher the melting point, the greater the degree of supercooling that is likely to exist at any given temperature of application, and thus the greater the likelihood of crystallization.

2.6.3 Characterization of Crystallinity

2.6.3.1 Intramolecular.   X-ray diffraction analysis can be used to characterize the crystallites themselves[17]. First, it is possible to obtain information about the chain conformation adopted by the polymer chains in the crystallites. Many polymers crystallize in helical conformations. These can be defined by the following helical parameters. The designation ρ_n tells how to rotate and simultaneously translate along an axis to generate the specified helix. First, a point of reference is rotated around the proposed axis by n/ρ times a complete rotation of 2π. Simultaneously, there should be a translation of n/ρ of the crystallographic repeat distance, a quantity that is also obtained from diffraction data. Repetition of this scheme gives a helix having ρ repeat units in n turns of the helix.

Helices, like threads on screws, have a handedness or chirality. The handedness can be determined by curving the fingers of the right hand so as to follow the chain in its rotation around the axis of a drawing or model of the proposed helix. If the thumb points in the direction of the translation, then the helix is right

Figure 2.29 A right-handed helix. Reprinted with permission from J. E. Mark, *Physical Chemistry of Polymers*, ACS Audio Course C-89, American Chemical Society, Washington, DC, 1986. Copyright 1986, American Chemical Society.

handed, as is illustrated in Figure 2.29. If, instead, the left hand has to be used to achieve this relationship between rotation and translation, then the helix is left handed.

2.6.3.2 Intermolecular.

The chains, now characterized in terms of intramolecular features, can also be characterized with regard to how they pack themselves into ordered arrangements in crystallites[17]. This is understood in terms of the "unit cell," which is the simplest volume by means of which the entire crystallite can be generated by translations in three directions. To characterize the unit cell requires a specification of the values of the lengths *a*, *b*, and *c* of the three sides, and the angles between them, if they are different from 90°. This is illustrated in Figure 2.30. Sliding the unit cell in the directions *a*, *b*, and *c* would then generate the macroscopic crystallite.

If the numbers and types of atoms present in the unit cell are also known, then it is possible to calculate the limiting or maximum density; that is, the density the material would have if it were 100% crystalline. In brief, the volume of the unit cell can be calculated from the geometric information mentioned earlier, and the weight of its contents can be obtained by summing the atomic weights of the atoms. The maximum density is then simply the ratio of this weight to the unit cell volume. This value of the density can be used to estimate the percentage crystallinity of an actual sample of the same polymer.

The simplest polymer morphology is obtained when polymer chains crystallize from dilute solution, circumstances under which they are not badly entangled. Crystallizations of this type have been carried out for a variety of organic polymers, most notably for polyethylene. The single crystals thus formed have a lamellar or "platelike" structure. These platelets are typically 100 Å thick and approximately 100,000 Å in the other two directions. Interestingly, the chains are found to be perpendicular to the plane of the platelet. Since the chains are much longer than the 100 Å thickness, they must be folded back and forth a large number of times. The regularity with which they do this has been a matter of contro-

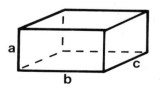

Figure 2.30 Sketch of a unit cell having a volume *abc* which, together with the weights of the atoms it contains, gives the crystallographic (maximum) density of the polymer. Reprinted with permission from J. E. Mark, *Physical Chemistry of Polymers*, ACS Audio Course C-89, American Chemical Society, Washington, DC, 1986. Copyright 1986, American Chemical Society.

versy for some time. Crystallizable inorganic polymers should also form such single crystals, but very little has been done to study them in this regard.

The morphology obtained by crystallizing an *undiluted* polymer can be considerably more complicated. The simplest model proposed consists of a system in which small crystallites exist, with amorphous chains attached to them like fringes on a colloidal aggregate. Such a system is called a *micelle*. This model is probably too simple except for a few polymers which from relatively few, very small crystallites.

It is much more common for melt-crystallized polymers to show a *spherulitic* type of structure. In this morphology, the crystallites are packed into spheres, except where a sphere encounters another sphere or an interface. The chains also exist in lamellae, but are folded perpendicular to the radii of the spherulites. The amorphous material is located *between* the radii as they diverge in progressing from the center of the spherulite to its surface.

2.6.4 Methods for Determining Percent Crystallinity

The four methods commonly used to determine the percent crystallinity of a partially crystalline polymer are dilatometry, x-ray crystallography, infrared spectroscopy, and calorimetry.

2.6.4.1 Dilatometry.   In dilatometry, specific volumes V or densities of materials are measured by use of a pycnometer, dilatometer, or density-gradient column. The data obtained are then interpreted through the equation

$$\% \text{ cryst} = \frac{100[V(\text{amorph}) - V(\text{partially cryst})]}{V(\text{amorph}) - V(\text{cryst})} \qquad \text{(Eq. 2.26)}$$

which utilizes the specific volumes of the totally amorphous polymer, the totally crystalline polymer, and the partially crystalline polymer being characterized. The specific volume of the totally amorphous polymer is easily measured if the polymer can be quenched (cooled rapidly from a moderate temperature) or can be prepared in a noncrystallizable (atactic) form. Otherwise, it is necessary to measure specific volumes at temperatures above the melting temperature T_m and then extrapolate these data to the lower temperature of interest. The specific volume of the totally crystalline polymer, as already mentioned, can be obtained from the unit cell data on the polymer.

The interpretation of eq. (2.26) is very simple. The denominator represents the maximum difference that can be observed for the polymer, namely, the difference between the values for the 100% amorphous polymer and the 100% crystalline polymer. The numerator is the difference observed between the 100% amorphous polymer and the actual, partially crystalline polymer. The ratio of this observed difference to the maximum possible difference is therefore the fraction crystallinity and, when multiplied by 100, is the % crystallinity.

2.6.4.2 Crystallography.   In the crystallographic method, the x-ray scattering intensity is plotted against diffraction angle. It is then necessary to separate the observed intensity peaks into amorphous and crystalline contributions. This can be done either by raising the temperature (to melt the crystallites in the normal manner) or by adding a diluent (which can melt the crystallites by depressing the melting point). The peaks that disappear must have been due to the crystalline regions and the ones that are relatively unaffected must have been due to the amorphous regions. The intensities I thus identified are then summed and used in the equation

$$\% \text{ cryst} = \frac{100 \sum I(\text{cryst})}{\sum I(\text{cryst}) + \sum I(\text{amorph})} \qquad \text{(Eq. 2.27)}$$

the interpretation of which is obvious.

2.6.4.3 Spectroscopy.   The infrared spectroscopic method is the same as the x-ray diffraction method, except that the infrared band absorptions are used instead of diffraction intensities. As in the x-ray analysis method, the dependence of the spectrum on temperature or on the presence of diluent can be used to determine which bands are due to the crystalline regions, and which to the amorphous.

2.6.4.4 Calorimetry.   In the calorimetric approach, it is necessary to know the heat of fusion of the totally crystalline polymer. This can be obtained from melting point depression measurements, as described in Section 2.7.5.4. The basic idea depends on the fact that the melting temperature is independent of the size of the system, since it is an intensive property. The extent to which it is depressed by the presence of solvent can be used to calculate a heat of fusion characteristic of the crystallites, irrespective of how many are present. This is the heat of fusion of the 100% crystalline polymer. The fractional crystallinity in an actual sample is then the ratio of its calorimetrically measured heat of fusion per gram to that of the 100% crystalline polymer. For example, if the actual polymer has a heat of fusion of 7 cal per gram, and the 100% crystalline polymer a heat of fusion of 10 cal per gram, then the fractional crystallinity is 0.7, and the percentage crystallinity is 70%.

2.6.5 Some Additional Information from X-ray Diffraction

2.6.5.1 Crystallite size.   The size of crystallites in a direction perpendicular to any chosen set of lattice planes can also be estimated from plots of x-ray scattering intensity against diffraction angle. Since infinitely sharp diffraction peaks occur only for crystallites of essentially infinite size, the width of a diffraction peak is an inverse measure of crystallite size. More specifically, the crystallite size perpendicular to the lattice planes giving rise to the diffraction can be obtained from the "half-width" of the corresponding peak (the range of scattering angle at half its maximum intensity).

2.6.5.2 Diffraction patterns, unoriented polymers.

The diffraction pattern obtained from an unoriented, partially crystalline polymer, or from a powdered crystalline material where the crystallites are unoriented, consists of a series of concentric circles. Such a pattern is illustrated in Figure 2.31.[18] Its exact pattern can be used to identify an unknown polymer. The origin of these circles can be described using the Bragg diffraction law. A diffraction spot will be placed on the film only where there is constructive interference, and this will occur only for those diffraction angles that obey the Bragg relationship $n\lambda = 2d \sin \theta$. Since the crystallites are completely unoriented about the incident beam, a cone of radiation is scattered for each of these permitted angles. The cone is intersected by the plane of the film perpendicular to its axis. Each intersection of a scattering cone by the film generates one of the concentric circles of the scattering pattern.

2.6.5.3 Diffraction patterns, oriented polymers.

The diffraction pattern obtained when the crystallites are oriented, as in a fiber, is a distinctive, symmetric pattern of arcs or layer lines. Such a pattern, obtained from a polyphosphazene, is shown in Figure 2.32.[19] The crystallites in the fiber have their axes nearly parallel to the fiber axis, but it is not possible to orient the other axes of the crystallites. Therefore, rotational disorder exists about the chain axes and,

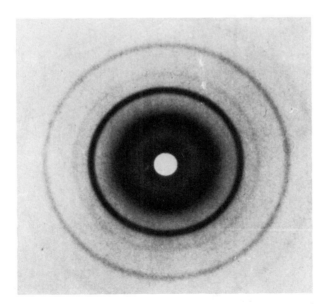

Figure 2.31 Illustrative diffraction pattern obtained from an unoriented polymer. Reprinted with permission from F. W. Billmeyer, Jr., *Textbook of Polymer Science*, 2nd ed.; Wiley-Interscience: New York, 1971. Copyright © 1971 John Wiley & Sons.

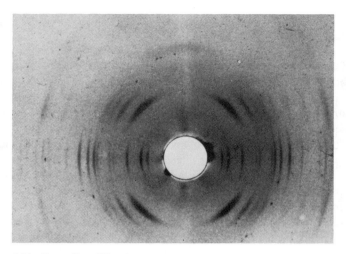

Figure 2.32 X-ray fiber diffraction pattern of poly(dichlorophosphazene). The layer-line separation distance corresponds to a polymer fiber axis repeating distance of 4.92 Å. Reprinted with permission from H. R. Allcock, *Phosphorous-Nitrogen Compounds;* Academic Press: New York, 1972.

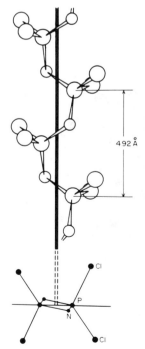

Figure 2.33 Distorted *cis-trans* planar conformation of poly(dichlorophosphazene). Reprinted with permission from E. Giglio, F. Pompa, A. Ripamonti, *J. Polym. Sci.* **1962,** *59*, 293, p. 299. Copyright © 1962 John Wiley & Sons.

thus, also about the fiber axis. Because of this disorder, a vertically mounted fiber will give cones of scattered radiation above and below the incident beam. Cutting such a pair of cones parallel to their common axis gives the observed, symmetrical pair of arcs in the scattering pattern. It is the separation between a pair of arcs that gives the crystallographic repeat distance mentioned earlier in this section. Models such as that shown in Figure 2.33[19, 20] are then postulated as possible structures for the polymer.

2.7 TRANSITIONS

2.7.1 Definitions

The simplest definition of a *first-order transition* is one in which heat flows into or out of the material with no change in temperature. Examples are melting and boiling and their reversals, crystallization, and condensation.

However, it is useful to provide a thermodynamic definition of a first-order transition. Specifically, it is one in which there is a discontinuity in a first derivative of the Gibbs free energy. The advantage of this definition is the guidance it provides for the experimental study of phase transitions. A useful expression for the free energy in this regard is

$$dG = Vdp - SdT \qquad \text{(Eq. 2.28)}$$

where the symbols have their usual significance. (G is the Gibbs free energy, V the volume, p the pressure, S the entropy, and T the absolute temperature.) It is called one of the fundamental equations of thermodynamics since it combines the first and second laws.

The constant-temperature first derivative with respect to pressure obtained from eq. (2.28) is

$$\left(\frac{\partial G}{\partial p}\right)_T = V \qquad \text{(Eq. 2.29)}$$

and is, thus, equal to the volume of the system. Therefore, studies of volume changes, called *dilatometry*, should be useful for characterizing first-order transitions. The constant-pressure first derivative with respect to temperature is given by

$$\left(\frac{\partial G}{\partial T}\right)_p = -S \qquad \text{(Eq. 2.30)}$$

and is equal to minus the entropy. Thus, calorimetric measurements should also be useful.

Second-order transitions are treated similarly. Properties that are second derivatives of the free energy show a discontinuity with changes in temperature. For example, the second derivative with respect to pressure and temperature is

$$\left[\frac{\partial(\partial G/\partial p)_T}{\partial T}\right]_p = \left(\frac{\partial V}{\partial T}\right)_p = V_{av}\alpha \qquad (Eq.2.31)$$

where α is the thermal expansion coefficient, $\alpha = (1/V_{av})(\partial V/\partial T)_p$. It is this coefficient, rather than the volume itself, that will show a discontinuity in a dilatometric study. The similarly obtained equation

$$\left[\frac{\partial(\partial G/\partial T)_p}{\partial T}\right]_p = -\left(\frac{\partial S}{\partial T}\right)_p = -\left(\frac{dq_{rev}}{TdT}\right)_p = -\frac{C_p}{T} \qquad (Eq.\ 2.32)$$

shows that the analogous calorimetric quantity is the heat capacity C_p.

In second-order transitions, no latent heat or discontinuous changes occur in volume or entropy.

2.7.2 Illustrative Representations

These conclusions are summarized graphically in Figure 2.34. The first-order transition of importance in polymer chemistry is the *melting point* T_m, and

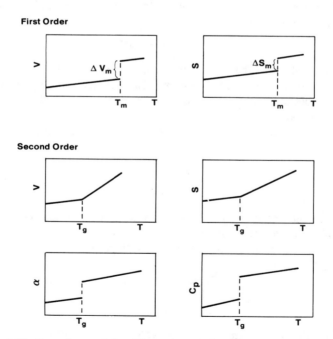

Figure 2.34 Some characteristics of first-order and second-order transitions. Reprinted with permission from J. E. Mark, *Physical Chemistry of Polymers*, ACS Audio Course C-89, American Chemical Society, Washington, DC, 1986. Copyright 1986, American Chemical Society.

the top two sketches show the discontinuities ΔV_m and ΔS_m in the volume and entropy, respectively.

The *glass transition temperature* T_g is thought by some to be a second-order transition, so some data relevant to it are shown in the middle two sketches. The volume and entropy merely show a change in slope at T_g. The relevant second derivatives are shown in the bottom pairs of sketches, with the expected discontinuities in the thermal expansion coefficient and heat capacity.

2.7.3 Dilatometric Results

Some typical, combined dilatometric results for a polymer are shown schematically in Figure 2.35. Since the volume is plotted against the temperature, there is a change in slope at T_g, where the polymer changes from a glass to a rubbery material, and nearly a discontinuity at T_m, where the crystalline regions melt to the liquid state. The melting point is generally unsharp, as shown, because of the presence of crystallites that are small or imperfectly formed.

2.7.4 Calorimetric Results

2.7.4.1 Bomb calorimetry.   Analogous calorimetric results obtained in a bomb calorimeter are shown in Figure 2.36. It is a very laborious process to obtain actual values of the specific heat as a function of temperature. At each temperature, a known amount of heat must be added to the polymer sample, in a sealed bomb calorimeter, and the temperature increase must be measured carefully. Because this is so time consuming, simpler techniques have been developed, and some are described in the following sections.

The polymer characterized in this figure was "quenched" so that it was

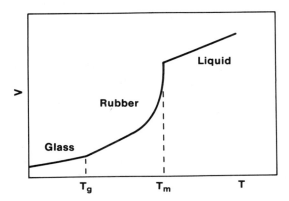

Figure 2.35 Typical dilatometric results for a partially crystalline polymer. Reprinted with permission from J. E. Mark, *Physical Chemistry of Polymers*, ACS Audio Course C-89, American Chemical Society, Washington, DC, 1986. Copyright 1986, American Chemical Society.

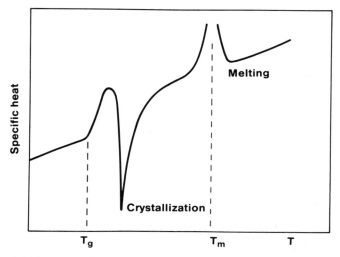

Figure 2.36 Calorimetry results obtained in a bomb calorimeter. Reprinted with permission from J. E. Mark, *Physical Chemistry of Polymers*, ACS Audio Course C-89, American Chemical Society, Washington, DC, 1986. Copyright 1986, American Chemical Society.

completely amorphous and glassy at the low-temperature start of the experiment. As the temperature is raised, the specific heat increases because of the more pronounced vibrations of the atoms in the polymer chains. (In atomic and molecular terms, the specific heat is a measure of the number of modes a system has for taking up energy and the efficiency with which this energy can be absorbed.)

At the glass transition temperature, the specific heat increases abruptly because the previously frozen large-scale molecular motions are now available to the chains for the uptake of thermal energy. The subsequent sharp downturn is due to crystallization of the chains, which is now possible because of their increased mobility. Latent heat of crystallization is being given off, meaning that less heat has to be added to bring about a given change in temperature. This has the effect of greatly decreasing the apparent heat capacity. After crystallization is complete, the specific heat increases slightly until melting starts. The apparent specific heat then increases dramatically because· a portion of the heat being added to the polymer goes into melting some of the crystallites. After melting is complete, the specific heat again decreases to a value typical of amorphous polymers.

2.7.4.2 Differential thermal analysis.   One of the simpler ways to obtain such information is called *differential thermal analysis (DTA)*, and a typical apparatus is described in Figure 2.37. Basically, the polymer sample P and an inert reference material R are heated from the same source. Thermocouples measure the temperature of the polymer and that of the reference, and the temperature difference $\Delta T = T_P - T_R$ is then plotted as a function of the temperature of the polymer.

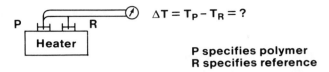

$$\Delta T = T_P - T_R = ?$$

P specifies polymer
R specifies reference

Figure 2.37 Differential thermal analysis equipment. Reprinted with permission from J. E. Mark, *Physical Chemistry of Polymers*, ACS Audio Course C-89, American Chemical Society, Washington, DC, 1986. Copyright 1986, American Chemical Society.

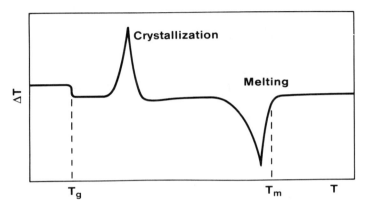

Figure 2.38 Typical DTA results. Reprinted with permission from J. E. Mark, *Physical Chemistry of Polymers*, ACS Audio Course C-89, American Chemical Society, Washington, DC, 1986. Copyright 1986, American Chemical Society.

Typical results are shown in Figure 2.38. As the temperature is increased, ΔT stays relatively constant until the glass transition temperature is reached. Then ΔT shifts downward; the temperature of the polymer has decreased because its specific heat has increased from the freeing of some of the molecular motions. The subsequent sharp upward peak in ΔT is due to crystallization, and the downward peak is a consequence of subsequent melting.

2.7.4.3 Differential scanning calorimetry.
A similar, simplified technique called *differential scanning calorimetry (DSC)* is described in Figure 2.39. In this approach, there are *separate* heating elements for the polymer and the reference material. The currents Q flowing into the two heating elements are adjusted to keep the temperature difference ΔT at zero. Changes in the difference

Figure 2.39 Differential scanning calorimetry apparatus. Reprinted with permission from J. E. Mark, *Physical Chemistry of Polymers*, ACS Audio Course C-89, American Chemical Society, Washington, DC, 1986. Copyright 1986, American Chemical Society.

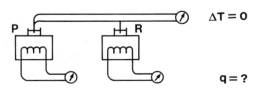

$$\Delta T = 0$$

$$q = ?$$

in current ΔQ are then used to locate the transitions. Values of ΔQ can also be used directly to obtain the heat of fusion of the polymer.

2.7.4.4 Some other techniques.   Another type of calorimetric technique is called *thermogravimetric analysis (TGA)*. It is the study of the weight of a material as a function of temperature. The method is used to evaluate the thermal stability from the weight loss caused by loss of volatile species. A final example, *thermomechanical analysis (TMA)*, focuses on mechanical properties such as modulus or impact strength as a function of temperature. Both types of analysis are essential for the evaluation of polymers that are to be used at high temperatures.

2.7.5 Crystallization and Melting

2.7.5.1 Thermodynamics.   For crystallization to occur, the Gibbs free energy of the system must decrease, as shown in the equation

$$\Delta G_{T,p} = \Delta H - T\Delta S < 0 \qquad \text{(Eq. 2.33)}$$

This leads to the inequality

$$T < \Delta H/\Delta S = T_m \qquad \text{(Eq. 2.34)}$$

which is the obvious requirement that the temperature has to be below the melting point for crystallization to occur. Thus, in the temperature range above T_m, crystallization is thermodynamically forbidden. Below T_m but not too close to T_g, crystallization can occur at a significant rate. However, near and below T_g, the chain motions are too sluggish for crystallization to occur. This can be very important if attempts are made to crystallize a polymer. Lowering the temperature too much introduces a kinetic complication: at temperatures close to T_g the chains should crystallize for thermodynamic reasons but cannot do so because of the very high viscosity of the medium. When this occurs, the polymer is said to be "quenched."

2.7.5.2 Annealing of polymers.   There is a rule of thumb that says that the temperature T_{max} for maximum rate of crystallization is given by

$$T_{max} = T_g + (\tfrac{2}{3})(T_m - T_g) \qquad \text{(Eq. 2.35)}$$

That is, measured from the T_g, it should be approximately two-thirds the temperature span between the glass transition temperature to the melting point. Predicted values are generally in sufficiently good agreement with theory to provide useful guidance for the development of annealing procedures carried out to develop crystallinity in a polymer.

2.7.5.3 Strain-induced crystallization.   The effect of stretching on the melting point of a polymer is shown in Figures 2.40 and 2.41. The upper sketch in Figure 2.40 shows the usual melting process for chains that can collapse into

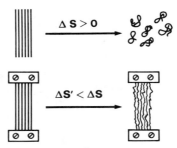

Figure 2.40 Effect of stretching on the melting point $T_m = \Delta H_m / \Delta S_m$. Reprinted with permission from J. E. Mark, *Physical Chemistry of Polymers,* ACS Audio Course C-89, American Chemical Society, Washington, DC, 1986. Copyright 1986, American Chemical Society.

random-coil configurations. As illustrated in the lower sketch, keeping the chains stretched by mechanical tension prevents their collapse into random coils on melting. This decreases their entropy of fusion and therefore increases their melting point. In a sense, the strain "induces" the crystallization by elevating the melting point. The force acts as a mechanical equivalent of the structural changes already mentioned, where the configurational entropy of a polymer was also decreased, but by the introduction of stiffening groups into the chain backbone.

The melting points of two polymers are shown as a function of elongation in Figure 2.41. An elastomer, such as "inorganic rubber," has a normal melting point T_m below room temperature. However, stretching it increases its melting point to above room temperature. The polymer is now supercooled, and crystal-

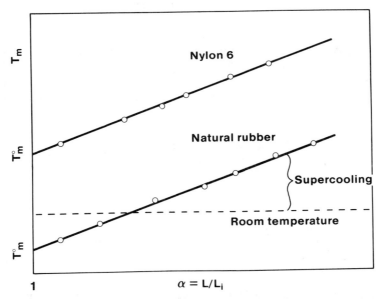

Figure 2.41 Effect of elongation α on high-melting fibers and low-melting elastomers. Reprinted with permission from J. E. Mark, *Physical Chemistry of Polymers*, ACS Audio Course C-89, American Chemical Society, Washington, DC, 1986. Copyright 1986, American Chemical Society.

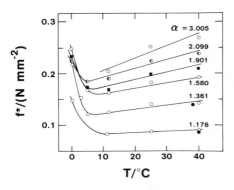

Figure 2.42 Stress-temperature data on "inorganic rubber," in which the upturns in the stress at low temperatures illustrate the reinforcing effects of strain-induced crystallization. Reprinted with permission from *Eur. Polym. J.* **1987**, *23*, Y.-H. Hsu and J. E. Mark, 829, p. 831. Copyright 1987, Pergamon Press PLC.

lization can occur. This can be very important since crystallites reinforce an elastomer, and thereby improve its mechanical properties. However, following removal of the stretching force, the melting point falls to its original value and the crystallites melt. The stress-temperature results for "inorganic rubber" networks presented in Figure 2.42 illustrate this type of reinforcement.[21]

A similar elevation of the melting point occurs for a polymer that has its normal melting point already above room temperature. An example would be a fibrous material such as one of the nylons. Stretching such a polymer can induce additional crystallization (and better orient the crystallites), but the crystallization is not lost on removal of the stretching force. This is also illustrated in Figure 2.41.

2.7.5.4 Polymer-diluent systems.

The addition of a diluent to a crystalline polymer depresses its melting point, as is shown schematically in Figure 2.43. The upper sketch again shows the standard reference case. In the lower sketch, solvent molecules are available to mix with the polymer chains once they

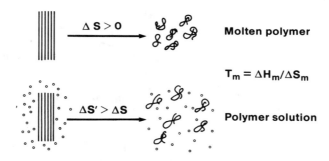

Figure 2.43 Mechanism for the solvent-induced depression of the melting point. Reprinted with permission from J. E. Mark, *Physical Chemistry of Polymers*, ACS Audio Course C-89, American Chemical Society, Washington, DC, 1986. Copyright 1986, American Chemical Society.

separate from the crystalline lattice. The final state is now a polymer solution, instead of a molten polymer. This additional disordering greatly increases the entropy change for the process and therefore decreases the melting point, frequently to the extent of 40 to 50°C.

Application of the Flory-Huggins theory[1-3] to the melting point depression gives the relationship

$$\left(\frac{1/T_m - 1/T_m^o}{v_1}\right) = \left(\frac{R}{\Delta H_m}\right)\left(\frac{V_2}{V_1}\right) - \left(\frac{R}{\Delta H_m}\right)\left(\frac{V_2}{V_1}\right)\chi_1 v_1 \quad \text{(Eq. 2.36)}$$

where the T's are melting points, with the zero superscript referring to the undiluted polymer, v_1 the volume fraction of solvent, and the V's molar volumes. As already mentioned, the temperatures and the heat of fusion ΔH_m do not depend on the amount of crystalline material present, and therefore pertain to the 100% crystalline material.

The entropy of fusion can be calculated from the equation

$$\Delta S_m = \frac{\Delta H_m}{T_m^o} \quad \text{(Eq. 2.37)}$$

and is frequently used as a direct measure of the flexibility of a chain.

Melting point depression data plotted in accordance with eq. (2.37) are shown schematically in Figure 2.44. The intercept gives the heat of fusion, and the slope gives the thermodynamic interaction parameter χ_1.

The melting point depression method can also be used to estimate melting points of polymers that degrade before they melt. The depressed melting points are simply plotted against the volume fraction v_1 of solvent in the solution and are extrapolated to the point where v_1 equals zero. The method, illustrated in Figure 2.45, has been used for a variety of polymers, particularly for cellulosic materials.

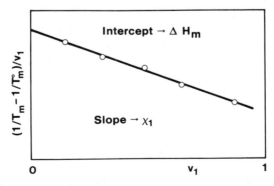

Figure 2.44 Typical melting point depression data. Reprinted with permission from J. E. Mark, *Physical Chemistry of Polymers*, ACS Audio Course C-89, American Chemical Society, Washington, DC, 1986. Copyright 1986, American Chemical Society.

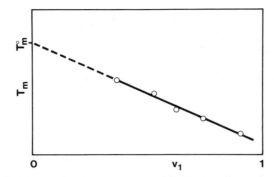

Figure 2.45 Extrapolation of melting point depression data to estimate the melting point of a polymer that degrades before it melts. Reprinted with permission from J. E. Mark, *Physical Chemistry of Polymers*, ACS Audio Course C-89, American Chemical Society, Washington, DC, 1986. Copyright 1986, American Chemical Society.

2.8 SPECTROSCOPY

The discussion of spectroscopic studies of polymers is relatively brief, in part because most of the uses are very similar to well-known, small-molecule applications.

2.8.1 Infrared and Ultraviolet Spectroscopy

The following applications are typical of those important in the field of polymers. The identification of unknown polymers proceeds through an analysis to identify various functional groups, as does the estimation of the ratios of co-monomers, end groups, unsaturation, and impurities. Studies of the chemical reactions of polymers are based on either the disappearance of one type of group or the appearance of another. An example would be the appearance of carbonyl absorption bands in an oxidized hydrocarbon polymer such as polyethylene. Stereochemical structure can also be studied by infrared (IR) or ultraviolet (UV) spectroscopy, although this is normally accomplished by nuclear magnetic resonance (NMR) spectroscopy. Finally, IR spectroscopy can be used to determine both the degree of crystallinity and the degree of chain orientation in a polymer.

2.8.2 NMR Spectroscopy

Illustrative applications include the determination of stereochemical structure and conformational preferences. Another application is the determination of chemical composition and chemical sequence distributions in copolymers. A final example is the study of relaxation processes and molecular motions in general, including the determination of transition temperatures from changes in resonance line widths.

2.8.3 Electron Paramagnetic Resonance Spectroscopy

Electron paramagnetic resonance (EPR) spectroscopy is also frequently used in polymer science. Applications involve the detection of free radicals in cross-linking, high-energy irradiation, photochemical degradation and oxidation, and mechanical fracture of polymer chains.

2.9 MECHANICAL PROPERTIES

This section contains two primary topics. The first is polymer rheology, which is concerned with how polymeric materials flow when they are placed under stress. Also of interest are the mechanical properties of polymers, particularly their elasticity. The combination of viscous effects with elastic phenomena is called viscoelasticity.

2.9.1 Elasticity

2.9.1.1 Continuous extension.   Figure 2.46 shows a stress-strain curve obtained for a typical partially crystalline, uncross-linked polymer, during elongation in the vicinity of room temperature. The data are typically taken on an Instron Tester, and no attempt is made to reach mechanical equilibrium. In this experiment, a strip of the polymer is mounted between two clamps which move apart at a constant rate. A typical rate would be 100% per minute which means that after 1 minute, a sample originally 1 centimeter long would increase to 2 centimeters. The *strain*, or relative length of the sample, is given the symbol γ and is plotted along the abscissa. The *stress* is given the symbol s and is taken to

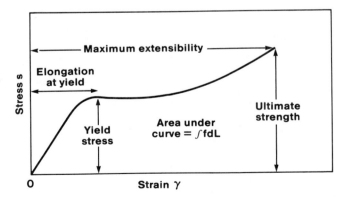

Figure 2.46 Typical nonequilibrium stress-strain curve in elongation. Adapted with permission from F. W. Billmeyer, Jr., *Textbook of Polymer Science*, 2nd ed.; Wiley-Interscience: New York, 1971. Copyright © 1971 John Wiley & Sons.

be the force per unit undeformed cross-sectional area (for example, Newtons mm^{-2}). Stress is shown plotted along the ordinate.

The ratio of any value of the stress to the corresponding value of the strain is called the *modulus*. It is a direct measure of the *hardness* of the material—how difficult it is to deform it. Only in the low-strain region of the curve are the stress and strain directly proportional to one another. Therefore, only in this region is the modulus a constant.

When the strain reaches a value called the "elongation at yield," the curve becomes very nonlinear. The stress at this point, the "yield stress," is sufficiently high to move the crystallites around in their very viscous surroundings and to cause them to melt and then recrystallize in new orientations that partly relieve the stress.

The curve then increases monotonically until rupture occurs. The strain at this point is called the "maximum extensibility" and the stress the "ultimate strength." The area under the curve up to the rupture point is also of interest. It corresponds to the integral of *fdL* and is therefore the work or energy required for rupture. It is the standard measure of toughness. The larger the area, the tougher the material.

Curves for typical polymeric materials are shown in Figure 2.47. The first

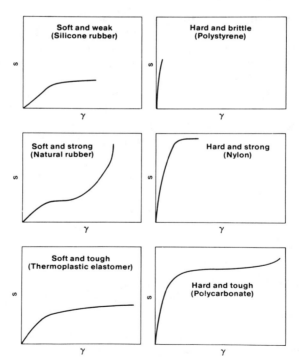

Figure 2.47   Nonequilibrium stress-strain curves for some types of polymers. Adapted with permission from F. W. Billmeyer, Jr., *Textbook of Polymer Science*, 2nd ed.; Wiley-Interscience: New York, 1971. Copyright © 1971 John Wiley & Sons.

is for a soft and weak material, such as an unfilled silicone rubber. "Soft" refers to the fact that the initial slope is small, which means a low value of the modulus. "Weak" refers to the low value of the ultimate strength. One doesn't have to be very strong to pull it apart.

The next type is hard and brittle, for example polystyrene. "Hard" refers to the fact that the initial slope and modulus are large, and "brittle" is another way of saying the maximum extensibility is very small.

These first two materials do not have large areas under their curves, and are therefore not very tough.

The curve shown by a soft and strong material like natural rubber is shown next. The small initial slope and modulus show the material to be soft. At higher elongations, however, strain-induced crystallization occurs and this reinforces the elastomer. As a result its ultimate strength is large and it is therefore quite strong. In other words, one has to be strong to pull it apart.

The fourth example chosen is hardness and strength, characteristics shown by many of the nylons. The hardness is demonstrated by the high initial slope and modulus and the strength by the large value of the ultimate strength.

Soft and tough might refer to one of the new thermoplastic elastomers. The initial modulus could be quite small, but high extensibility would give a large area under the curve. The high degree of toughness means it would take a lot of work or energy to cause it to rupture.

Polymers that are both hard and tough are exemplified by some of the polycarbonates. The high values of the ultimate strength and maximum extensibility combine to give unusually large areas under the curve, and thus considerable toughness.

2.9.1.2 Hookean elasticity.

The simplest type of elasticity, Hookean elasticity, is defined by the equation

$$\frac{s}{\gamma} = \text{constant} \equiv G \qquad \text{(Eq. 2.38)}$$

A polymer is said to exhibit Hookean elasticity if the ratio of stress to strain is a constant. In this case, the ratio is called *Young's modulus*, and is given the symbol G.

2.9.1.3 Rubberlike elasticity.

A good operational definition of rubberlike elasticity[22] is high deformability with essentially complete recoverability. The high deformability can be remarkably high, with some rubbery materials extending up to 15 times their original lengths.

In the simplest molecular theory now available, there are two major assumptions. The first is that *inter*molecular interactions are unimportant; that is, they do not change with deformation. The second is that the deformation is "affine," which means that the molecular deformation is the same as the macroscopic deformation. More precisely, the assumption is that the cross-link positions vary in a simple linear manner with the macroscopic dimensions.

The results for elongation given by the simplest theory are embodied in what is called an *equation of state*, that is, an equation that interrelates the properties required to characterize the system. It is given by

$$f^* \equiv \frac{f}{A^*} = \nu k T(\alpha - \alpha^{-2}) \qquad \text{(Eq. 2.39)}$$

where f^* is the nominal stress, f the equilibrium force, A^* the undeformed cross-sectional area, ν the number density of network chains, k the Boltzmann constant, T the absolute temperature, and $\alpha = L/L_i$ the elongation or relative length of the sample.

This equation is very similar to the ideal gas law, given by

$$p = NkT\left(\frac{1}{V}\right) \qquad \text{(Eq. 2.40)}$$

where p is the pressure of the gas, N the number of gas molecules, and V the volume of the sample. In the case of the gas, the deformation, $1/V$, is simpler than the strain function $(\alpha - \alpha^{-2})$. The α^{-2} term comes from the fact that elongation occurs at essentially constant volume, so that the width and thickness of the sample must decrease proportionally as the length increases. This causes some of the chains to be compressed instead of elongated, and the compression is the origin of the subtractive α^{-2} term.

In general, the force is not directly proportional to the absolute temperature, as specified by eq. (2.39). A more refined molecular theory permits the use of precise force-temperature or thermoelastic data to determine the energy change ΔE resulting from the deformation.

Theory also predicts the reduced stress or modulus $[f^*]$, defined as the ratio of the nominal stress to the strain function $(\alpha - \alpha^{-2})$, to be independent of the elongation α. Experimentally, however, the modulus is found to change with α, generally decreasing linearly with decreasing reciprocal elongation. For this reason, stress-strain isotherms are frequently represented by the semiempirical Mooney-Rivlin relationship[22]

$$[f^*] = 2C_1 + 2C_2\alpha^{-1} \qquad \text{(Eq. 2.41)}$$

where $2C_1$ and $2C_2$ are constants. In the most recent theories, the constant $2C_2$ is considered to be due to the deformation becoming increasingly nonaffine as the elongation increases.

2.9.2 Viscosity

The viscosity η of a fluid may be defined by the equation

$$s = \eta \frac{d\gamma}{dt} = \eta\dot{\gamma} \qquad \text{(Eq. 2.42)}$$

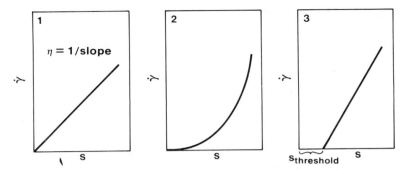

Figure 2.48 Newtonian (1), non-Newtonian (2), and retarded flow (3), where $\dot{\gamma}$ is the rate of flow, s the stress, and η the viscosity. Adapted with permission from F. W. Billmeyer, Jr., *Textbook of Polymer Science*, 2nd ed.; Wiley-Interscience, New York: 1971. Copyright © 1971 John Wiley & Sons.

where s is the force per unit area required to impart a rate of deformation or rate of shear $\dot{\gamma} = d\gamma/dt$ to the fluid.

In the case of a fluid, the deformation γ itself is not important since a fluid (by definition) will flow to take the shape of the vessel that contains it. Only the time derivative of γ is important, and this quantity, also called the rate of flow or the rate of shear, is represented by the symbol $\dot{\gamma}$.

If a shearing stress s gives rise to a rate of deformation $\dot{\gamma}$, then the proportionality factor is called the viscosity η. If this factor is a constant then the flow is said to be "Newtonian"; if not, it is non-Newtonian.

These two types of viscous flow are illustrated in the first two sketches of Figure 2.48, where the reciprocal of the slope of the curve gives the viscosity. In the first sketch, the slope and viscosity are constant, so the fluid is Newtonian. In the second sketch the viscosity starts out high (low slope) and then decreases significantly as the stress increases. This type of behavior is shown by liquids that have a temporary gel-like structure. "Nondrip" paints are in this category. The shearing by the paintbrush breaks up the structure long enough to get the paint onto the surface. Once the brush is removed, the network sets again and the paint can no longer flow. Retarded flow, illustrated in the third sketch, is just an extreme case of type 2 behavior. The viscosity is infinite (zero slope) until a threshold value of the stress is reached. The flow that occurs after this point can be either Newtonian, as illustrated, or non-Newtonian.

2.9.3 Viscoelasticity

With the information given, it becomes possible to combine viscous characteristics with elastic characteristics to describe the viscoelasticity of polymeric materials. The two simplest ways of combining these features are shown in Figure 2.49, where a spring having a modulus G models the elastic response. The viscous response is modeled by what is called a "dashpot." It consists of a piston moving in a cylinder containing a viscous fluid of viscosity η. If a downward force is

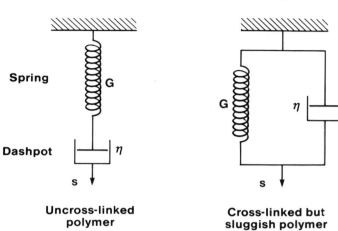

Figure 2.49 Two viscoelasticity models. Adapted with permission from F. W. Bill-meyer, Jr., *Textbook of Polymer Science*, 2nd ed.; Wiley-Interscience: New York, 1971. Copyright © 1971 John Wiley & Sons.

applied to the cylinder, more fluid flows into it, whereas a upward force causes some of the fluid to flow out. The flow is retarded because of the high viscosity, and this element thus models the retarded movement and flow of polymer chains.

The combination of spring and dashpot in series is called the Maxwell model, and was in fact first investigated by the same Maxwell famous for his work on gases and molecular statistics. It is used to model the viscoelastic behavior of uncross-linked polymers. The spring is used to describe the recoverability of the chains that are elongated, and the dashpot the permanent deformation or creep resulting from the uncross-linked chains irreversibly sliding by one another.

The parallel arrangement, also shown in Figure 2.49, is called the Voigt model. It is used to model the behavior of a cross-linked but sluggish polymer, such as one of the polyacrylates. Since the spring and dashpot have to move in parallel, both the deformation and the recoverability are retarded.

The general differential equation which describes the Maxwell model is

$$\dot{\gamma} = \frac{s}{\eta} + \left(\frac{1}{G}\right)\left(\frac{ds}{dt}\right) \qquad \text{(Eq. 2.43)}$$

and is obtained simply from the defining relationships for the viscosity and modulus. In physical terms, it says that the flow of the polymer has two, simply additive parts. The first is from the movement of the chains, and the second is from the effect of a changing stress on the elastic part of the deformation. This equation does not have a general solution, but it can be solved for the two situations of interest to polymer chemists, namely, stress relaxation and creep.

Experiment

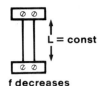

f decreases

Figure 2.50 The stress-relaxation experiment. Reprinted with permission from J. E. Mark, *Physical Chemistry of Polymers*, ACS Audio Course C-89, American Chemical Society, Washington, DC, 1986. Copyright 1986, American Chemical Society.

The stress-relaxation experiment, described in Figure 2.50, involves stretching the uncross-linked sample to a length L and then measuring the force as it decreases at this constant length as a consequence of the polymer chains undergoing slippage past each other. Because the length is constant, $\dot\gamma$ is zero, and this simplifies eq. (2.43) in that the left-hand side disappears. The solution to this equation is

$$s = s_o e^{-t/\tau} \qquad \text{(Eq. 2.44)}$$

where s_o is the stress at the initial $t = 0$ and $\tau = \eta/G$ is the relaxation time (the time required for the deformation to decrease to $1/e$ of its original value).

A plot of eq. (2.44), including the definition of the relaxation time, is shown in Figure 2.51. As can be seen, the stress eventually declines to zero if the chains have enough time to undergo the required amount of slippage.

The application of the same Maxwell model to creep data is described in Figures 2.52 and 2.53. As shown in Figure 2.52, the force on the sample is fixed by suspending a constant weight at the end of the sample. The length of the sample is then measured as a function of time. Since the force is constant, $ds/dt = 0$ and

Results

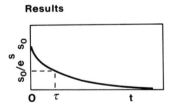

Figure 2.51 Typical stress-relaxation results, where τ is the relaxation time. Reprinted with permission from J. E. Mark, *Physical Chemistry of Polymers*, ACS Audio Course C-89, American Chemical Society, Washington, DC, 1986. Copyright 1986, American Chemical Society.

Experiment

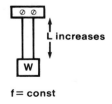

f = const

Figure 2.52 The creep experiment. Reprinted with permission from J. E. Mark, *Physical Chemistry of Polymers*, ACS Audio Course C-89, American Chemical Society, Washington, DC, 1986. Copyright 1986, American Chemical Society.

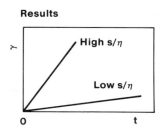

Results

Figure 2.53 Typical creep results either for two polymers having different viscosities η, or the same polymer under two different stresses s. Reprinted with permission from J. E. Mark, *Physical Chemistry of Polymers*, ACS Audio Course C-89, American Chemical Society, Washington, DC, 1986. Copyright 1986, American Chemical Society.

the general differential equation again simplifies to a solvable form, namely,

$$\dot{\gamma} = \frac{d\gamma}{dt} = \frac{s}{\eta} \tag{Eq. 2.45}$$

Its solution is

$$\gamma = \left(\frac{s}{\eta}\right) t \tag{Eq. 2.46}$$

and states that the deformation increases linearly with time, without limit, and that the proportionality constant is the ratio of stress to viscosity. Plots of eq. (2.46) are shown schematically in Figure 2.53. A high slope means that the sample pulls apart rapidly, and this takes place when s/η is large. Thus, this occurs when the stress on the sample is high, or when the viscosity of the polymer is low. The process is slower when s/η is low, that is, when the stress is small or the viscosity of the sample is high.

The use of the Voigt model to characterize cross-linked but sluggish polymers is illustrated in Figure 2.54. The corresponding equation in this case is

$$s = G\gamma + \eta\dot{\gamma} \tag{Eq. 2.47}$$

which arises because the stress is partitioned between the two parallel arms of the model. The first term comes from the definition of the modulus of the spring, and the second from the definition of the viscosity of the polymer. This differential equation has a general solution, which is given by

$$\gamma = \left(\frac{s}{G}\right)(1 - e^{-t/\tau}) \tag{Eq. 2.48}$$

when $t = 0$, the exponential is unity and the deformation γ is zero. When t goes to infinity, the exponential is zero, and the deformation reaches its maximum value, specifically the value s/G dictated by the values of the stress and the modulus of the spring.

Removal of the stress makes the left-hand side of eq. (2.48) zero. This simplified equation can be solved to show how the deformation decreases back to zero. The solution is given by

$$\gamma = \gamma_o e^{-t/\tau} \tag{Eq. 2.49}$$

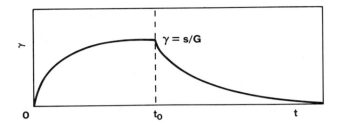

Figure 2.54 Stress-time behavior of a cross-linked but sluggish elastomer. Reprinted with permission from J. E. Mark, *Physical Chemistry of Polymers*, ACS Audio Course C-89, American Chemical Society, Washington, DC, 1986. Copyright 1986, American Chemical Society.

where γ_o is the deformation immediately after removal of the stress s. The sketch in Figure 2.54 shows the asymptotic increase in deformation given by eq. (2.48), followed by the asymptotic decrease given by eq. (2.49).

In this general approach to viscoelasticity, appropriate models are constructed for the interpretation of the stress-strain-time behavior of a polymer. Then, values of Young's modulus G of the elastic elements and the viscosities η of the viscous elements are used to characterize and predict the general behavior of the material.

Such analyses are essential in the design of polymeric materials that have predictable mechanical properties.

2.10 REFERENCES

1. General information on the subject of this chapter can be found in any of the standard books on polymers. These include
   (a) Flory, P. J. *Principles of Polymer Chemistry*, Cornell University: Ithaca, NY, 1953.
   (b) Allcock, H. R.; Lampe, F. W. *Contemporary Polymer Chemistry*, 2nd ed.; Prentice Hall: Englewood Cliffs, NJ, 1990.
   (c) Hiemenz, P. C. *Polymer Chemistry. The Basic Concepts*; Marcel Dekker: New York, 1984.
   (d) Billmeyer, Jr., F. W. *Textbook of Polymer Science*, 3rd ed.; Wiley-Interscience: New York, 1984.
   (e) Sperling, L. H. *Introduction to Physical Polymer Science*; Wiley-Interscience: New York, 1986.
   (f) *Comprehensive Polymer Science*; Allen, G., Ed.; Pergamon: Oxford, 1989.
   (g) Munk, P. *Introduction to Macromolecular Science*; Wiley-Interscience: New York, 1989.
2. More detailed information of this type can be found in the following books, which more specifically address characterization techniques:
   (a) *Newer Methods of Polymer Characterization*; Ke, B., Ed.; Interscience: New York, 1964.

(b) *Characterization of Macromolecular Structure*; McIntyre, D., Ed.; National Academy of Sciences: Washington, DC, 1968.

(c) McCaffery, E. M. *Laboratory Preparation for Macromolecular Chemistry*; McGraw-Hill: New York, 1970.

(d) Collins, E. A.; Bares, J.; Billmeyer, Jr., F. W. *Experiments in Polymer Science*; Wiley-Interscience: New York, 1973.

(e) *Determination of Molecular Weight*; Cooper, A. R., Ed.; Chemical Analysis; Wiley-Interscience: New York, 1989; Vol. 103.

3. Mark, J. E. *Physical Chemistry of Polymers*; Manual, Audio Course; American Chemical Society: Washington, DC, 1987.

4. Hagnauer, G. L. *J. Macro. Sci.-Chem.* **1981,** *A16*, 385.

5. West, R. *J. Organometallic Chem.* **1986,** *300*, 327.

6. *Inorganic and Organometallic Polymers*; Zeldin, M.; Wynne, K. J., Allcock, H. R., ACS Symposium Series; American Chemical Society; Washington, DC. 1988; Vol. 360.

7. *Thermoplastic Elastomers. A Comprehensive Review*; Legge, N. R.; Holden, G.; Schroeder, H. E., Eds.; Hanser Publishers (Oxford): New York, 1987.

8. Ibemesi, J.; Gvozdic, N.; Kuemin, M.; Lynch, M. J.; Meier, D. J. *Preprints, Div. of Polym. Chem., Inc.* **1985,** *26(2)*, 18; Ibemesi, J.; Gvozdic, N.; Kuemin, M., Tarshiani, Y.; Meier, D. J. in *Polymer Based Molecular Composites*, Schaefer, D. W.; Mark, J. E., Eds., Materials Research Society: Pittsburgh, PA, 1990.

9. Flory, P. J. *Statistical Mechanics of Chain Molecules*; Wiley-Interscience: New York, 1969.

10. Hopfinger, A. J. *Conformational Properties of Macromolecules*; Academic: New York, 1973.

11. Burkert, U.; Allinger, N. L.; *Molecular Mechanics*; ACS Monograph 177; American Chemical Society: Washington, DC, 1982.

12. Damewood, Jr., J. R.; West, R. *Macromolecules* **1985,** *18*, 159.

13. Welsh, W. J.; DeBolt, L.; Mark, J. E. *Macromolecules* **1986,** *19*, 2978.

14. Van Krevelen, D. W. *Properties of Polymers. Their Estimation and Correlation with Chemical Structure*, 2nd ed.; Elsevier: Amsterdam, 1976.

15. Mandelkern, L. *Crystallization of Polymers*; Wiley-Interscience: New York, 1964.

16. Wunderlich, B. *Macromolecular Physics*; Academic: New York, 1973.

17. Tadokoro, H. *Structure of Crystalline Polymers*; Wiley-Interscience: New York, 1979.

18. Billmeyer, Jr., F. W. *Textbook of Polymer Science*, 2nd ed.; Wiley-Interscience: New York, 1971.

19. Allcock, H. R. *Phosphorous-Nitrogen Compounds*; Academic: New York, 1972.

20. Giglio, E.; Pompa, F.; Ripamonti, A. *J. Polym. Sci.* **1962,** *59*, 293.

21. Hsu, Y.-H.; Mark, J. E. *Eur. Polym. J.* **1987,** *23*, 829.

22. Mark, J. E.; Erman, B. *Rubberlike Elasticity. A Molecular Primer*; Wiley-Interscience: New York, 1988.

3

Polyphosphazenes

3.1 INTRODUCTION

The polyphosphazenes comprise by far the largest class of inorganic macromolecules. At least 300 different polymers of this type have been synthesized, with a range of physical and chemical properties that rivals those known hitherto only for synthetic organic macromolecules.

Polyphosphazenes have the general molecular structure shown in Structure 3.1. The polymer backbone consists of alternating phosphorus and nitrogen atoms,

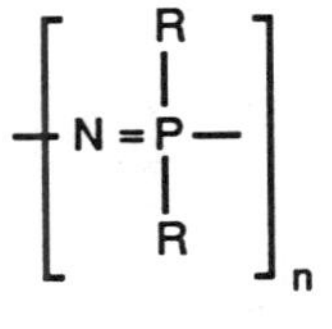

$$\left[\begin{array}{c} R \\ | \\ -N = P- \\ | \\ R \end{array} \right]_n$$

with two side groups, R, being attached to each phosphorus. The side groups may be organic, organometallic, or inorganic units. Each macromolecule typically contains 15,000 or more repeating units linked end to end, which means that the molecular weights are in the range of 2 million to 10 million. The bonding structure in the backbone is formally represented as a series of alternating single and double bonds. However, this formulation is misleading. Structural measurements suggest that all the bonds along the chain are equal or nearly equal in length, but without the extensive conjugation found in organic polyunsaturated molecules. This anomaly will be discussed later.

Perhaps the most important feature of polyphosphazene chemistry is the method of synthesis that allows the side groups, R, to be varied over a very broad range. Different side groups generate different properties such that the characteristics may vary from those of elastomers to glasses, from water-soluble to hydrophobic polymers, from bioinert to bioactive materials, and from electrical insulators to conductors. This versatility underlies the technological developments that have occurred in this area. For example, certain variants of Structure 3.1 are solvent-resistant elastomers that are used in demanding engineering applications (Figure 3.1). Others are used as flame-resistant, heat-, electrical-, or sound-insulation materials (Figure 3.2). Yet other variations are under development as nonburning textile fibers (Figure 3.3) or as biomaterials, controlled drug delivery systems, hydrogels, and membranes.

In the rest of this chapter we will describe how this field developed, how polyphosphazenes are synthesized, how the system provides almost unprecedented opportunities for the design of new macromolecules, and how the molecular structure-property relationships have been developed to produce a wide range of advanced materials.

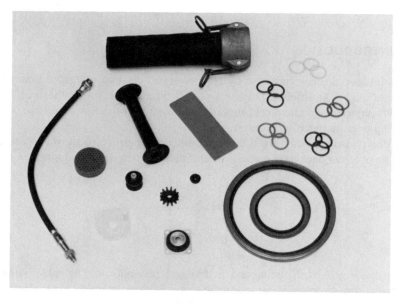

Figure 3.1 Polyphosphazene fluoroalkoxy elastomers of general formula, $[NP(OCH_2CF_3)(OCH_2(CF_2)_xCF_2H)]_n$, fabricated into fuel lines, O-rings, gaskets, and other hydrocarbon-resistant devices. (Courtesy of the Firestone Tire and Rubber Company and Ethyl Corporation.)

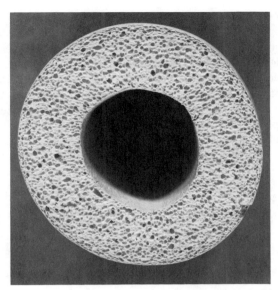

Figure 3.2   Cross section of an expanded foam rubber heat- and sound-insulation tube of a poly(aryloxyphosphazene). (Courtesy of Ethyl Corporation.)

Figure 3.3   Textile fibers of $[NP(OCH_2CF_3)_2]_n$, a nonburning polymer. (Courtesy of Ethyl Corporation.)

3.2 HISTORY

The beginnings of this field can be traced back more than 150 years to the observation by Wohler[1] and Rose[2] in 1834 that phosphorus pentachloride reacts with ammonia to yield a stable, white, crystalline solid. Only sporadic interest in this product apparently existed during the next 50 years. Papers by Gerhardt,[3] Laurent,[4] Gladstone and Holmes,[5] and Wichelhaus[6] addressed the question of the nature of this compound, and it was concluded that the formula was $(NPCl_2)_3$ (Structure 3.2). It is now known that the reaction is as shown in eq. (3.1).

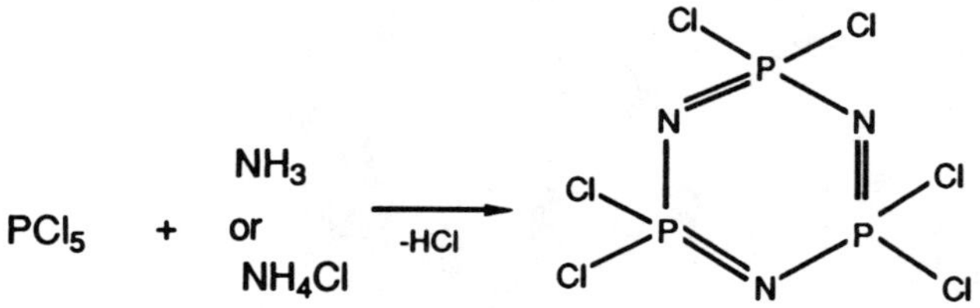

3.2

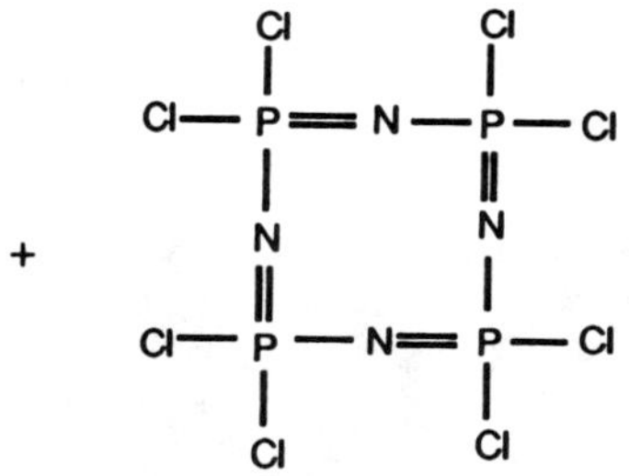

$$PCl_5 \;+\; \begin{array}{c} NH_3 \\ \text{or} \\ NH_4Cl \end{array} \xrightarrow[\text{-HCl}]{} \quad \textbf{3.2}$$

$$+ \quad \textbf{3.3} \quad + \; (NPCl_2)_5 \; \cdots \cdots \qquad \text{(Eq. 3.1)}$$

However, from the viewpoint of inorganic polymer chemistry, the principal early contributor was an American chemist, H. N. Stokes, who, at the turn of the century, first suggested the cyclic structure of 3.2, identified cyclic homologs such as 3.3 up to the species $(NPCl_2)_7$, and reported that chlorophosphazenes, when heated, were transformed into an elastomeric material known subsequently as "inorganic rubber."[7-10] Stokes also described how inorganic rubber decomposed to reform the cyclic compounds when heated to high temperatures under reduced pressure. Considering the laboratory conditions under which Stokes probably worked, it is remarkable that he achieved so much. All the compounds he worked with are sensitive to a moist atmosphere, in which they hydrolyze to ammonium phosphate and hydrochloric acid.

The reading of Stokes's papers (see Figure 3.4) nearly 100 years after their publication is an uncanny experience. He could not have known that "inorganic rubber" was a high polymer with thousands of repeating units linked end to end. Nor could he have foreseen the later development of the substitution chemistry of poly(dichlorophosphazene), because his material was insoluble in all solvents and was probably highly cross-linked. Nevertheless, one senses that he understood the uniqueness of the transformation of a molten small-molecule system to an insoluble, rubbery, elastomeric solid. This was a new phenomenon that would have to wait for another 70 years before it could be fully understood or utilized for the synthesis of a broad range of useful polymers.

During the next 40 years inorganic rubber was mentioned sporadically as a laboratory curiosity, but was largely ignored by the mainstream scientists of the day. The slow and controversial acceptance of Staudinger's idea[11] that long-chain macromolecules could exist, and the demonstration by Meyer and Mark[12] in the 1920s and 1930s that natural rubber is a linear macromolecule, stimulated a brief resurgence of interest in "inorganic rubber." X-ray diffraction experiments by Meyer, Lotmar, and Pankow[13] in 1936 strongly suggested that this material contained linear high polymeric chains with a repeating structure of the type shown in Structure 3.4. However, once again it was the insolubility of the polymer in all

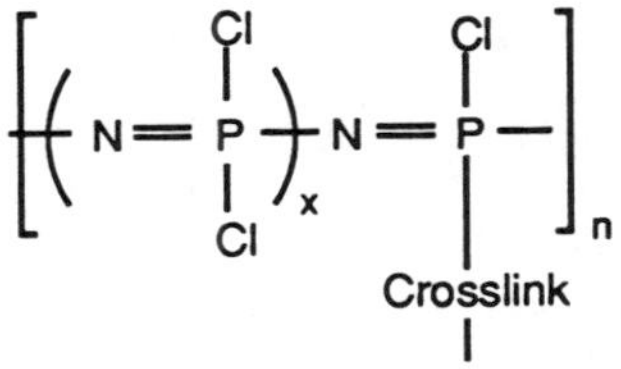

$$\left[\left(N = \underset{\underset{Cl}{|}}{\overset{\overset{Cl}{|}}{P}} \right)_x N = \underset{\underset{\underset{\underset{|}{Crosslink}}{|}}{|}}{\overset{\overset{Cl}{|}}{P}} - \right]_n$$

3.4

known solvents and its hydrolytic instability in the atmosphere that discouraged a more serious interest in this material.

This situation persisted until the mid-1960s when a series of three papers by Allcock, Kugel, and Valan[14-16] were published. The theme of these papers was as follows:

1. The principal defect of "inorganic rubber"—its hydrolytic instability—might be utilized as its principal advantage. The hydrolytic sensitivity implied a high reactivity of the P—Cl bonds, a characteristic that might be translated into reactions that could be used to replace chlorine atoms by hydrolytically stable organic groups. Thus, the halogenophosphazene high polymer might be used as a macromolecular reactive intermediate, providing access to a very broad range of stable organic derivative polymers.

2. Before this could be accomplished, a method had to be found to prepare poly(dichlorophosphazene) in a form that was soluble in organic solvents. It was known that Stokes's polymer swelled in organic solvents such as benzene, but it did not dissolve. This is a characteristic of a cross-linked polymer. Reactions carried out on this swollen polymer provided encouragement for the idea of chlorine replacement reactions, but the substitutions were never complete because of the insolubility. Thus, a critical need existed for a polymerization process that would yield an uncross-linked polymer. In fact, it was during a mechanistic study of the

ON THE CHLORONITRIDES OF PHOSPHORUS. (II).[1]

BY H. N. STOKES.

In a former article[2] I have shown that in addition to the phosphonitrilic chloride,[3] $P_3N_3Cl_6$, discovered by Liebig, there exists another, $P_4N_4Cl_8$, of similar properties, which is formed at the same time, but in smaller quantity. The opinion was expressed that these bodies belong to a series of polymers, $(PNCl_2)_n$, the existence of other members of which was indicated by the formation, in small amount, of a liquid of the same empirical composition.[4] The yield of this secondary product, only 2 per cent. of the theoretical or 1 per cent. of the pentachloride used, was too small to allow of its preparation in quantities large enough to admit of the isolation of its supposed constituents, but a fractional distillation of the few grams at my disposal showed that it contained crystalline substances of higher boiling-points than those of the two bodies thus far known.

The method of preparation then employed consisted in distilling phosphorus pentachloride with a large excess of ammonium chloride in a retort, at atmospheric pressure; it offered but little prospect of obtaining the higher members. The total yield of phosphonitrilic chloride was but 15 per cent. of the theoretical, most of the pentachloride being converted into "phospham" by the excess of ammonium chloride, while only those members could be obtained which distil unchanged at atmospheric pressure. Decreasing the amount of ammonium chloride resulted only in a loss of pentachloride by volatilization, without increasing the yield of the bodies sought after.

The following method has been found to give entirely satisfactory results; several new bodies have been obtained, and the simpler phosphonitrilic chlorides, at least, are now easily accessible substances: If equal molecular weights of phosphorus pentachloride and ammonium chloride be heated in a

1 Published by permission of the Director of the United States Geological Survey.
2 This JOURNAL, 17, 275 (1895): Ber. d. chem. Ges., 28, 437.
3 I propose in future to use the term *phosphorus chloronitride* to denote any body composed of phosphorus, nitrogen, and chlorine, the name *phosphonitrilic chloride* being reserved for chloronitrides belonging to the series $(PNCl_2)_n$.
4 This JOURNAL, 17, 277, 280, 290.

Figure 3.4 Two pages from a paper by H. N. Stokes in the *American Chemical Journal*, Vol. 19, pp. 782 and 783 (1897), in which the first preparation of poly(dichlorophosphazene) (cross-linked) was described. Reproduced by permission of the American Chemical Society.

Chloronitrides of Phosphorus.   783

sealed tube, there results a mixture of chloronitrides, which is partly crystalline and soluble in gasoline, but for the greater part liquid and insoluble in this solvent, and of a high degree of complexity. This may be distilled almost without residue, the distillate being a crystalline mass, impregnated with an oil, and composed almost wholly of a mixture of members of the series $(PNCl_2)_n$ in nearly theoretical amount, containing about 50 per cent. $P_3N_3Cl_6$, and 25 per cent. $P_4N_4Cl_8$, the remainder consisting of the higher homologues. From this distillate the new bodies, with one exception, have been isolated.

The series, as at present known, consists of the following :[1]

		Melting-point.	Boiling-point. 13 mm.	760 mm.
Triphosphonitrilic chloride,	$(PNCl_2)_3$	114°	127°	256.5°[2]
Tetraphosphonitrilic chloride,	$(PNCl_2)_4$	123.5°	188°	328.5°[3]
Pentaphosphonitrilic chloride,	$(PNCl_2)_5$	40.5–41°	223–224.3°	Polymerizes
Hexaphosphonitrilic chloride,	$(PNCl_2)_6$	91°	261–263°	Polymerizes
Heptaphosphonitrilic chloride,	$(PNCl_2)_7$ Liquid at –18°		289–294°	Polymerizes

Polyphosphonitrilic chloride, $(PNCl_2)x$ Below red heat. Depolymerizes on distillation.

There were obtained, further, a liquid residue of the same empirical composition, of a *mean* molecular weight corresponding nearly to $(PNCl_2)_{11}$, and a small amount of a chloronitride, $P_6N_7Cl_9$, not belonging to the above series. The absence of the lower members, $PNCl_2$ and $(PNCl_2)_2$, is remarkable, and theoretically significant. Indications of a trace of a substance more volatile than the compound $(PNCl_2)_3$, and of similar but stronger odor, were observed, but there is no evidence that it consists of one of the missing bodies.

One of the most remarkable properties of the phosphonitrilic chlorides is that each member of the series is converted by heat into the rubber-like *polyphosphonitrilic chloride*, a body, or mixture of bodies, of very high molecular weight, which is highly elastic and insoluble in all neutral solvents, but which swells enormously in benzene, and which, on distilling at a higher temperature, breaks down into a mixture of all the lower members mentioned above, which can then be separated by appropriate means. In this way it is possible to convert any phosphonitrilic chloride quantitatively into any other by

[1] The melting- and boiling-points are corrected.
[2] 183.8° at 100 mm.   [3] 242° at 100 mm.

Figure 3.4   (*Continued*)

polymerization of hexachlorocyclotriphosphazene that the answer was found. Careful control of the time, temperature, and trimer purity, and termination of the reaction before it reached a stage of 70% polymerization, yielded an essentially linear high polymer (Structure 3.5) that dissolved completely in organic solvents such as benzene, toluene, or tetrahydrofuran. Further heating of this polymer caused cross-linking of the chains and yielded the insoluble "inorganic rubber" described by Stokes (eq. 3.2).

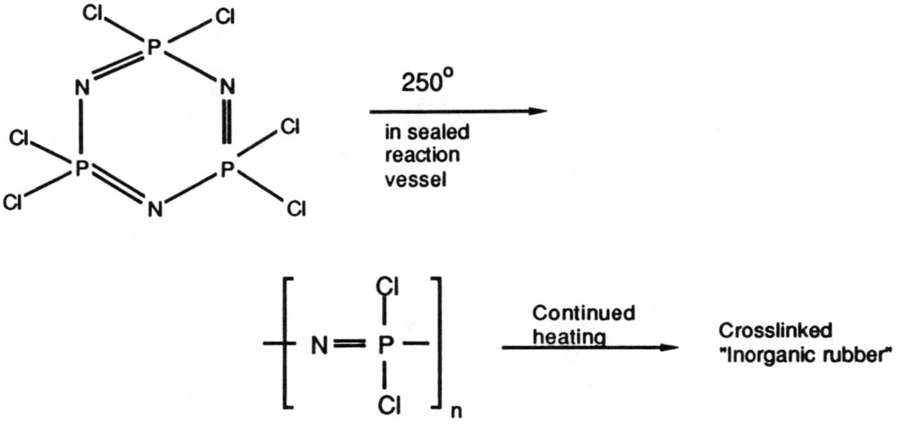

$$3.5 \qquad\qquad (\text{Eq. } 3.2)$$

3. When dissolved in a suitable solvent, poly(dichlorophosphazene) (3.5) was found to behave as a remarkable macromolecular reactant. Treatment with organic nucleophiles such as the sodium salts of alcohols or phenols, or with primary or secondary amines, brought about total replacement of the chlorine atoms by the organic units. These derivative polymers proved to be hydrolytically stable and to possess a broad range of unusual, interesting, and useful properties. In later years, this synthesis route has been used to prepare several hundred different types of polyphosphazenes. It is used today as a manufacturing process.

3.3 THE THREE SYNTHESIS ROUTES

3.3.1 The Macromolecular Substitution Route

The synthesis method mentioned in the previous section is the most extensively developed route to the preparation of poly(organophosphazenes). It is summarized in Scheme 3.1.

Hexachlorocyclotriphosphazene (Structure 3.2) is prepared on an industrial scale by the interaction of phosphorus pentachloride with ammonium chloride in an organic solvent such as chlorobenzene or tetrachloethane. This compound, after

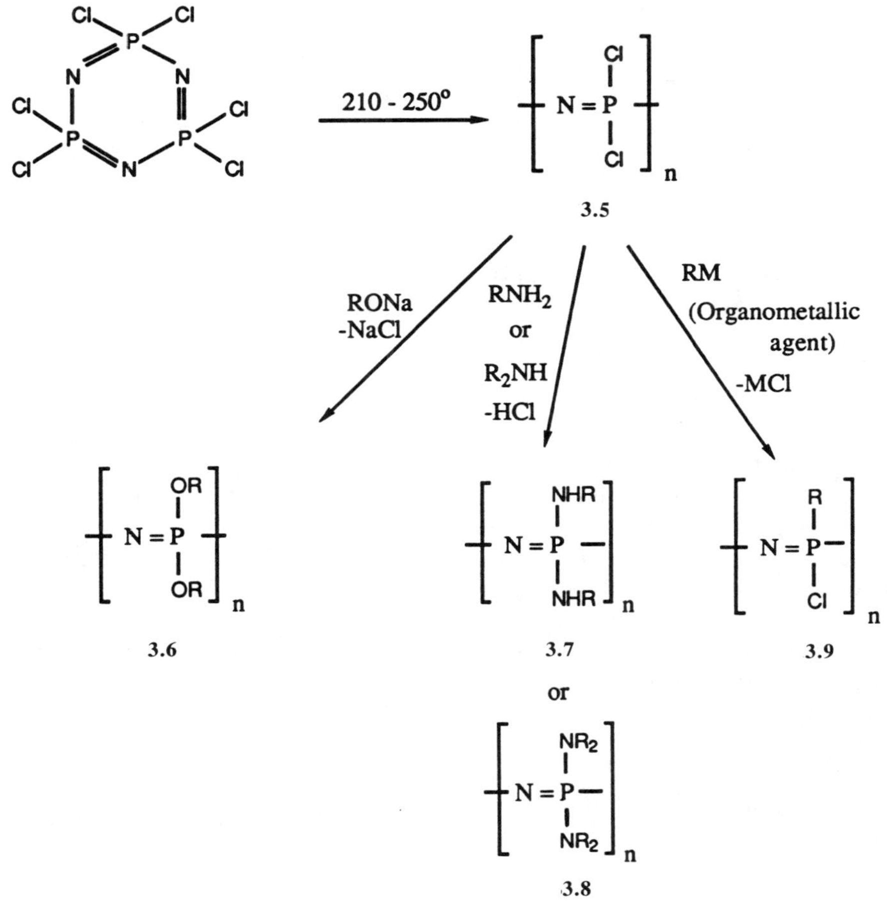

Scheme 3.1

careful purification and protection from moisture, is heated in the molten state at temperatures between 210 and 250°C to induce polymerization.[14-17]

Solutions of poly(dichlorophosphazene) (Structure 3.5) in benzene, toluene, or tetrahydrofuran react rapidly and completely with nucleophiles such as sodium trifluoroethoxide to yield derivative polymers such as poly[bis(trifluoroethoxy)phosphazene] (Structure 3.10)[14, 15] An impetus for this substitution is the precipitation of sodium chloride from solution, a process that drives the reaction to completion. Polymer 3.10 was the first stable poly(organophosphazene) to be synthesized. It is still one of the most interesting (see later). A wide range of other alkoxy or aryloxy groups can be introduced, such as the ethoxy groups in Structure 3.11 or the phenoxy groups in Structure 3.12. One of the few limitations to this process is the ability of certain bulky organic side groups to slow the replacement of nearby chlorine atoms by steric hindrance. Hence, the introduction

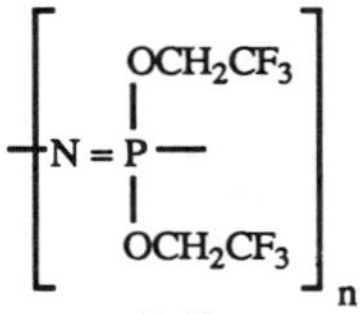

$$\left[\begin{array}{c} OCH_2CF_3 \\ | \\ -N=P- \\ | \\ OCH_2CF_3 \end{array} \right]_n \qquad \left[\begin{array}{c} OC_2H_5 \\ | \\ -N=P- \\ | \\ OC_2H_5 \end{array} \right]_n \qquad \left[\begin{array}{c} O-\bigcirc \\ | \\ N=P- \\ | \\ O-\bigcirc \end{array} \right]_n$$

3.10 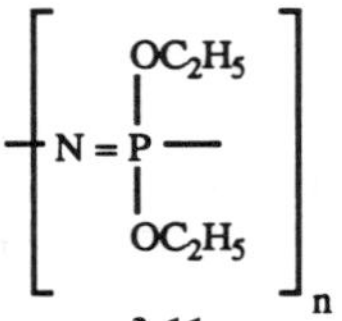   3.11 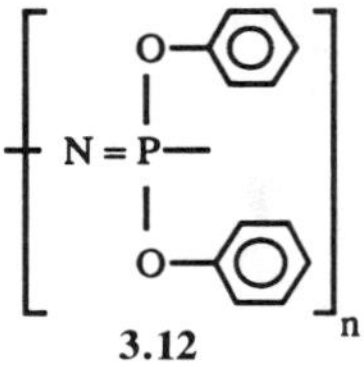   3.12

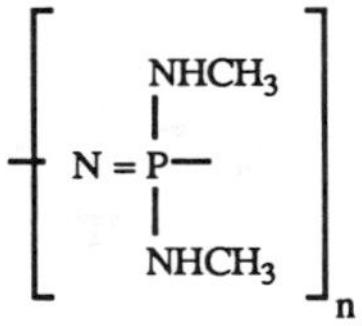

$$\left[\begin{array}{c} NHCH_3 \\ | \\ N=P- \\ | \\ NHCH_3 \end{array} \right]_n \qquad \left[\begin{array}{c} NHC_4H_9 \\ | \\ -N=P- \\ | \\ NHC_4H_9 \end{array} \right]_n \qquad \left[\begin{array}{c} NH-\bigcirc \\ | \\ -N=P- \\ | \\ NH-\bigcirc \end{array} \right]_n$$

3.13 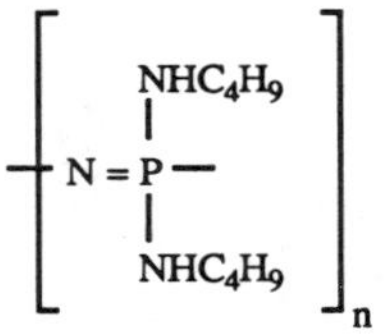   3.14 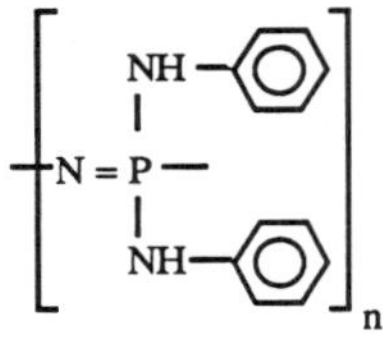   3.15

of multiring- or ortho-substituted aryloxy units may require forcing (high-temperature) conditions.

Aminolysis of poly(dichlorophosphazene) also takes place readily to yield poly(aminophosphazenes) (Structures 3.7 and 3.8) with specific examples shown as the polymers in 3.13, 3.14, or 3.15.[16] Here too, steric hindrance effects may slow the replacement of the last 25% of the remaining chlorine atoms, thus requiring forcing conditions toward the end of the reaction.

Inherent in the macromolecular substitution method is the possibility that two or more different organic groups can be introduced either simultaneously or sequentially. Steric hindrance effects that slow the reaction rate after a bulky side group has been introduced allow the controlled introduction of a second set of side groups, as shown in Scheme 3.2.

Thus, the treatment of $(NPCl_2)_n$ with diethylamine (even in excess) results in the replacement of only one chlorine atom per phosphorus.[17] However, the

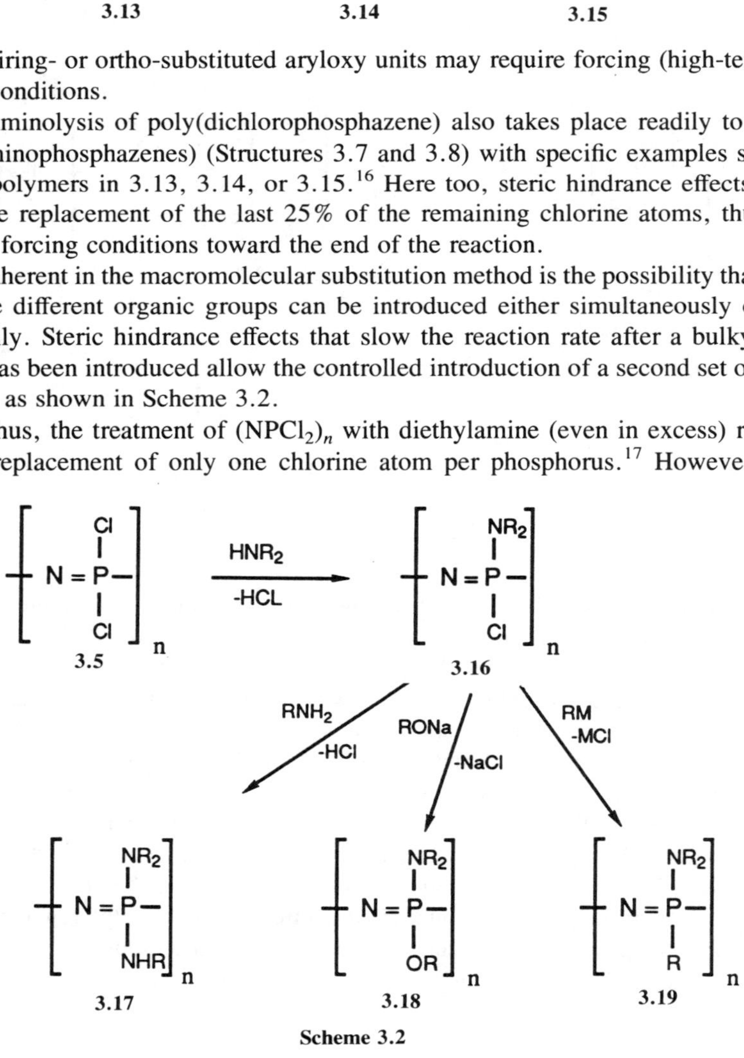

$$\left[\begin{array}{c} Cl \\ | \\ N=P- \\ | \\ Cl \end{array} \right]_n \xrightarrow[-HCL]{HNR_2} \left[\begin{array}{c} NR_2 \\ | \\ N=P- \\ | \\ Cl \end{array} \right]_n$$

3.5   3.16

$$\xrightarrow[-HCl]{RNH_2} \qquad \xrightarrow[-NaCl]{RONa} \qquad \xrightarrow[-MCl]{RM}$$

$$\left[\begin{array}{c} NR_2 \\ | \\ N=P- \\ | \\ NHR \end{array} \right]_n \qquad \left[\begin{array}{c} NR_2 \\ | \\ N=P- \\ | \\ OR \end{array} \right]_n \qquad \left[\begin{array}{c} NR_2 \\ | \\ N=P- \\ | \\ R \end{array} \right]_n$$

3.17   3.18   3.19

Scheme 3.2

remaining chlorine can be replaced by treatment with a less hindered nucleophile such as methylamine, short-chain alkoxides, or, in a few cases, alkyl or aryl organometallic reagents.

This principle can be utilized for a number of different initial nucleophiles. Phenoxy and substituted phenoxy side groups replace less than 100% of the chlorine atoms under mild reaction conditions, but the remaining chlorine atoms react with a less hindered nucleophile such as trifluoroethoxide.

The third pathway shown in Scheme 3.1 illustrates the interaction of poly(dichlorophosphazene) with organometallic nucleophiles such as Grignard or organolithium reagents. Unlike their oxo- or nitrogen-nucleophile counterparts, organometallic reagents generate more complicated reactions.[18,19] For example, the interactions of $(NPCl_2)_n$ with RMgX or RLi usually follow two concurrent and conflicting pathways. Replacement of chlorine by the group R certainly occurs, but this is accompanied by (or followed by) cleavage of the phosphorus-nitrogen bonds in the skeleton. This sequence of events is summarized in Scheme 3.3.

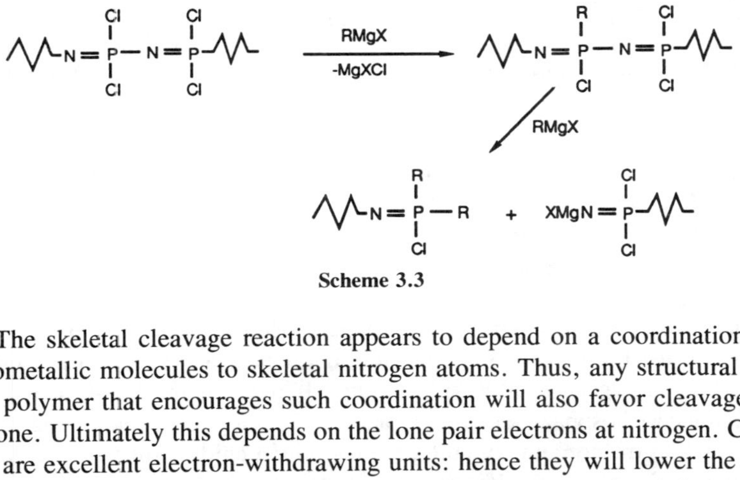

Scheme 3.3

The skeletal cleavage reaction appears to depend on a coordination of the organometallic molecules to skeletal nitrogen atoms. Thus, any structural feature of the polymer that encourages such coordination will also favor cleavage of the backbone. Ultimately this depends on the lone pair electrons at nitrogen. Chlorine atoms are excellent electron-withdrawing units: hence they will lower the density of the lone pair electrons at skeletal nitrogen. This protects the skeleton against cleavage. But, if the chlorine atoms are replaced by alkyl or aryl groups, this protective effect is lost. Thus, as an organometallic substitution reaction proceeds, an increased probability will exist that phosphorus-nitrogen bond cleavage will occur rather than halogen replacement.

Because fluorine is a more electron-withdrawing unit than chlorine, it should be expected that fluorophosphazenes might be more resistant to skeletal cleavage than their chlorophosphazene counterparts.

This is illustrated by the reaction profiles shown in Figure 3.5, which compare the chain lengths of polymers obtained by treatment of the polymers $(NPCl_2)_n$ and $(NPF_2)_n$ with phenyllithium.[19,20] When $(NPCl_2)_n$ is the reaction substrate, phosphorus-nitrogen skeletal cleavage is detectable almost from the start of the reaction. However, chain cleavage is not noticeable with $(NPF_2)_n$ until roughly 75% of the fluorine atoms have been replaced by the organic groups.

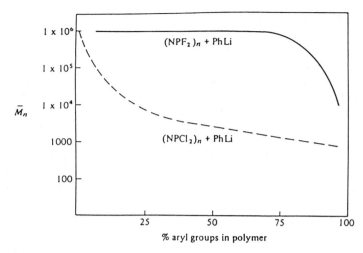

Figure 3.5 Comparison of the variation in average molecular weight for $[NP(C_6H_5)_x(OCH_2CF_3)_y]_n$ versus the percentage of phenyl groups introduced via the reaction of $(NPF_2)_n$ or $(NPCl_2)_n$ with phenyllithium. The trifluoroethoxy groups were introduced later to facilitate the molecular-weight measurements. (Reprinted from T. L. Evans and H. R. Allcock, *J. Macro. Sci.—Chem.* **1981**, *A16(1)*, 409, by courtesy of Marcel Dekker Inc.)

Incidentally, $(NPF_2)_n$ is prepared by polymerization of the trimer, $(NPF_2)_3$, which itself is obtained from $(NPCl_2)_3$ by treatment with sodium fluoride. In principle, $(NPF_2)_n$ is a very useful macromolecular intermediate for the reasons just discussed. However, in practice, its insolubility in all solvents except specialized fluorocarbon media limits its usefulness.

One additional technique has been developed to allow organometallic reagents to be used as reactants for the introduction of alkyl, aryl, carboranyl, or transition metal organometallic side groups. If most of the halogen atoms in a poly(dihalogenophosphazene) are replaced by electron-withdrawing organic groups, such as trifluoroethoxy units, the remaining chlorine or fluorine side units can be replaced without significant skeletal cleavage.[21] This is a useful technique for the preparation of mixed-substituent polymers.

However, in spite of these restrictions, the overall macromolecular substitution route allows access to an enormous range of different polymeric structures. Moreover, the length of the polymer chain is usually unaffected by the replacement of halogen by organic side groups. Thus, it is possible to alter the side group structure in polyphosphazenes without changing any other structural features, and this is of vital importance for the assessment of structure-property relationships.

3.3.2 Polymerization of Organo- or Organometallo-Substituted Cyclic Phosphazenes

The macromolecular substitution route just described is a powerful method for the synthesis of a broad range of new phosphazene polymers. Yet that approach has limitations if the target polymers are to contain organic side groups linked to the skeleton through carbon-phosphorus bonds. As discussed, organometallic macromolecular substitution reactions allow the introduction of alkyl, aryl, or organometallic groups, but often at the expense of skeletal cleavage or incomplete halogen replacement. The methods discussed in this and the following section provide alternative approaches that avoid the interaction of organometallic reagents with high polymeric phosphazenes.

It is usually easier to carry out organometallic substitution reactions with small-molecule chloro- or fluoro-cyclic phosphazenes than with the analogous high polymers. For one thing, the consequences of skeletal cleavage reactions are less severe in small-molecule chemistry than for high polymers. Consider the following example. Suppose that the interaction of an organometallic reagent (say, an organolithium reagent) with a chlorophosphazene results in an average of 1 skeletal cleavage reaction for every 12 halogen replacement reactions. At the high polymer level this situation would result in a chain cleavage once for every six repeating units along the chain. Thus, the products would on the average be only six repeating units long. These molecules would have none of the properties of high polymers. So the yield of polymer would be zero. On the other hand, the same reaction carried out with a cyclic trimeric phosphazene would cleave the ring of only one molecule out of every two. Hence, the yield of the substituted cyclic phosphazene would be 50%. For this reason it is much easier to prepare organosubstituted cyclic trimeric phosphazenes by organometallic substitution chemistry than it is to prepare the analogous high polymers.

Thus, the second approach to the synthesis of poly(organophosphazenes) involves the introduction of the organic (or organometallic) side groups at the cyclic trimer level, followed by ring-opening polymerization of the substituted cyclic trimer to the high polymer (Scheme 3.4).[22]

This approach has both advantages and limitations. For example, although cyclic trimers that bear one or two organic or organometallic side groups and five

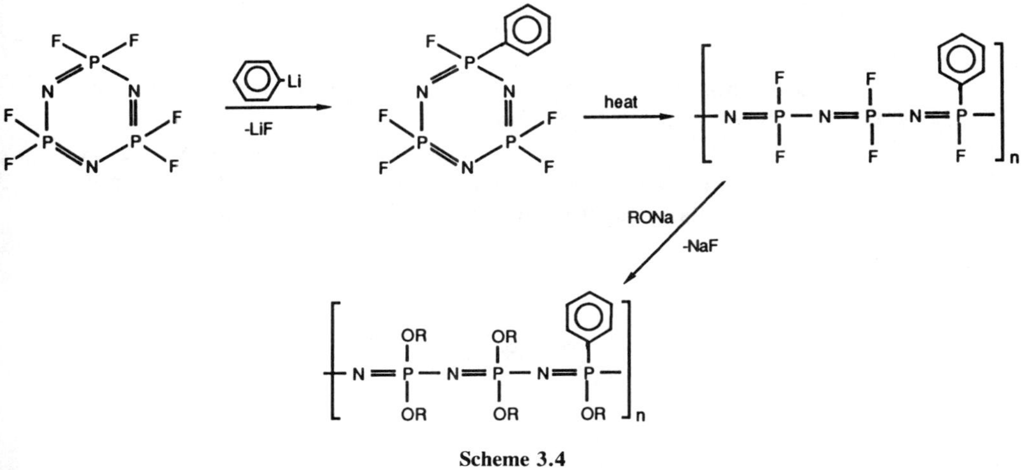

Scheme 3.4

or four chlorine or fluorine atoms usually polymerize almost as easily as $(NPCl_2)_3$ or $(NPF_2)_3$, the tendency for polymerization declines as more and more halogen atoms in the trimer are replaced by organic groups.[23] The cyclic trimer that bears six methyl groups, $(NPMe_2)_3$ (Structure 3.20), undergoes ring-ring equilibration to the eight-membered cyclic tetramer (Structure 3.21) when heated, but it does not polymerize to the high polymer.[24] The same is true if the trimeric ring bears six phenyl groups or six trifluoroethoxy units. Possible reasons for this will be discussed in a later section.

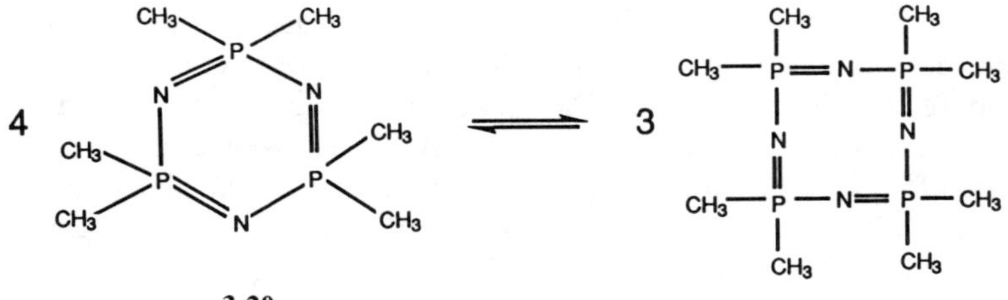

$$\text{(Eq. 3.3)}$$

3.20   **3.21**

However, this restriction does not hold if the phosphazene ring is subjected to ring strain by the presence of a transannular ferrocenyl group, as in Structure 3.22.[25] Under these circumstances, polymerization takes place to give Structure 3.23 even if no halogen atoms are attached to the phosphorus atoms, although the reaction is accelerated by the presence of catalytic quantities of $(NPCl_2)_3$.

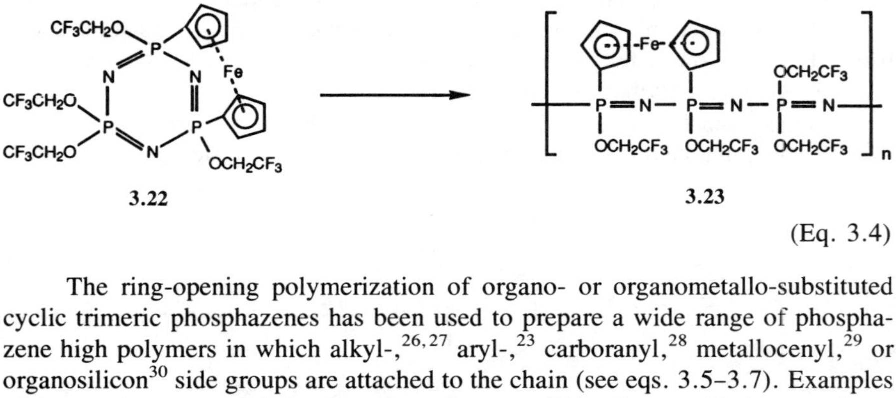

3.22   **3.23**

$$\text{(Eq. 3.4)}$$

The ring-opening polymerization of organo- or organometallo-substituted cyclic trimeric phosphazenes has been used to prepare a wide range of phosphazene high polymers in which alkyl-,[26,27] aryl-,[23] carboranyl-,[28] metallocenyl-,[29] or organosilicon[30] side groups are attached to the chain (see eqs. 3.5–3.7). Examples of these polymers, and their properties and uses, will be discussed in later sections.

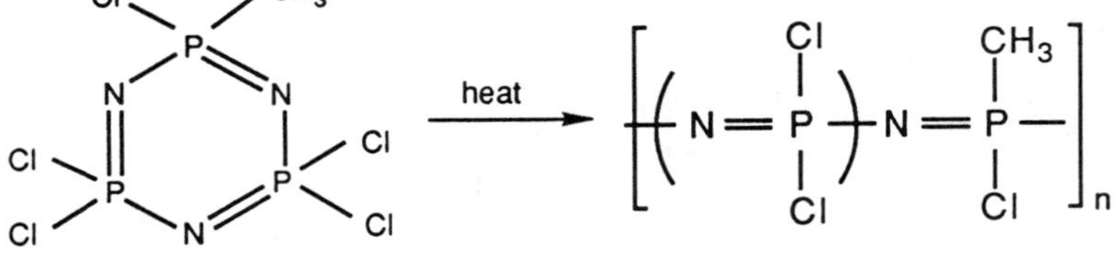

$$\text{(Eq. 3.5)}$$

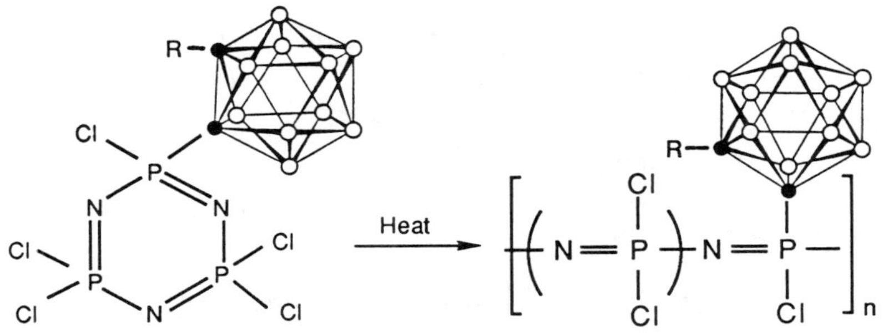

(Eq. 3.6)

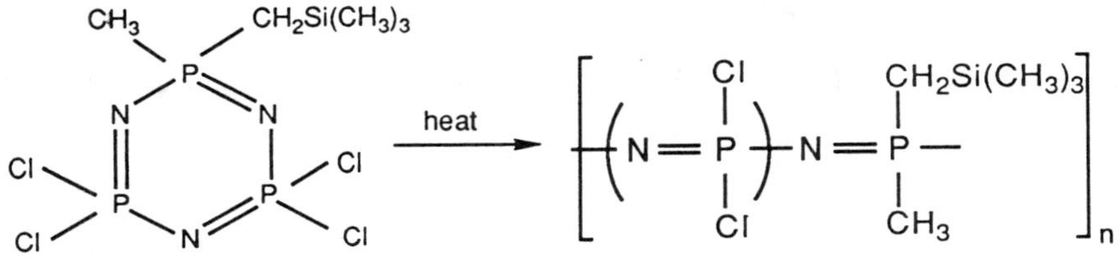

(Eq. 3.7)

3.3.3 Synthesis of Poly(organophosphazenes) by Condensation Reactions

Condensation reactions form the basis of the synthesis of $(NPCl_2)_3$. The reaction between phosphorus pentachloride and ammonia proceeds in a stepwise fashion by elimination of hydrogen chloride to form first a monomer (Structure 3.24), then a linear dimer (Structure 3.25), trimer (Structure 3.26), tetramer, and so on. Cyclization could occur to give cyclic chlorophosphazenes at any stage beyond the dimer (Scheme 3.5).

Similar condensation processes are used to prepare cyclic arylphosphazenes, starting from Cl_2PPh_3 and ammonium chloride. These processes strongly favor the formation of small-molecule cyclic phosphazenes rather than the high polymers.

However, for reasons that are not fully understood, one specific condensation reaction preferentially gives polymer rather than cyclic oligomers. This reaction, developed by Neilson and Wisian-Neilson,[31-33] is shown in Scheme 3.6.

The small molecule precursor (Structure 3.28) is heated in vacuum to drive off $Me_3SiOCH_2CF_3$ and leave $(NPMe_2)_n$ (Structure 3.29) behind. Although this synthesis is somewhat restricted, in the sense that the side groups that can be incorporated are limited to two methyl groups per repeating unit, a methyl group and a phenyl group, and a few others, the polymers obtained by this route are precisely those that are so difficult to produce by the macromolecular substitution approach. Moreover, polymers such as Structure 3.29 undergo lithium-hydrogen

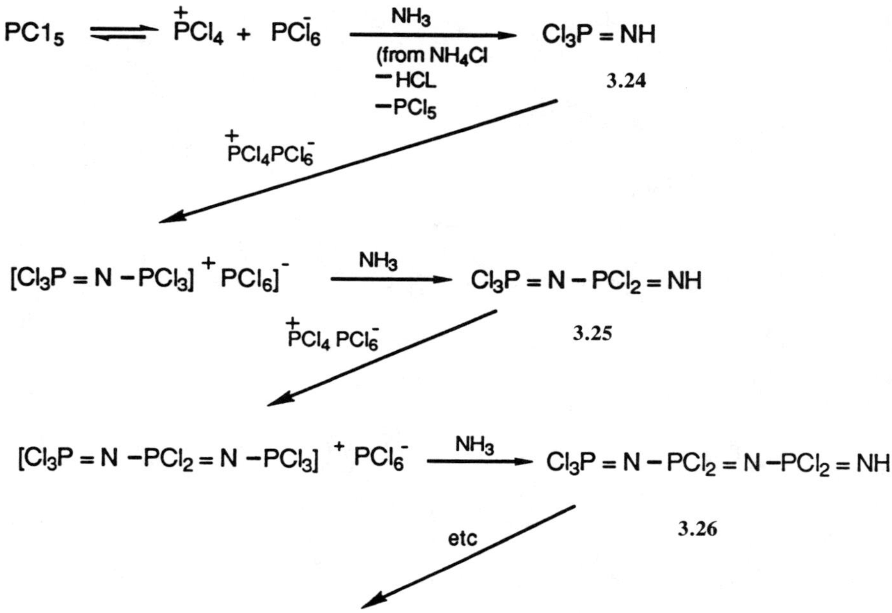

$$PCl_5 \; \rightleftharpoons \; \overset{+}{P}Cl_4 \; + \; \overset{-}{P}Cl_6 \; \xrightarrow[\substack{(from\ NH_4Cl) \\ -HCL \\ -PCl_5}]{NH_3} \; Cl_3P = NH$$

3.24

$$\overset{+}{P}Cl_4 PCl_6^-$$

$$[Cl_3P = N - PCl_3]^+ PCl_6^- \; \xrightarrow{NH_3} \; Cl_3P = N - PCl_2 = NH$$

$$\overset{+}{P}Cl_4\, PCl_6^-$$

3.25

$$[Cl_3P = N - PCl_2 = N - PCl_3]^+ PCl_6^- \; \xrightarrow{NH_3} \; Cl_3P = N - PCl_2 = N - PCl_2 = NH$$

etc

3.26

Scheme 3.5

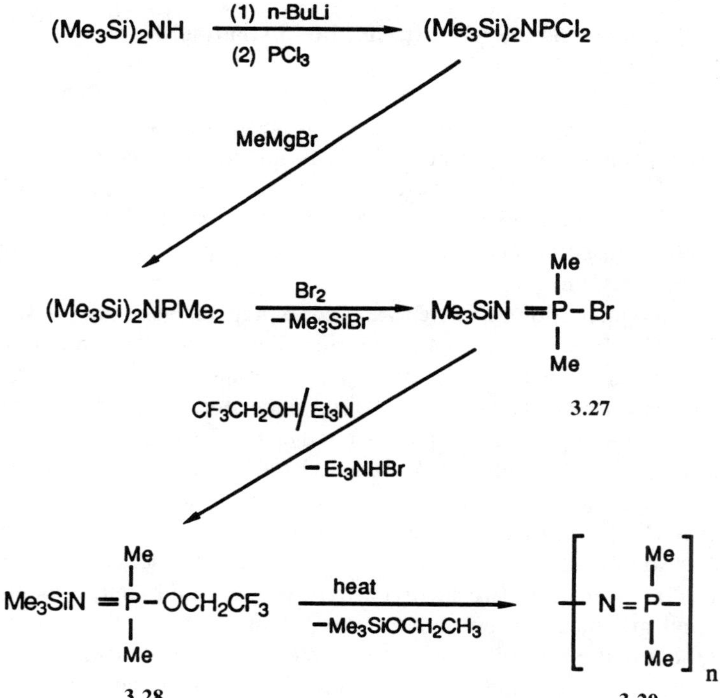

$$(Me_3Si)_2NH \; \xrightarrow[(2)\ PCl_3]{(1)\ n\text{-}BuLi} \; (Me_3Si)_2NPCl_2$$

MeMgBr

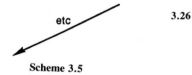

$$(Me_3Si)_2NPMe_2 \; \xrightarrow[-Me_3SiBr]{Br_2} \; Me_3SiN \overset{Me}{\underset{Me}{=\!\!\!\!\!\overset{|}{\underset{|}{P}}\!\!-}} Br$$

3.27

$$CF_3CH_2OH / Et_3N$$

$$-Et_3NHBr$$

$$Me_3SiN \overset{Me}{\underset{Me}{=\!\!\!\!\!\overset{|}{\underset{|}{P}}\!\!-}} OCH_2CF_3 \; \xrightarrow[-Me_3SiOCH_2CH_3]{heat} \; \left[-N = \overset{Me}{\underset{Me}{\overset{|}{\underset{|}{P}}}} - \right]_n$$

3.28   3.29

Scheme 3.6

$$\left[\begin{array}{c} CH_3 \\ | \\ -N=P- \\ | \\ CH_3 \end{array} \right]_n \xrightarrow[-RH]{R\,Li} \left[\begin{array}{c} \overset{-}{C}H_2Li^+ \\ | \\ -N=P- \\ | \\ CH_3 \end{array} \right] \xrightarrow[-LiCl]{R'Cl} \left[\begin{array}{c} R' \\ | \\ CH_2 \\ | \\ -N=P- \\ | \\ CH_3 \end{array} \right]_n$$

3.29   3.30   3.31

(Eq. 3.8)

exchange reactions to give anionic species (Structure 3.30), and these react with organic or organometallic halides to give derivative polymers such as Structure 3.31.[34]

It is interesting that the condensation reaction gives high polymer when the leaving group is OCH_2CF_3, but yields cyclic oligomers only when the leaving group is bromine (Structure 3.27).

The condensation process also provides an alternative route to fluoroalkoxyphosphazene polymers such as Structure 3.10. For example, the pyrolysis of $Me_3Si-N=P(OCH_2CF_3)_3$ gives Structure 3.10,[35] and this reaction can be catalyzed by the presence of Bu_4NF.[36]

Another condensation route to polyphosphazenes has been reported in the patent literature.[37,38] In this method, short-chain linear chlorophosphazenes undergo end-group condensation to generate longer chain species. The medium-molecular-weight polymers formed can then be used as macromolecular intermediates for the preparation of organophosphazene derivatives.

3.4 THE RING-OPENING POLYMERIZATION MECHANISM

Synthesis methods 1 and 2 discussed earlier both rely on the ring-opening polymerization of small-molecule cyclic phosphazenes. Thus, the reaction mechanism followed in these polymerizations is of considerable interest. By understanding this mechanism, it may be possible to induce the polymerization of cyclic trimers, such as $(NPMe_2)_3$ or $(NPPh_2)_3$, that so far have resisted polymerization.

The initial thinking on this topic was focused on the polymerization of $(NPCl_2)_3$, $(NPCl_2)_4$, and the fluoro analog, $(NPF_2)_3$. The following experimental observations were made.

1. Polymerization appears to require the presence of several halogen atoms attached to the ring phosphorus atoms. Replacement of all the halogen atoms by methyl, phenyl, or OCH_2CF_3 groups blocks the polymerization process (but not the ring-expansion equilibration).

2. Ions appear to participate in the polymerization process. The ionic conductivity of molten $(NPCl_2)_3$ remains low until the trimer is heated into the

temperature range where polymerization occurs. At this point the conductiv-ity rises dramatically.[39]

3. Traces of impurities, especially water or BCl_3, exert a catalytic effect on the polymerization of $(NPCl_2)_3$.[40,41]

4. The temperature required for the onset of polymerization of $(NPF_2)_3$ (350°C) is much higher than that required for polymerization of $(NPCl_2)_3$ (210–250°C).[42-44]

These observations were used as a basis for proposing the polymerization mech-anism shown in Scheme 3.7.[39]

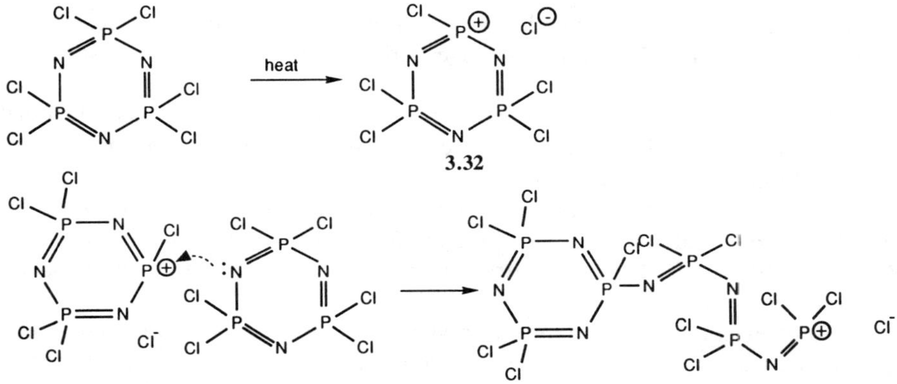

Scheme 3.7

In this mechanism it is assumed that polymerization is induced by ionic spe-cies, such as Structure 3.32, generated by the ionization of P—Cl bonds to form a cyclic (or linear) phosphazenium ion. The phosphazenium ion would then act as a cationic initiator by attack on the skeletal nitrogen atom of an $(NPCl_2)_3$ molecule, inducing ring opening and chain propagation by a cationic mechanism.

Traces of water might function as a catalyst by assisting the separation of chloride ion, or could serve as co-catalytic species. Boron trichloride, commonly used as a catalyst in the commercial process,[41] would also be expected to promote the separation of Cl^- from phosphorus, and would be stabilized as the BCl_4^- ion. Moreover, this mechanism would explain why $(NPF_2)_3$ requires a higher temper-ature for polymerization than $(NPCl_2)_3$, since it is generally assumed that more energy is required to thermally cleave a P—F bond than a P—Cl bond. The mech-anism would also explain why replacement of all the halogen atoms in $(NPCl_2)_3$ or $(NPF_2)_3$ by organic groups blocks the polymerization process, since it would be exceedingly difficult to induce the thermal ionization of a methyl or phenyl group from phosphorus.

However, this cannot be the complete story. As mentioned earlier, ring-strained phosphazenes that lack halogen side groups do polymerize, and organo-substituted trimers undergo ring expansion without forming high polymers. Either two independent mechanisms are involved or another factor entirely must be con-sidered.

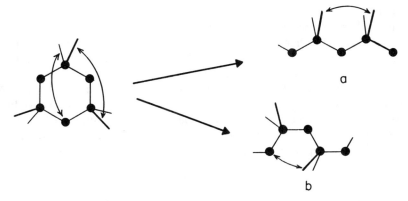

Figure 3.6 Schematic representation of the effect of steric hindrance generated by bulky side groups on a cyclic trimeric and a high polymeric phosphazene. Depolymerization of a high polymer to a cyclic trimer relieves the intramolecular crowding.

Another factor that appears to exert a powerful influence on the ability of a cyclic phosphazene to polymerize is the amount of intramolecular steric hindrance generated by the side groups.

Figure 3.6 illustrates that, irrespective of the chain conformation assumed by the high polymer, the side groups are always closer to their neighbors along a chain than they are when attached to a cyclic trimer. Thus, if the groups are large, they will interfere with each other and experience more repulsions in the polymer than in the cyclic trimer. Hence, the polymer will be thermodynamically less stable than the cyclic trimer. This means that, even if a polymerization mechanism is accessible, the trimer will not polymerize. The system would be above its "ceiling temperature," the temperature at which a polymer is always unstable relative to small-molecule species.

Conversely, if the polymer could be made by some other route (for example, by macromolecular substitution), it might be stable at moderate temperatures where the rate of depolymerization is very slow, but would depolymerize to the cyclic trimer or tetramer when heated to higher temperatures. In fact, this behavior is found for polymers such as $[NP(OPh)_2]_n$ that appear to be kinetically stabilized at moderate temperatures, but are sufficiently destabilized thermodynamically by the bulky aryloxy side groups that they depolymerize when heated above 150–200°C.

3.5 SMALL-MOLECULE MODELS

Most chemists begin their training by learning about small molecules rather than polymers. The reasons for this are both traditional and practical. Small molecules are often easier to synthesize, purify, and characterize than are polymers. Moreover, in phosphazene chemistry it is easier to study small-molecule reactions, reaction mechanisms, and molecular structures than it is to obtain comparable information at the high polymer level.

For these reasons, small molecules have played a critical role in the development of phosphazene high-polymer chemistry.[45] In particular, the substitution reactions, reaction mechanisms, NMR spectroscopy, and x-ray diffraction analysis of small-molecule cyclic phosphazenes, such as Structure 3.2 or 3.3 have provided information that could not be obtained directly from the high polymers.

For example, one of the first clean reactions carried out between $(NPCl_2)_3$ and an alkoxide ion was the one shown in eq. (3.9), that is, the reaction of $(NPCl_2)_3$ with sodium trifluoroethoxide.[46,47] It was the isolation and study of this exceedingly stable derivative (Structure 3.33) that pointed the way to the use of the same nucleophile for the replacement of the chlorine atoms in $(NPCl_2)_n$ by fluoroalkoxy groups.[14,15]

3.33

(Eq. 3.9)

The model compound approach is also useful for studying reaction mechanisms. For example, as chlorine atoms in $(NPCl_2)_3$ are replaced by OCH_2CF_3 groups, what is the pattern of halogen replacement? Does the second fluoroalkoxy group enter the molecule at the same phosphorus atom as the first, or at an unsubstituted phosphorus atom (at the so-called nongeminal site)? Does this pattern persist when the same reaction is carried out on the high polymer? If not, what influence does a small-molecule ring or a linear chain have on the mechanism? Many of these questions have not yet been answered for most phosphazene polymer reactions, but they are critically important. Small-molecule reaction mechanism studies provide one of the few ways to answer these questions.[48]

It is exceedingly difficult to determine the molecular structure of a synthetic macromolecule. X-ray diffraction—the ultimate structural tool for small-molecule studies—yields only limited information for most synthetic high polymers, and critical data about bond lengths and bond angles are difficult to obtain.[49] However, that same information can be obtained relatively easily from single-crystal x-ray diffraction studies of cyclic trimers, tetramers, and short-chain linear phosphazene oligomers. The information obtained may then be used to help solve the structures of the high polymeric counterparts.

Of course, cyclic trimers and tetramers are not always good reaction or structural models for high polymers. The constraints of the ring may alter reactivities, bond angles, and bond lengths. But the model data provide a good starting point. And generally it is found that if a certain reaction cannot be carried out with a phosphazene cyclic trimer or tetramer, it will not work for the high polymer.

3.6 MOLECULAR STRUCTURE OF POLYPHOSPHAZENES

Molecular structure determination for high polymers is a more complicated matter than is the corresponding process for small molecules. For one thing, different polymer molecules within the same sample have different chain lengths. In addition, if more than one type of side group is attached to the same polymer chain, the organization of those groups may differ from one point along the chain to another. For example, some regions of the chain may bear only one type of side group (a block structure), while another region may contain both groups organized in a random or an alternating sequence. If the two are disposed in a regular, repetitive fashion, the arrangement may be isotactic, syndiotactic, or heterotactic.[49]

Once the primary framework structure of the polymer has been determined, the question of chain conformation remains to be solved. Small molecules have access to only a limited range of molecular conformations, but polymer molecules, with their thousands of skeletal bonds and side group units, can adopt millions of different conformations—ranging from a totally random coil to extended chain or regular helical arrangements. Three physical tools have proved to be particularly useful for the structure determination of synthetic polymers—nuclear magnetic resonance (including magic angle spinning NMR), Fourier transform infrared spectroscopy, and fiber-type x-ray diffraction. Polyphosphazenes can be studied by all three methods, although the presence of several NMR-active nuclei (^{31}P, ^{13}C, ^{1}H, and often ^{19}F and ^{28}Si) makes nuclear magnetic resonance a particularly valuable technique. X-ray diffraction is a useful method for conformational studies.

In the following sections, four different aspects of polyphosphazene structure will be reviewed briefly—side group disposition, chain conformation, skeletal bonding, and skeletal flexibility.

3.6.1 Side Group Disposition

The properties of a polymer depend not only on the nature of the skeleton and the types of side groups attached to it, but also on the way in which different side groups are sequenced along the same chain. For example, isotactic or syndiotactic sequencing may favor efficient intermolecular chain packing and microcrystallite formation. On the other hand, a random distribution of two or more different side groups will discourage crystallization, and may perhaps favor the appearance of elastomeric character.

Because most poly(organophosphazenes) are synthesized by a macromolecular substitution route, the disposition of side groups usually depends on the steric and electron-directing characteristics of the incoming groups and on the side groups already present. Clearly, the order in which two or more different side groups are introduced will also affect the outcome.

The first step in the structural investigation of a polyphosphazene is to determine the side group ratios by microanalysis and by NMR spectroscopy. The second step is to attempt to deduce the sequencing and configuration by a detailed analysis of the NMR spectra.

The ^{31}P NMR spectra shown in Figures 3.7 and 3.8 illustrate how the arrangement of side groups can be determined at the simplest level.

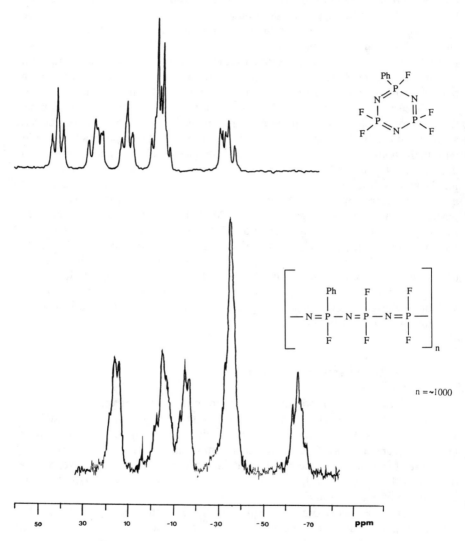

Figure 3.7 ^{31}P NMR spectra of cyclic trimeric and high polymeric phenyl-fluoro-phosphazenes. Note (1) the shift in the whole spectrum that occurs in moving from a cyclic small molecule phosphazene to a related high polymer and (2) the chemical shift and splitting pattern that results from phosphorus coupling to the two fluorine atoms or to one fluorine. (Spectra provided by W. D. Coggio.)

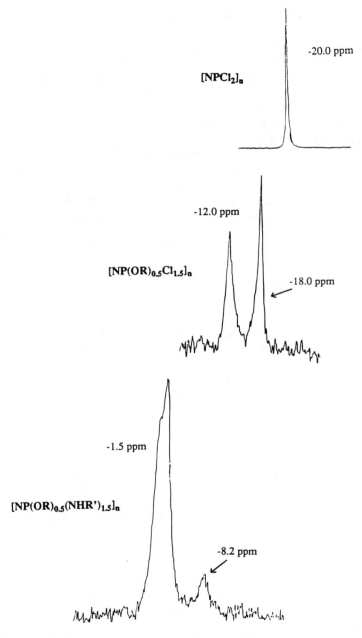

Figure 3.8 ^{31}P NMR spectral changes that occur when 25% of the chlorine atoms in $(NPCl_2)_n$ are replaced first by CF_3CH_2O- groups and the remainder by $RNH-$ groups. The signal at ~ -8.2 ppm in the spectrum of $[NP(OCH_2CF_3)_{0.5}(NHR)_{1.5}]_n$ results from blocks of $NP(OCH_2CF_3)_2$ units. (Spectra provided by W. D. Coggio.)

3.6.2 Macromolecular Conformations

The chain conformation of a macromolecule is determined by the torsional angles assumed by the backbone bonds.[50] By convention, the angles 0°, 0° are used to define a *trans-trans* planar conformation as shown in Figure 3.9a. Torsion (rotation) of bonds 2 and 4 in Figure 3.9a by 180° generates the *cis-trans* planar conformation (Figure 3.9b). Other torsional angle values can give rise to a broad range of spiral helices or even (if the torsional angles are not sequenced in some regular way) to a random coil.

In most polymers the torsional angles assumed by the molecule depend on two factors: (1) repulsions or attractions between nearby side groups on the same chain and (2) the forces associated with the packing of many chains into a micro-crystalline domain. If organic-type double bonds are present in the chain, an additional influence will be exerted by the *cis-trans* barrier imposed by the $p_\pi - p_\pi$ double bond.

For most polymers it is difficult to predict the outcome of a balancing of all

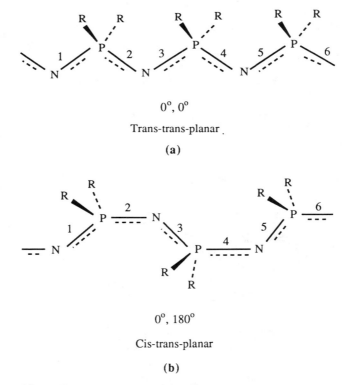

$0°, 0°$

Trans-trans-planar

(a)

$0°, 180°$

Cis-trans-planar

(b)

Figure 3.9 (a) *Trans-trans* planar conformation of a polyphosphazene. (b) *Cis-trans* planar conformation, the form detected for most of the phosphazene polymers studied to date.

these influences. Hence, the prediction of chain conformations is a considerable challenge. However, for polyphosphazenes, the situation is simpler than for most classes of macromolecules.

First, in polyphosphazenes the side groups are attached to every other skeletal atom, rather than to every skeletal atom in the chain. This contrasts with the situation in most organic polymers. Second, although the structure of these polymers is written as a sequence of alternating single and double bonds, in fact the bonding is not of the classical organic $p_\pi - p_\pi$-type, and it does not impose a skeletal torsional barrier (see later). It is as if all the backbone bonds behave like single bonds, at least so far as their torsional characteristics are concerned. Thus, the conformation assumed by a polyphosphazene can be understood mainly in terms of the repulsions or attractions between side groups attached to nearby phosphorus atoms. It can be seen from Figure 3.9 that the *cis-trans* planar conformation allows the side groups to move as far away from each other as possible. Hence, this conformation should minimize the repulsions and generate the lowest energy. Molecular mechanics calculations tend to confirm this supposition.[51,52]

Of course, calculations of this type are interesting but not conclusive proof of the conformational preferences of a polymer. For one thing, they neglect the forces between neighboring polymer molecules. Experimental proof can be obtained by x-ray fiber diffraction experiments. A beam of x-rays passed through a stretched ("oriented") fiber of a microcrystalline polymer will yield a diffraction pattern. A diffraction pattern obtained for poly(dichlorophosphazene)[53,54] is shown in Figure 3.10.

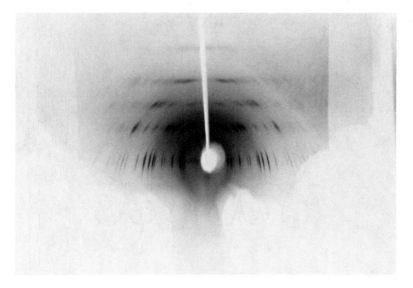

Figure 3.10 X-ray diffraction pattern obtained from a fiber of poly(dichlorophosphazene). The pattern of diffraction arcs is consistent with a near *cis-trans* planar conformation of the polymer chains.

The separation of the horizontal layer lines in such a photograph gives an indication of the conformational repeating distance along the polymer chain. For virtually all polyphosphazenes studied so far (including those with F, Cl, OCH_2CF_3, and OPh side groups), the repeating distance is close to 4.9 Å, which is the value expected if the chains assume a near *cis-trans* planar conformation. Moreover, the presence of a meridional reflection on the $l = 2$ layer line is an indication that there are two monomer units per turn of the "helix." This too is consistent with a near-*cis-trans*-planar arrangement of the chain.

The conformation of complex side groups that consist of more than one atom is more difficult to assess from x-ray fiber diagrams, and much work remains to be done to obtain detailed structural data for such polyphosphazenes.

3.6.3 Bonding in Polyphosphazenes

As mentioned earlier, the skeletal bonds in phosphazenes are unlike their counterparts in classical organic polymers. To understand the differences, it is necessary to consider the disposition of the valence electrons in a short segment of the chain. Each phosphorus atom provides five valence electrons per repeating unit, and each nitrogen contributes an additional five. If two of the electrons from nitrogen are confined to a lone pair orbital, and electron pairs are assigned to the sigma bond framework, two electrons are left unaccounted for—one from phosphorus and one from nitrogen (Structure 3.34).

$$-\overset{\displaystyle \cdot}{\underset{\displaystyle \cdot\cdot}{N}} - \overset{\displaystyle |}{\underset{\displaystyle |}{\overset{\displaystyle \cdot}{P}}} -$$

3.34

These electrons do not remain unpaired. It is believed that the electron on nitrogen is accommodated in a $2p_z$ orbital, and the one from phosphorus in a $3d$ orbital to generate an arrangement of the type shown in Structure 3.35.[55] Thus,

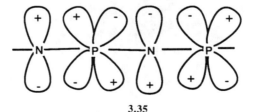

3.35

although the pi-bonds are delocalized over three atoms, they are not broadly delocalized over the whole chain because of the orbital mismatch and nodes that occur at every phosphorus. Moreover, because each phosphorus can use as many as five $3d$-orbitals, torsion of a P—N bond can bring the nitrogen p-orbital into an overlapping position with a d-orbital at virtually any torsion angle. Hence, the inherent torsional barrier is much smaller than in a $p_\pi - p_\pi$ double bond of the type found in organic molecules. Calculations suggest that the inherent torsional barrier in the backbone bonds may be as low as 0.1 kcal per bond.

An "island" pi-bond structure of this type may explain why most polyphosphazenes are colorless rather than colored materials, and are insulators rather than electronic conductors. Exceptions do exist, but the exceptions are for polymers that have chromophores in the side groups or which bear electroactive side units.

3.6.4 Skeletal Flexibility

A close connection exists between the presence of a flexible polymer skeleton and the flexibility of the bulk material. Macromolecular flexibility is often defined in terms of the glass transition temperature, T_g.[56] Below this temperature, the polymer is a glass, and the backbone bonds have insufficient thermal energy to undergo significant torsional motions. As the temperature is raised above the T_g, an onset of torsional motion occurs, such that individual molecules can now twist and yield to stress and strain. In this state the polymer is a quasi-liquid (an elastomer) unless the bulk material is stiffened by microcrystallite formation. Thus, a polymer with a high T_g is believed to have a backbone that offers more resistance to bond torsion than a polymer with a low T_g.

Polyphosphazenes are highly unusual macromolecules because specific polymers have some of the lowest T_g's known in polymer chemistry. For example, $(NPCl_2)_n$ has a T_g of $-66°C$, $(NPF_2)_n$ has a value of $-96°C$, $[NP(OCH_3)_2]_n$ of $-74°C$, species $[NP(OC_2H_5)_2]_n$ of $-84°C$, $[NP(OC_3H_7)_2]_n$ of $-100°C$, and $[NP(OCH_2CH_2OCH_2CH_2OCH_3)_2]_n$ of $-84°C$. Note that the low T_g values quoted are for polyphosphazenes that bear very small side groups (F, Cl, or OCH_3) or side groups that are themselves very flexible ($OCH_2CH_2OCH_2CH_2OCH_3$).

However, if the side groups are large and inflexible, they generate steric interference with each other as the skeletal bonds attempt to undergo twisting motions. Thus, as a general rule, large, rigid side groups impose their own restrictions on the flexibility of the macromolecule, even though the backbone bonds have a very low barrier to torsion. For example, if the side groups are OPh the T_g is $-8°C$. If the side units are $-OC_6H_4C_6H_5-p$, the T_g is $+93°C$.

It will now be clear why the polyphosphazene system is one of the most versatile polymer systems known. By the choice of appropriate side groups, polymers can be tailored to be low-temperature elastomers, flexible, microcrystalline film- and fiber-forming materials, or high melting glasses. The choice of side group also affects properties such as solubility, refractive index, chemical stability, hydrophobicity or hydrophilicity, electrical conductivity, nonlinear optical activity, and biological behavior. These effects are illustrated in the following sections.

3.7 STRUCTURE-PROPERTY RELATIONSHIPS

Enough is now known about the effect of different side groups attached to a polyphosphazene chain to allow some general structure-property relationships to be understood. To a limited extent, these relationships allow the prediction of the properties of polymers not yet synthesized. Some general relationships will be described in the following sections, but specific properties associated with certain side groups are summarized in Table 3.1.

3.7.1 Crystalline Versus Amorphous Polymers

As mentioned in Chapter 1 and earlier in this chapter, the presence of microcrystalline domains in an amorphous (random coil) polymer matrix has the effect of stiffening the material, generating opalescence rather than transparency, and raising the temperature at which the material can be used before it undergoes liquidlike flow.

Crystallization is a consequence of molecular symmetry. A macromolecule with a precise, regular sequence of side groups arrayed along the chain will be more prone to pack tightly with neighboring molecules than will a polymer that has an irregular disposition of side groups.

The conditions for microcrystallinity in polyphosphazenes can be met in only one set of circumstances—when only one type of side group is present and when those side groups are small or fairly rigid. When two or more different side groups are present, the macromolecules usually lack the necessary symmetry and regularity. This is because the substitutive mode of synthesis employed for the majority of polyphosphazenes does not allow sufficient control over the stereoregularity. Hence, most mixed-substituent polyphosphazenes are amorphous. Microcrystallinity has been detected in single-substituent polymers when the side groups are F, Cl, CH_3, OCH_2CF_3, OC_6H_5, and various substituted phenoxy groups. Most aminophosphazene polymers are amorphous, perhaps because of intra- and intermolecular hydrogen bonding.

3.7.2 Hydrophobic Versus Hydrophilic or Water Soluble

The phosphazene backbone itself appears to be hydrophilic, due mainly to the presence of the nitrogen lone pair electrons and their ability to form hydrogen bonds to water molecules. However, the overall hydrophilic or hydrophobic character is determined by the side groups and by the degree to which they shield the skeleton.

For example, side groups such as $-NHCH_3$, $-OCH_2CH_2OCH_2CH_2OCH_3$, glucosyl, and glyceryl, which are themselves hydrophilic, generate solubility of the polymer in water. Methyl side groups are sufficiently small that they do not shield the backbone nitrogen atoms: thus, methylphosphazene polymers are hy-

TABLE 3.1 Physical Properties of Selected Polyphosphazenes

Formula	Properties	T_g (°C)	T_m (°C)
$[NP(OC_6H_4C_6H_5\text{-}p)_2]_n$	Microcrystalline thermoplastic; high refractive index	+93	>350
$[NP(NHC_6H_5)_2]_n$	Glass	+91	—
$[N_3P_3(OCH_2CF_3)_4C_5H_4FeC_5H_4]_n$,[a]	Amber-colored glass	+61	—
$[NP(OC_6H_5)(OC_6H_4C_6H_5\text{-}p)]_n$	Glass	+43	—
$[NP(NHC_2H_5)_2]_n$	Glass; soluble in aqueous acid	+30	—
$[NP(OC_6H_5)(OC_6H_4C_6H_5\text{-}o)]_n$	Glass	+24	—
$[NP(NHCH_3)_2]_n$	Glass; water soluble	+14	—
$[NP(OC_6H_4CH_3)(OC_6H_4CHO)]_n$	Thermoplastic	+11	
$[NP(OC_6H_4COOEt)_2]_n$	Microcrystalline thermoplastic (films)	+7.5	127
$[NP(OC_6H_4COOH)_2]_n$	Glass; soluble in aqueous base	−4.7	—
$[NP(OC_6H_5)_2]_n$	Microcrystalline thermoplastic (films, fibers)	−8	+390
$[NP(OC_6H_5)(OC_6H_4C_2H_5)]_n$	Elastomer	−10[b]	c
$(NPBr_2)_n$	Leathery material; hydrolytically unstable	−15	—
$[NP(O(CH_2)_8CH_3)_2]_n$	Elastomer	−56	—
$[NP(OCH_2CF_3)(OCH_2(CF_2)_x\text{—}CF_2H)]_n$	Elastomer	−60[d]	—
$[N_3P_3(OCH_2CF_3)_x(CH_2SiMe_3)]_n$	Elastomer	−61	—
$[N_3P_3(OCH_2CF_3)_5(CH_3)]_n$	Elastomer	−63	—
$[NP(OCH_2CF_3)_2]_n$	Microcrystalline thermoplastic (films, fibers)	−66	+242 −7.2
$(NPCl_2)_n$	Elastomer; hydrolytically unstable	−66	(+39)[e]
$[NP(OCH_3)_2]_n$	Elastomer	−76	—
$[NP(OC_2H_5)_2]_n$	Elastomer	−84	—
$[NP(OCH_2CH_2OCH_2CH_2\text{—}OCH_3)_2]_n$	Water-soluble elastomer	−84	—
$(NPF_2)_n$	Elastomer, hydrolytically unstable	−96	−68 −40
$[NP(OC_3H_7)_2]_n$	Elastomer	−100	—
$[NP(OCH_2CH_2CH_2CH_3)_2]_n$	Elastomer	−105	—

[a]Ferrocenyl polymer (Structure 3.69), where OR = OCH_2CF_3.

[b]Varies with ratio of side groups.

[c]Complex melting phenomena.

[d]Varies with values of x and ratio of side groups.

[e]For the stretched polymer.

drophilic without being water soluble. On the other hand, side groups such as $-OCH_2CF_3$ or $-OC_6H_5$, which are both hydrophobic and large enough to shield the skeleton, generate strong water repellency and confer solubility in specific organic solvents. For example, the polymer with trifluoroethoxy side groups is soluble in acetone, tetrahydrofuran, or ethyl acetate. The phenoxy derivative is soluble in hot, aromatic hydrocarbons. The ability to tailor the hydrophobic or hydrophilic character is vital if the polymers are to be used in biomedical applications (see later), while insolubility in hydrocarbons is important if the material is to be used for seals or O-rings in aircraft or automobiles.

3.7.3 Water Stable Versus Water "Erodible"

Until recently, polymers that were unstable to water were considered to be unworthy of further consideration. Today it is recognized that such materials can be immensely valuable for use in biomedical applications, especially if the products of hydrolysis are nontoxic.

Most poly(organophosphazenes) are stable to water. Only a selected few are not. These include polymers with amino acid ester side groups and species with imidazolyl, glyceryl, or glucosyl side units. These will be discussed later.

3.7.4 High-T_g Versus Low-T_g Polymers

The side group characteristics needed to move the T_g up or down the temperature scale have been mentioned earlier. The data in Table 3.1 illustrate these principles for a number of different side groups.

3.7.5 Materials Structure Imposed by Side Group Stacking

In crystals that are formed from small molecules, it is often found that large, flat molecules stack together like a pile of plates or saucers. This stacking imposes considerable order and can generate unusual optical or electrical properties.

If a large, flat side group unit is attached to a flexible polymer chain, the same phenomenon may occur. In fact the tendency of the side units to form stacks may overwhelm the preference of the chain to adopt its own preferred conformation (Structure 3.36). In practice, a flexible "spacer group" is usually required to decouple the motions of the side groups from the thermal motions of the chain. In

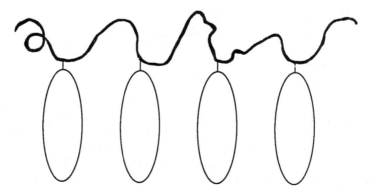

3.36

some systems, side groups on neighboring chains may interleaf to form intermolecular stacks. The phenomenon of polymer liquid crystallinity can arise when side group stacking or colinear orientation occurs.

It will be shown in later sections that side groups such as aromatic azo or biphenyl units generate liquid crystallinity, metal phthalocyanines and tetracyanoquinodimethane generate electroactive domains, and polyaromatic units such as naphthyl or anthracene groups alter the physical properties markedly as they attempt to align their molecular axes or form stacks of side groups.

In the following sections, examples will be given of how structure-property correlations of these types have been used to make materials that are useful in technology or medicine.

3.8 ADVANCED ELASTOMERS

From information presented earlier in this chapter, it will be clear that three types of molecular structures in polyphosphazenes are known to give rise to rubbery or elastomeric properties. These are summarized in Table 3.1. First, if the side groups are single atoms, such as fluorine, chlorine, or bromine, the inherent flexibility of the backbone dominates the materials' flexibility and gives rise to elastomeric character. Unfortunately, the polymers $(NPF_2)_n$, $(NPCl_2)_n$, and $(NPBr_2)_n$ are slowly hydrolyzed in contact with atmospheric moisture. Hence, they are of interest as reaction intermediates but not as usable materials.

The second group of elastomeric polyphosphazenes are those in which the organic side groups are sufficiently flexible that they can readily undergo "avoidance" motions as the skeletal bonds twist. Thus, the side groups impose little or no additional barrier to the twisting motions of the skeleton. Linear side groups such as $-OCH_3$, $-OC_2H_5$, $-OC_3H_7$, $-OC_4H_9$, $-OCH_2CH_2OCH_3$, and $-OCH_2CH_2OCH_2OH_2OCH_3$ fall into this category. Side groups such as these probably assist polymer chain motions in another way also. Being themselves capable of assuming different conformations and shapes, they have a low tendency to fill the available space in the most efficient manner. Thus, molecular voids of "free volume" exist, and this space provides freedom for movement of the polymer chains. The greater the free volume, the greater the material's flexibility. The elastomers in this class are, in general, stable to water.

One phosphazene elastomer in this group has received considerable attention because of its molecular flexibility and its application to battery technology. This is poly[bis(methoxyethoxyethoxy)phosphazene)] (Structure 3.37), also known by the acronym "MEEP."[57,59-61]

$$\left[\begin{array}{c} \text{OCH}_2\text{CH}_2\text{OCH}_2\text{CH}_2\text{OCH}_3 \\ | \\ -\text{N} = \text{P} - \\ | \\ \text{OCH}_2\text{CH}_2\text{OCH}_2\text{CH}_2\text{OCH}_3 \end{array} \right]_n$$

3.37

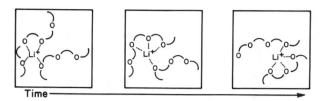

Time ————————————————————————————————▶

Figure 3.11 Ionic electrical conductivity for solutions of lithium triflate in solid poly[bis(methoxyethoxyethoxy)phosphazene] (MEEP) is believed to occur following coordination of the etheric side groups to Li+ ions, cation-anion separation, and ion transfer from one polymer to another as the polymer chains and side groups undergo extensive thermal motions. Reprinted with permission from D. F. Shriver and G. C. Farrington, *Chem. & Eng. News* **1985**, 42. Copyright 1985 American Chemical Society.

Considerable interest exists in flexible polymers that possess etheric units in the backbone or the side groups. Such polymers often function as *solid* solvents for salts such as lithium triflate ($LiSO_3CF_3$) or silver triflate. The interest in polymer/dissolved salt systems is a consequence of their ability to function as ionic conductors of electricity. Solvation of the ions (particularly the cations) by the oxygen atoms, favors ion-pair separation. This is illustrated in Figure 3.11. Moreover, if the polymer chains are sufficiently flexible (and if sufficient free volume exists), the cations can be transferred from chain to chain as the macromolecules undergo thermal motion. If an electric current is applied, the cations will migrate toward the negative electrode and the anions toward the positive electrode.

This principle is made use of in the design and construction of experimental lightweight, high-energy-density rechargeable lithium batteries, as illustrated in Figure 3.12. In theory, there is almost no limit to the size of batteries constructed

Li negative electrode

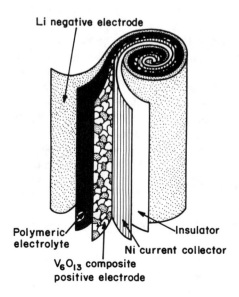

Polymeric electrolyte

V_6O_{13} composite positive electrode

Ni current collector

Insulator

Figure 3.12 Design for a rechargeable lithium battery based on the conductivity of lithium triflate in solid poly[bis(methoxyethoxyethoxy)phosphazene]. Reprinted with permission from D. F. Shriver and G. C. Farrington, *Chem. & Eng. News* **1985**, 42. Copyright 1985 American Chemical Society.

in this way since the solid electrolyte component can be solution cast as a film. Until recently, poly(ethylene oxide), $(CH_2CH_2O)_n$, was the prototype polymer for such applications. However, poly(ethylene oxide) is microcrystalline, and ionic conduction in the amorphous regions is interrupted by the microcrystalline domains. For this reason, this polymer must be heated to nearly 100°C before a high ionic conductivity can be obtained.

On the other hand, polyphosphazene (Structure 3.37) is noncrystalline. The ionic conductivity of its lithium triflate complexes at room temperature is 1000 times greater than that of (polyethylene oxide). For battery-type applications, it must be cross-linked lightly to prevent slow liquidlike flow, but this can be accomplished without lowering the conductivity. A similar type of polymer, with a polysiloxane backbone and oligoether side groups, is being studied for similar applications.

The third group of polyphosphazene elastomers includes species in which two or more different alkoxy, aryloxy, or organosilicon units are attached to the polymer chains.[62-68] The principle behind the design of elastomers of this type is that the random distribution of two or more different side groups reduces the tendency for interchain alignment and crystallinity, and also (if the side groups are of different dimensions) generates free volume that assists macromolecular reorientation. Three main classes of elastomers that fulfill these requirements are known— mixed-substituent fluoroalkoxyphosphazenes, mixed-substituent aryloxyphosphazenes, and polymers that contain both organosilicon and alkoxy or aryloxy cosubstituent groups. These will be considered in turn.

The first polyphosphazenes to be developed commercially were polymers with two different types of fluoroalkoxy side groups attached to the same chain.[62-65] A typical example is shown in Structure 3.38.

$$\left[-N = P - \begin{array}{c} OCH_2CF_3 \\ | \\ | \\ OCH_2(CF_2)_xCF_2H \end{array} \right]_n$$

3.38

This polymer 3.38 has a more complicated structure than is implied by the illustration. The polymer is synthesized by the reaction of a mixture of CF_3CH_2ONa and $CHF_2(CF_2)_xCH_2ONa$ with poly(dichlorophosphazene). This means that the two nucleophiles *compete* for replacement of the available chlorine atoms, a process that could lead to a random distribution of side groups, but could also generate repeating units in which the phosphorus atoms bear two of the same side groups.

Elastomers of this type are usually cross-linked during fabrication, and often contain "fillers" such as carbon black or iron oxide to reduce the "compliance" of the elastomer (i.e., to provide a greater resistance to deformation). Such materials are depicted in Figure 3.1. They are used in technology because of their flexibility and elasticity at low temperatures ($-60°C$); their resistance to hydro-carbon solvents, oils, and hydraulic fluids; and their flame resistance.[67] For these reasons, they are utilized in aerospace and advanced automotive applications. Some interest exists in their development as inert biomaterials, mainly because of their surface hydrophobicity and consequent biocompatibility.

The mixed-substituent aryloxyphosphazenes are typified by the structure shown in Structure 3.39, where R is an alkyl group.[66] Again, the polymers prob-

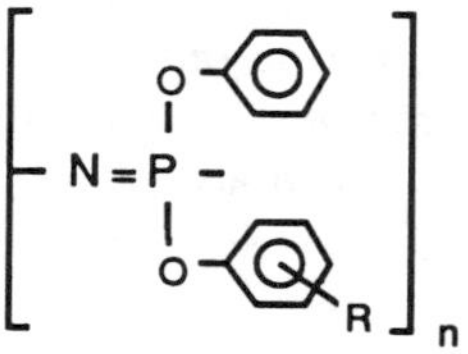

3.39

ably contain some repeating units with a single type of side group. The glass tran-sition temperatures of these materials are not as low as for the fluoroalkoxy deriv-ative. These polymers are of interest mainly as nonflammable, sound-, heat-, and electrical-insulating materials, often in the form of an expanded foam rubber.

The final class of mixed-substituent elastomers is still at the stage of labo-ratory investigation, rather than manufacturing. They comprise a group of poly-mers such as Structures 3.40 or 3.41 that contain organosilicon side groups as

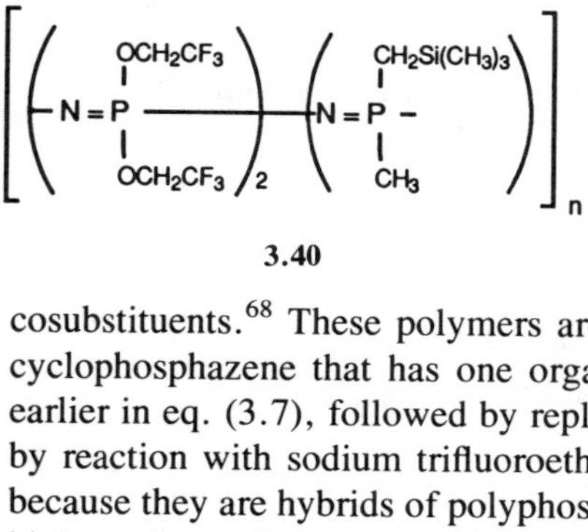

3.40 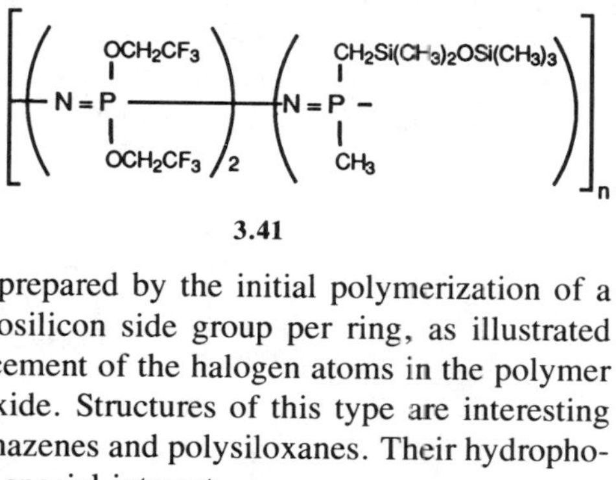   3.41

cosubstituents.[68] These polymers are prepared by the initial polymerization of a cyclophosphazene that has one organosilicon side group per ring, as illustrated earlier in eq. (3.7), followed by replacement of the halogen atoms in the polymer by reaction with sodium trifluoroethoxide. Structures of this type are interesting because they are hybrids of polyphosphazenes and polysiloxanes. Their hydropho-bicity and membrane properties are of special interest.

3.9 FIBER- AND FILM-FORMING POLYPHOSPHAZENES

The main requirements for good textile fiber- or film-forming polymers are a T_g below room temperature (to provide flexibility at normal temperatures), together with microcrystallinity, to provide strength, orientability, and (sometimes) fabri-cation by melting. The classical organic polymers (nylons, polyesters, and poly-olefins) generally have these characteristics.

As discussed earlier, microcrystallinity in polyphosphazenes can generally be achieved only with single-substituent polymers such as those with OCH_2CF_3, OPh, or OC_6H_4R-p side groups,[14, 15, 69–71] although some mixed-substituent derivatives are also crystalline.[69] Many of these polymers are film- and fiber-forming materials. They can be fabricated more easily by solution casting or solution spinning than by melt-fabrication techniques. This is because the microcrystallites melt at temperatures that are so high (sometimes over 250°C) that the polymers begin to depolymerize as they are melt fabricated.

In general, the fiber, film, and coatings technology of polyphosphazenes is at an early stage of development, with laboratory studies more evident than manufacturing.

3.10 POLYPHOSPHAZENES AS BIOMEDICAL MATERIALS

3.10.1 Background

The use of synthetic polymers in medicine is a subject of wide interest. Polymers are used or are being considered for use in replacement blood vessels, heart valves, blood pumps, dialysis membranes, intraocular lenses, and surgical sutures and in a variety of targeted, controlled drug delivery devices. One class of inorganic-organic polymers, the poly(organosiloxanes), have been used for many years as inert prostheses and heart valves. Biomedical materials based on polyphosphazenes are being considered for nearly all the uses mentioned earlier.

A significant advantage of poly(organophosphazenes) over nearly all other biomedical polymers is the ease with which different organic side groups can be linked to the polyphosphazene chain. Thus, by changing the types of side groups present, it is possible to generate virtually any combination of properties needed for a particular biomedical application, including the ability to hydrolyze ("erode") in the body to nontoxic small molecules that can be metabolized or excreted. In the following sections an overview of this topic will be given, organized according to the design and synthesis of the following types of materials: solid, bioinert polymers; amphiphilic polymers and membrane materials; hydrogels; and water-soluble bioactive polymers.

3.10.2 Solid, Bioinert Polyphosphazenes

Some of the most useful biomaterials in use at the present time are inert, hydrophobic organic polymers, such as polyethylene or poly(tetrafluoroethylene) (Teflon or GoreTex). Much of the utility of these materials stems from their hydrophobic surface character which minimizes the "foreign body" interactions that normally occur when nonliving materials are implanted in contact with living tissues, including blood. Three different types of poly(organophosphazenes) are being

Figure 3.13 Illustration of the extreme hydrophobicity of the surface of poly-[bis(trifluoroethoxy)phosphazene. In this example, a cotton textile material has been impregnated with the polymer. Water droplets do not wet the surface.

studied as bioinert materials, and these are typified by the polymers shown earlier as Structures 3.10 and 3.12, and as Structures 3.38 and 3.41.

Surface hydrophobicity is estimated by the contact angle formed when a drop of water is placed on the surface of a polymer. Low-contact-angle values are found when the droplet interacts with and wets the surface. High contact angles indicate a hydrophobic surface (Figure 3.13). All the polymers shown have high contact angles in the region of 100–107°. Poly(tetrafluoroethylene) has a value of 107°. The fluoroalkoxy and aryloxy derivative (Structures 3.10 and 3.12) have been studied by means of implantation tests in animals and have shown minimal tissue response.[72] Thus, polyphosphazenes of these types are good candidates for use in cardiovascular replacements (heart pumps, heart valves, or blood vessel prostheses), or as coatings for pacemakers or other implantable devices.

3.10.3 Solid Polymers with Hydrophilic or Bioactive Surfaces

Not all biomaterials have hydrophobic surfaces. Some are deliberately designed to stimulate tissue adhesion or infiltration or to generate a biochemical response. Such polymers are hydrophilic and chemically inert, or they have bioactive agents on the polymer surface.

3.10.3.1 Amino- and alkoxy-phosphazenes. First, many amino-phosphazene polymers are intrinsically hydrophilic because the amino side groups

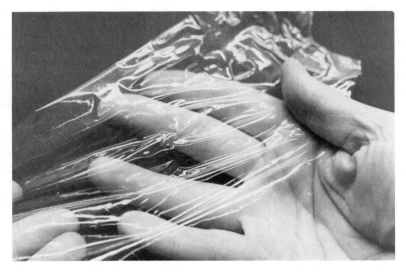

Figure 3.14 A thin film of poly[bis(butylamino)phosphazene] fabricated by solution casting for membrane experiments.

can form hydrogen bonds with water molecules. Examples include polymers with side groups such as CH_3NH-, C_2H_5NH-, C_3H_7NH-, and so on[73] (Figure 3.14). In addition, certain alkoxy side groups such as $-OC_2H_5$ or $-OCH_2CH_2OCH_3$, or aryloxy units with OH or COOH substituent groups, also generate hydrophilic surface character. In some of these cases the polymers are so hydrophilic that they may dissolve in water, but this tendency can be counteracted by cross-linking or by the presence of hydrophobic cosubstituent units.

3.10.3.2 Sulfonic acid groups.

Biological compatibility to blood or other tissues often depends on surface charge or the presence of ionic groups. One way in which such surface structure can be generated is by sulfonation of the surface of an aryloxyphosphazene by treatment with concentrated sulfuric acid (Scheme 3.8).[74]

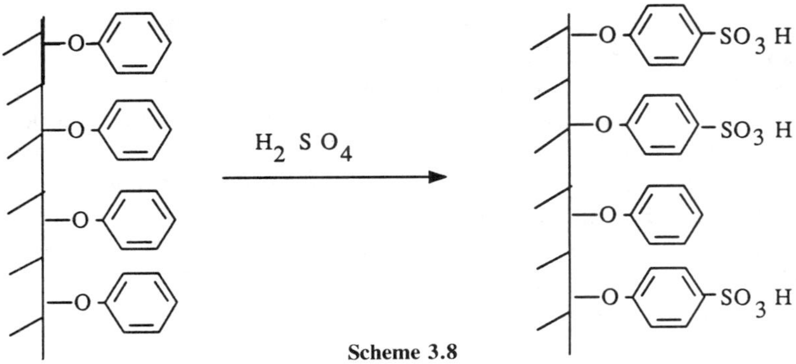

Scheme 3.8

The reaction conditions must be controlled carefully to prevent dissolving of the polymer in strong acid and subsequent chain cleavage. But, under the right conditions, an outer layer of sulfonated aryloxyphosphazene units is formed. This change converts a hydrophobic surface (contact angle, 101°) to a hydrophilic surface (contact angle near 0°) in one simple step. The influence of this transformation on blood and tissue compatibility is still being investigated, but it is known that similar changes in the surface character of organic polymers bring about a dramatic improvement in blood compatibility.

3.10.3.3 Heparinized surfaces. A second approach to the surface modifications of polyphosphazenes is illustrated in Scheme 3.9. Here, the objec-

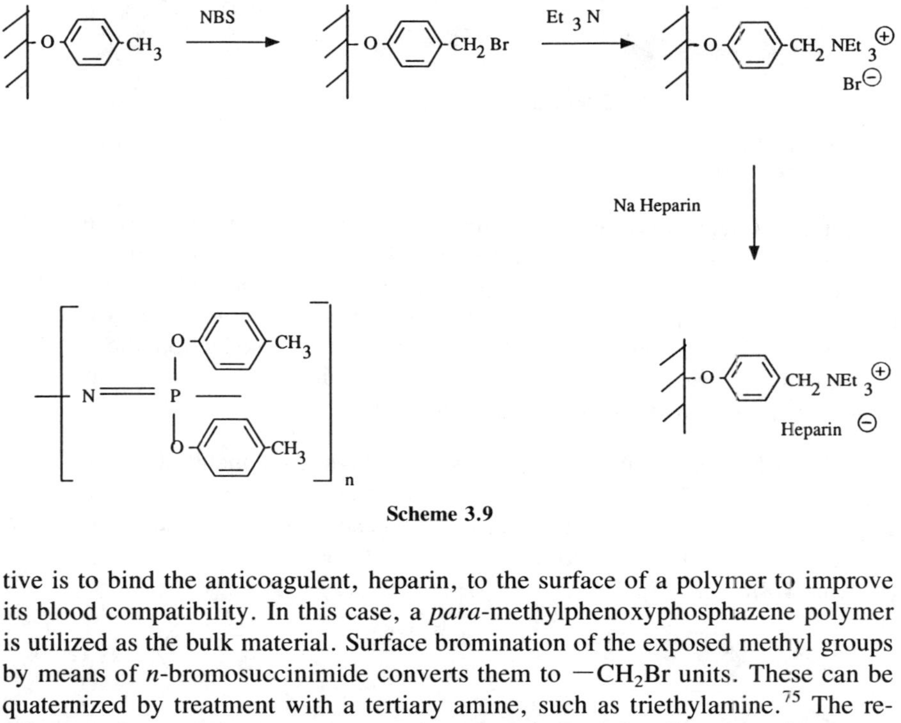

Scheme 3.9

tive is to bind the anticoagulent, heparin, to the surface of a polymer to improve its blood compatibility. In this case, a *para*-methylphenoxyphosphazene polymer is utilized as the bulk material. Surface bromination of the exposed methyl groups by means of *n*-bromosuccinimide converts them to $-CH_2Br$ units. These can be quaternized by treatment with a tertiary amine, such as triethylamine.[75] The resultant surface, when exposed to an aqueous solution of sodium heparin, undergoes anion exchange, formation of sodium bromide which is removed, and ionic binding of the heparin anion to the surface. Polymer surfaces treated in this way show a dramatic (fivefold) improvement in blood compatibility as measured by blood clotting times. However, over a long period of time the heparin may be leached from the surface and the beneficial properties are lost. This illustrates a key distinction between the behavior of bioactive species that are ionically bound or even adsorbed on a surface and those that are linked to it through covalent bonds. The following two examples show how covalent binding of biologically active species can be accomplished.

3.10.3.4 Immobilized enzymes.

This example illustrates how enzymes can be linked covalently to the surface of a poly[bis(aryloxy)phosphazene].[76] Numerous reasons exist for attempting to immobilize enzymes on the surfaces of polymers. For one thing, surprisingly perhaps, immobilization often enhances the length of time over which the protein maintains its catalytic activity, compared to the same enzyme free in solution. Second, immobilization facilitates the separation of the enzyme from the products of an enzyme-induced reaction, for example, by using the polymer/enzyme conjugate as the stationary phase in a continuous-flow reactor. Third, it may be possible to immobilize several different enzymes on the same surface so that the products of a reaction induced by one enzyme become the starting materials for a transformation catalyzed by the second, and so on. For all such processes it is usually advantageous to use a form of the immobilization polymer that provides the greatest possible surface area. For this reason, small particles of the polymer are normally used.

In the system to be described, the high surface area was ensured by depos-

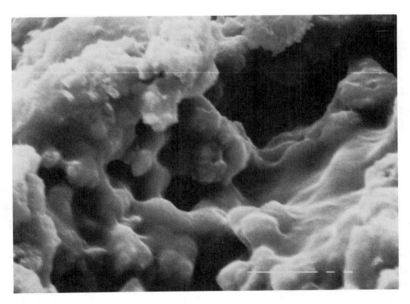

Figure 3.15 Scanning electron micrograph of the surface of a porous alumina particle covered with poly[bis(phenoxy)phosphazene]. The polymer was then surface nitrated and reduced, and the reactive sites were used for coupling to enzymes. Reprinted from H. R. Allcock and S. Kwon, *Macromolecules* **1986,** *19* 1502. Copyright 1986 American Chemical Society.

iting a thin film of poly[bis(phenoxy)phosphazene] (Structure 3.12) on to the surface of highly porous small particles of alumina powder. Provided the deposition of the polymer is controlled carefully so as not to block the microscopic pores, an enormous surface area of polymer can be generated (see Figure 3.15). Surface chemistry on the polymer was then carried out according to the reactions shown in Scheme 3.10.

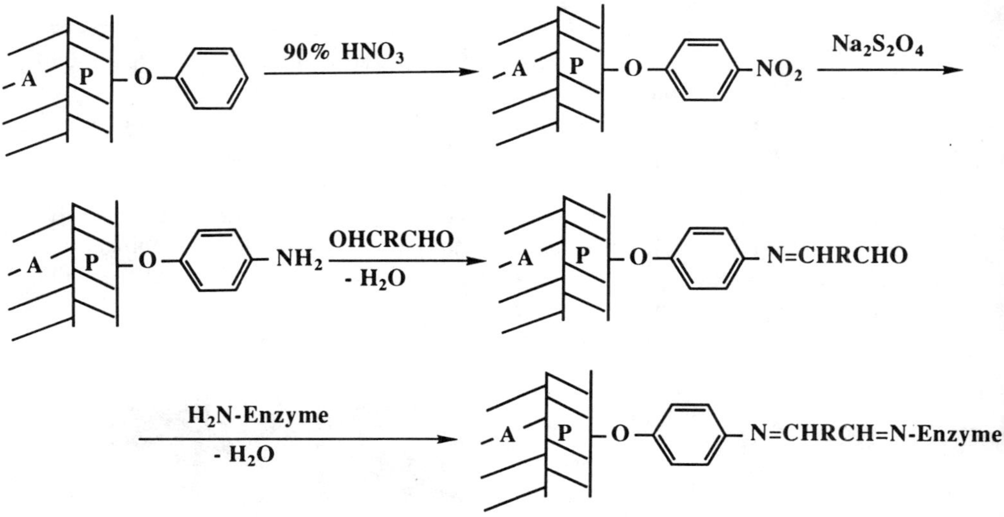

A = Alumina
P = Poly[bis(phenoxy)phosphazene]
R = (CH$_2$)$_3$

Scheme 3.10

Surface nitration generated nitrophenoxy units at the interface. Reduction with dithionite then converted the nitro groups to amino functions. At this stage a variety of methods were used to couple the enzymes trypsin or glucose 6-phosphate dehydrogenase to the surface. The most effective method was to use glutaric dialdehyde [OHC(CH$_2$)$_3$CHO] as a coupling agent. This reagent reacts with both the amino groups on the surface and amino groups on the enzyme to complete the immobilization process.

A large part of the enzymic activity was retained after immobilization, as illustrated in Figure 3.16. Moreover, packing of the conjugate particles into a chromatography-type column allowed the construction of a continuous-flow reactor. Figure 3.16 illustrates that the enzyme molecules continued to catalyze their reactions for long periods of time. Practical uses of immobilized enzymes can be

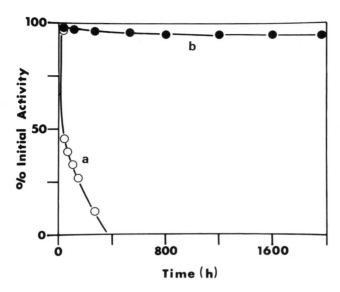

Figure 3.16 Plots showing the activity of the enzyme, glucose-6-phosphate dehydrogenase as a function of time for (○) the free enzyme in solution and (●) the enzyme bound to the surface of poly[bis(phenoxy)phosphazene]. The polyphosphazene is an excellent surface substrate because of the stability of the polymer backbone to nitration and reduction, and the ease with which the hydrophobicity or hydrophilicity of the surface can be changed. From H. R. Allcock and S. Kwon, *Macromolecules* **1986,** *19* 1502. Copyright 1986 American Chemical Society.

foreseen not only in the biochemical and fermentation industries, but also in medical diagnostic equipment and, ultimately, in the construction of the artificial liver.

3.10.3.5 Immobilized dopamine.
The last example involves a series of reactions that lead to the immobilization of catecholamines, such as dopamine.[77] Again, an aryloxyphosphazene was chosen as the substrate material, but in this case the functional groups were introduced at the polymer synthesis stage. As shown in Scheme 3.11, both phenoxy and *para*-nitrophenoxy groups can be linked to the polyphosphazene chain during the macromolecular substitution step. The nitro groups were then catalytically reduced to amino groups, and these were diazotized to give the diazonium salt. Treatment with dopamine then generated the polymer/dopamine system via a diazo coupling reaction. All these reactions were carried out with the polymer in solution. The polymer was then cast into thin films by solvent evaporation. The surface of these films bore immobilized dopamine units as shown in Structure 3.42.

Biochemical tests showed that cultured rat pituitary cells in contact with the films underwent the same biochemical responses as did similar cells immersed in

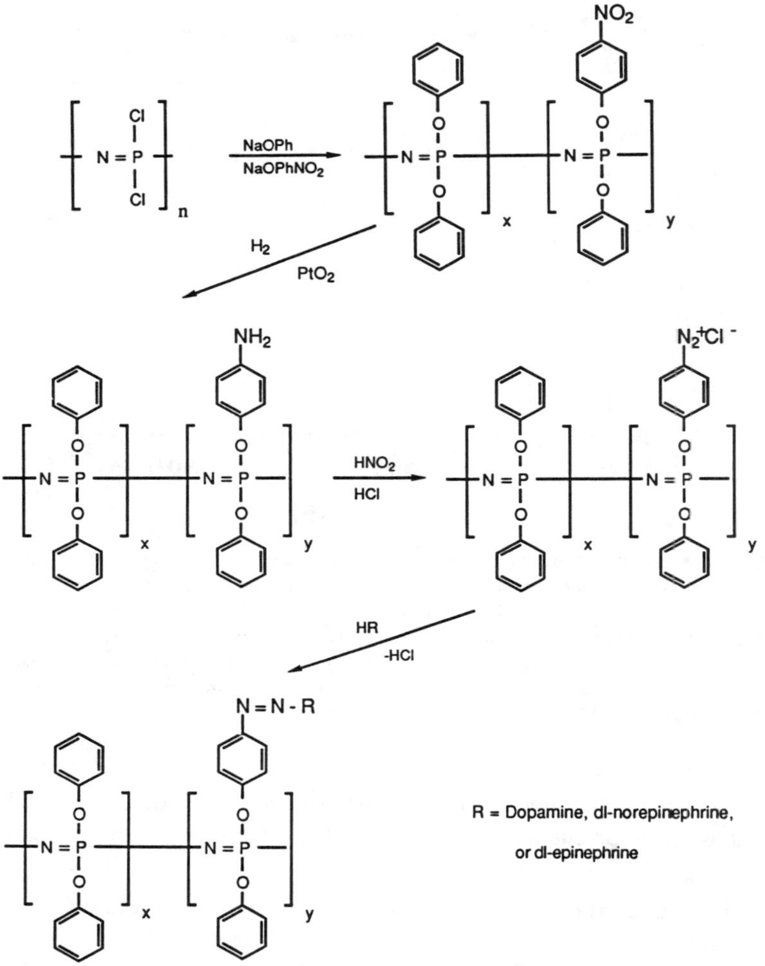

Scheme 3.11

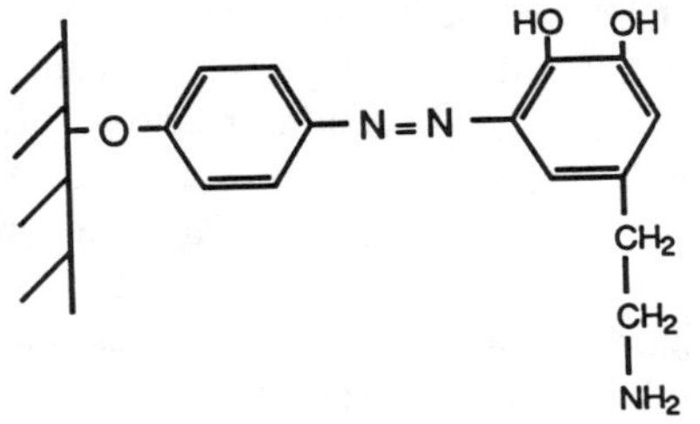

R = Dopamine, dl-norepinephrine, or dl-epinephrine

3.42

solutions of dopamine. Thus, it appears that the immobilized dopamine is capable of interacting with the cell surface receptors of pituitary cells even though the dopamine is held close to the polymer surface by a relatively short and rigid spacer group. The development of this principle is of great interest in medicine since it would allow the stimulation of biochemical responses in a patient, and the possibility that the response could be turned off simply by removal of the implanted device. The utilization of such mechanisms to stimulate antigen production is an important prospect.

3.10.4 Bioerodible Polyphosphazenes

As mentioned, one of the main advantages of polyphosphazenes over most classical organic polymers is the fact that a few specific polymers can be designed to decompose slowly in water at the pH of many body fluids (pH 7.5) to give products that are nontoxic. The erosion process converts a macromolecular system that has all the attributes of a polymer (strength, elasticity, ease of fabrication, ability to act as a reservoir for other molecules) to small molecules that are soluble in water and can be removed by metabolism or excretion. Bioerodible polymers are of interest as matrices for the controlled release of drug molecules at a specific site in the body (Figure 3.17) and as absorbable surgical sutures that avoid the necessity for the secondary surgery normally needed to remove the nonabsorbable (catgut) variety.

Because of the potential importance of this subject in the development of both medicine and polyphosphazene chemistry, considerable research has been carried out in this area. The seven systems discussed next provide an overview of the possibilities.

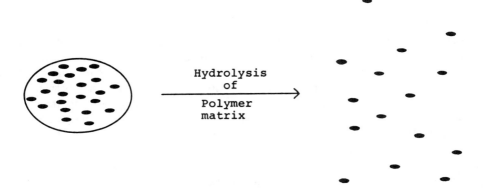

Figure 3.17 Bioerodible polymers can be used for the controlled release of pharmaceutical molecules (black ellipses). Ideally, hydrolysis of the implanted matrix polymer should occur at the polymer surface only so that the drug molecules are released at a constant rate—a so-called zero-order release profile.

3.10.4.1 Amino acid ester derivatives.   The first class of bioerodible polyphosphazenes to be considered are derivatives with amino acid ester side groups. They are prepared by the reaction of poly(dichlorophosphazene) with the ethyl esters of amino acids such as glycine, phenylalanine, and so on (Scheme 3.12).[78] The *ethyl ester* must be used as the nucleophile in this reaction for two

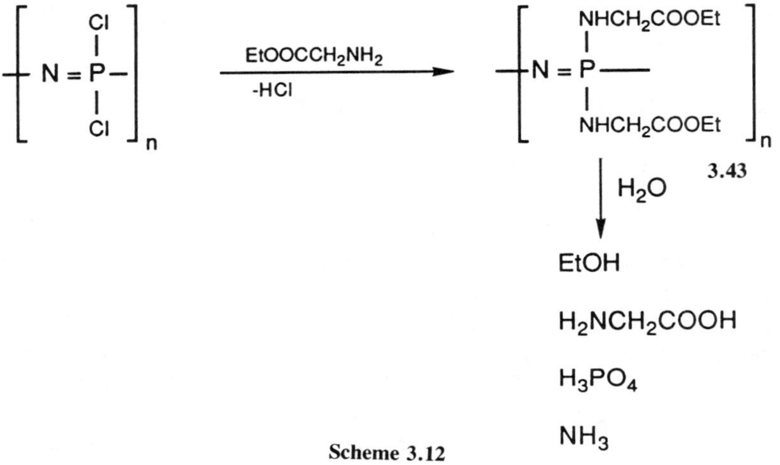

Scheme 3.12

reasons. First, a free carboxylic acid unit would provide a second nucleophilic site that could lead to cross-linking of the polymer chains. Thus, the ethyl ester function serves as a protecting group. Second, even if the amino terminus were the only site to react, the pendent free carboxylic acid unit would induce rapid decomposition of the polymer. In fact, this is what happens once the side unit is deprotected during hydrolysis. Hence the ester function stabilizes the polymer until it is exposed to an aqueous environment.

The utility of polymers such as Structure 3.43 is that they are sensitive to hydrolysis at body pH and that, on hydrolysis, they yield ethanol, amino acid, and phosphate, which can be metabolized, and small amounts of ammonia, which can be excreted.[78] Moreover, the rate of hydrolysis depends on the type of amino acid ester used. For example, phenylalanine derivatives hydrolyze more slowly than do glycine derivatives. The mechanism of hydrolysis is quite complex and changes with variation in pH.[79] In near-neutral aqueous media, the ester function appears to be hydrolyzed first to deprotect the carboxylic acid units. Hydrolytic cleavage of a side group P—N bond occurs next to release glycine. This leaves a hydroxyl group attached to phosphorus in place of the organic side group. Hydroxyphosphazenes are intrinsically unstable, and the skeletal bonds are broken as degradation to phosphate and ammonia occurs.

These polymers show a great deal of promise as biomedical materials. In subcutaneous tissue compatibility experiments, the polymers decompose without evidence of toxicity, irritation, or giant cell formation. They have been studied in detail for the controlled release of the antitumor agent, L-phenylalanine mustard (melphan).[80] A development from this chemistry is the use of the bioerosion process to release drug molecules that are bound to the polymer skeleton as substituents or cosubstituents. For example, the two steroidal structures shown in Structures 3.44 and 3.45 have been used, along with ethyl glycinate groups, as a means for the controlled release of the steroid.[81]

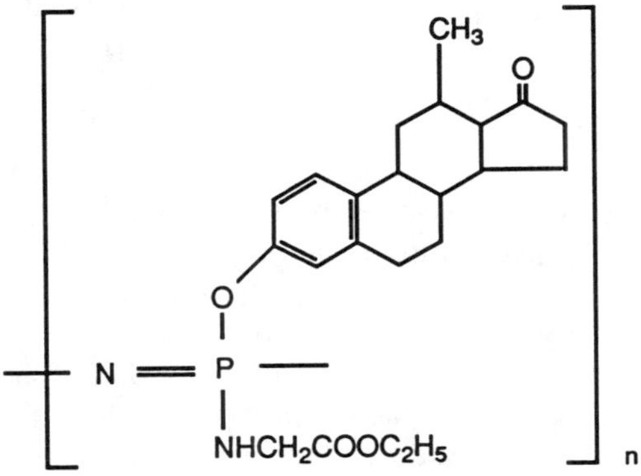

3.44

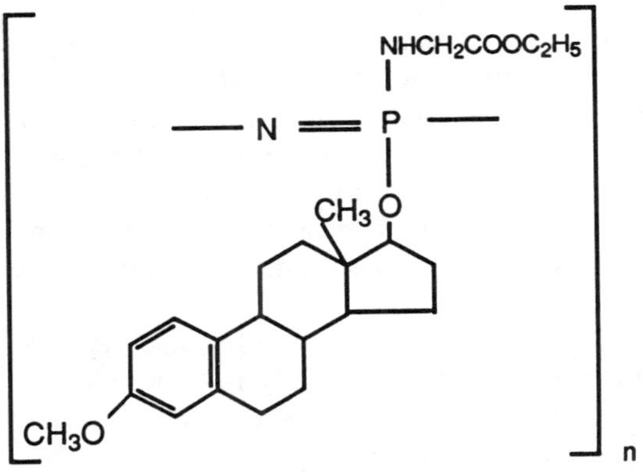

3.45

In a similar fashion, a controlled release system has been developed based on polyphosphazenes that bear amino acid ester side groups and covalently bound naproxen [($^+$)-2-(6-methoxy-2-naphthyl)propionic acid].[82]

3.10.4.2 Imidazolylphosphazene polymers.

Amino acid ester side groups are not the only units that sensitize the system to hydrolysis. The *imidazolyl* group has an even greater effect.[79,83] For example, the polymer of Structure 3.46, prepared by the reaction of poly(dichlorophosphazene) with imidazole, is so unstable hydrolytically that it decomposes in moist air to imidazole, phosphate, and

                    3.46   3.47

ammonia. This is too high a sensitivity for most biomedical applications. Hence, an emphasis has been placed on the study of polymers such as Structure 3.47 in which a hydrophobic cosubstituent group, such as aryloxy, is present to reduce the rate of erosion.[84] The aryloxy substituent serves a second purpose also. For drug release applications, it is normally preferred that the device should erode steadily *from the outside* only. Polymers such as Structure 3.46 are so hydrophilic that water absorption followed by catastrophic breakdown of the capsule may occur, with the result that the drug is released precipitously in a massive dose. This is a situation to be avoided. The hydrophobic groups in Structure 3.47 prevent rapid water penetration and, in so doing, limit the hydrolysis to the surface layers.

The polymer of 3.47 has been evaluated as a matrix for the controlled release of progesterone.[84] It was first shown that the rate of release of this steroid and of bovine serum albumin can be controlled by variations in the ratio of aryloxy to imidazolyl side groups attached to the polyphosphazene chain. In vitro and in vivo studies were conducted to examine the release rate of labeled steroid from devices implanted subcutaneously in rats. Typical data are shown in Figure 3.18. The biocompatibility of this system, at least in rats, was found to be good.

3.10.4.3 Schiff's base linkages.   Another approach to the use of polyphosphazene chemistry in pharmacology is to design a polymer in which a drug molecule is linked to the side groups via a hydrolyzable linkage. Hydrolysis either at the surface of a solid polymeric film or capsule or hydrolysis from individual polymer molecules in solution would then bring about release of the drug in a specific region of the body. (Solutions of a polymeric drug would be injected into a compartment of the body or into a tumor, and the diffusion of the carrier polymer from that site would be restricted by the inability of polymer molecules to pass through semipermeable membranes.)

A simple prototype approach to a system of this type is illustrated by the synthesis of Structure 3.48.[85] An aldehydic group at the terminus of an aryloxy side unit is allowed to form a Schiff's base linkage to a biologically active amine. Schiff's base linkages are hydrolytically unstable. Hence, exposure to water cleaves this bond and brings about release of the bioactive agent. Amino-4-picoline, sulfadiazine, and dopamine have been immobilized and released from polyphospha-

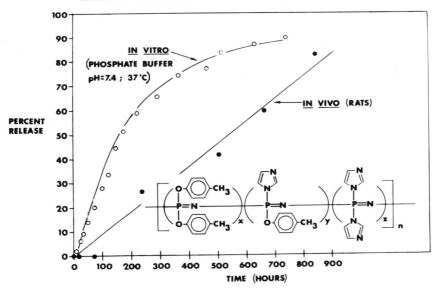

Figure 3.18   Release rate of radioactive progesterone from an erodable matrix of a poly(imidazolyl-aryloxy)phosphazene polymer. From work by C. Laurencin, H. J. Koh, T. X. Neenan, H. R. Allcock, and R. S. Langer. See also (same authors), *J. Biomed. Mater. Res.* **1987**, *21*, 1231.

zenes by this mechanism. As in the previous example, the rate of release can be controlled by the ratio of phenoxy to substituted phenoxy units in the polymer.

3.10.4.4 Bioactive aminophosphazene polymers (local anesthetics).
A related approach is to synthesize a polyphosphazene with an amino drug linked directly to phosphorus through a side group phosphorus-nitrogen bond. The procaine derivative shown in Structure 3.49 is a typical example designed to

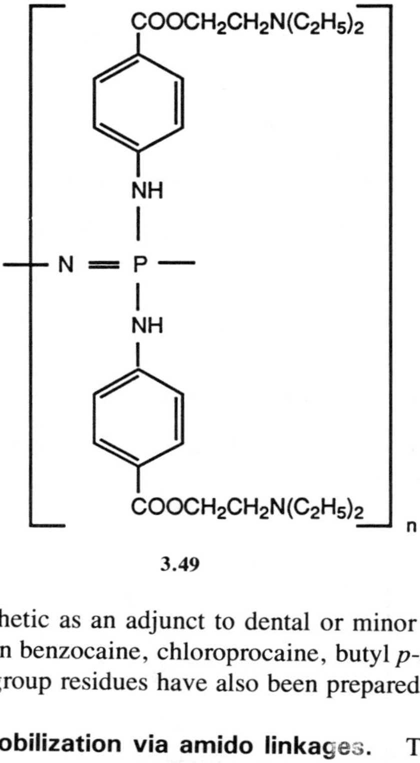

$$\left[\begin{array}{c} COOCH_2CH_2N(C_2H_5)_2 \\ \\ NH \\ | \\ -N=P- \\ | \\ NH \\ \\ COOCH_2CH_2N(C_2H_5)_2 \end{array} \right]_n$$

3.49

release the local anesthetic as an adjunct to dental or minor surgical procedures. Derivatives that contain benzocaine, chloroprocaine, butyl p-amino-benzoate, and 2-aminopicoline side group residues have also been prepared.[86]

3.10.4.5 Immobilization via amido linkages.
The amido linkage is relatively labile in an aqueous environment. Hence, it is a promising connector group to link bioactive molecules with carboxylic acid units to polyphosphazenes that have amino groups in the side chain structure. An example is shown in Structure 3.50.

By this means, oligopeptides, N-acetyl-DL-penicillamine, p-(dipropyl-sulfamoyl)benzoic acid, and nicotinic acids can be linked to a polyphosphazene structure.[87] Such species can be water insoluble but bioerodable, or water soluble, depending on the cosubstituent groups.

3.10.4.6 Glyceryl phosphazenes.
Glycerol is a trifunctional reagent. Attempts to use such a molecule as a nucleophile for reaction with poly(dichlorophosphazene) would lead to cross-linking of the system. Hence, the

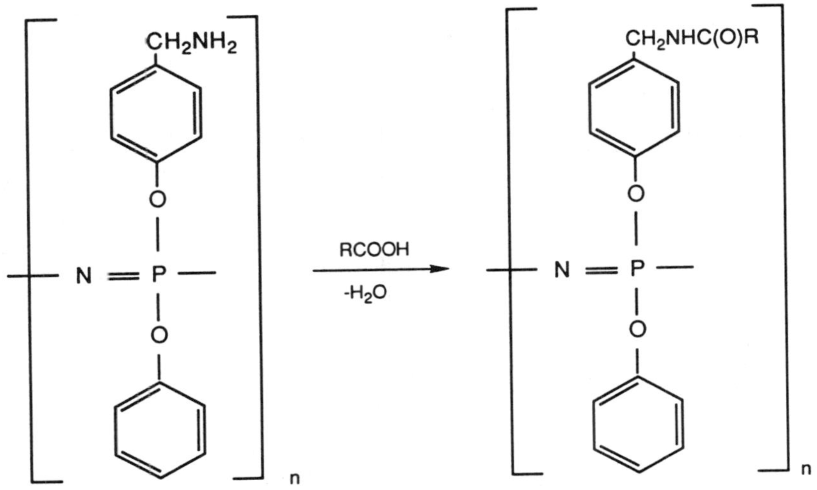

3.50

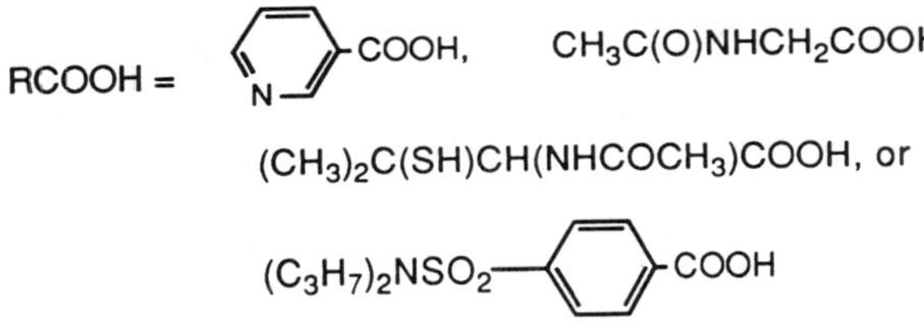

$$RCOOH = \quad \text{(pyridine)}-COOH, \quad CH_3C(O)NHCH_2COOH,$$

$$(CH_3)_2C(SH)CH(NHCOCH_3)COOH, \text{ or}$$

$$(C_3H_7)_2NSO_2-\text{(benzene)}-COOH$$

linkage of glycerol to the polymer chain must be preceded by protection of two of the hydroxyl groups. After the macromolecular substitution reaction has been carried out, the protected groups must be deprotected. The chemistry is outlined in Scheme 3.13.[88]

Poly[bis(glyceryl)phosphazene] (Structure 3.51) is soluble in water. However, it is hydrolyzed slowly to glycerol, which can be metabolized readily, and to phosphate and ammonia. A further advantage of this polymer is that the pendent hydroxyl groups are sites for the attachment of drug molecules. If a water-insoluble polymer is needed, this can be accomplished by the use of hydrophobic cosubstituents or by cross-linking of the chains.

3.10.4.7 Glucose-substituted polyphosphazenes.

Polyphosphazenes that bear glucosyl side (Structure 3.52) groups are promising biomedical polymers. For one thing, they are soluble in water. When hydrolyzed they yield glucose, phosphate, and ammonia. The presence of four free hydroxyl groups per side unit offers many opportunities for the linkage of bioactive agents. Moreover, the synthesis of polymers with varying ratios of glucosyl to other side groups allows the system to be fine-tuned with respect to water solubility, surface character, and hydrolytic stability or instability. In many respects, glucosyl polyphospha-

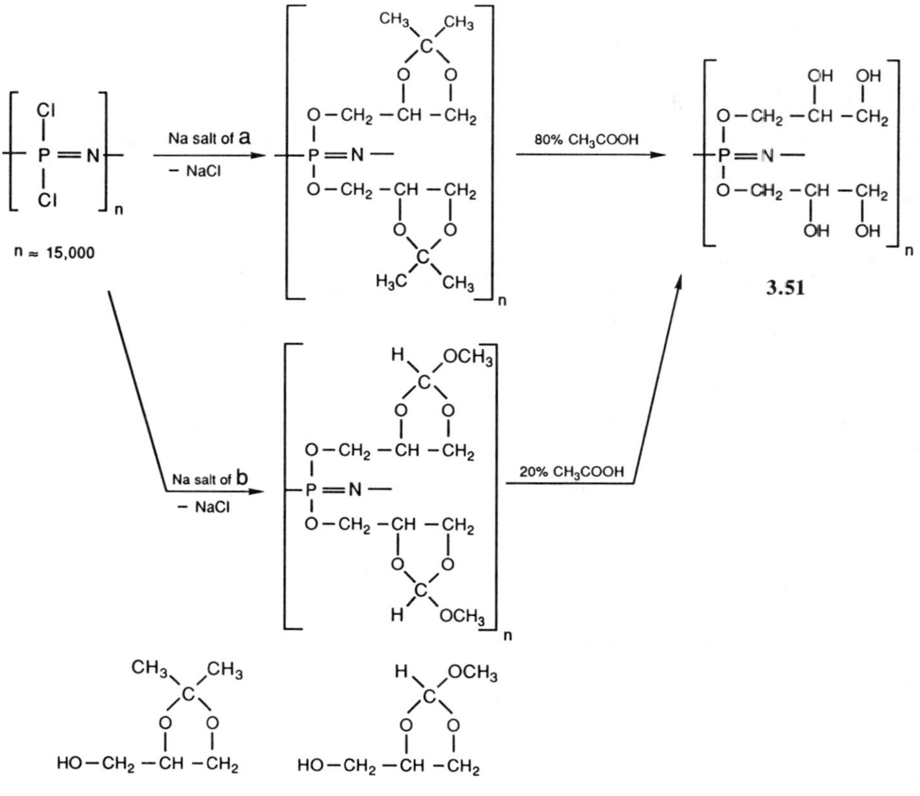

Scheme 3.13

zenes can be viewed as "inorganic polysaccharides," with close similarities being evident to an amorphous cellulose (with a built-in flame retardant!), and with all the side group reactions that are available for cellulose.

Obviously, the synthesis of such polymers requires a hydroxyl group protection-deprotection cycle to prevent a gross cross-linking of the system during the macromolecular substitution step. The synthesis process is summarized in Scheme 3.14.[89]

First, four of the hydroxyl groups of glucose must be protected by the formation of diacetone glucose. This is converted to its monosodium salt, and this species is allowed to react with poly(dichlorophosphazene). The diacetone glucoxide units are quite bulky species, and the replacement of every single chlorine atom along the polyphosphazene chains is not easy. However, the use of a second, less hindered nucleophile to complete the reaction assists this process. The pendent hydroxyl groups must then be deprotected by careful treatment with trifluoroacetic acid. Excessive exposure to this reagent causes backbone cleavage.

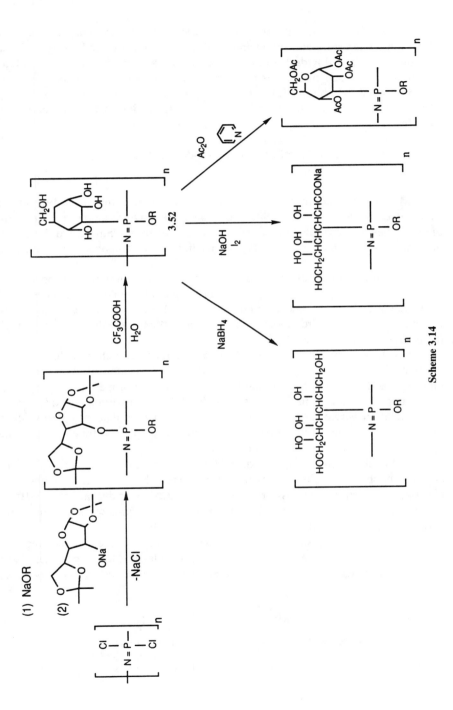

Scheme 3.14

111

In spite of the difficulty of synthesis, glucosyl polyphosphazenes are some of the most interesting and promising derivatives. They are moderately resistant to hydrolysis but will hydrolyze to glucose, phosphate, and ammonia over a long period of time. When cosubstituents are present, their properties can be fine-tuned to solve a number of biomedical problems—not only in the area of controlled drug delivery, but also in the design of materials for tissue ingrowth and blood compatibility.

3.10.5 Amphiphilic Polymers and Membrane Materials

Membranes are thin polymeric films that may permit the faster diffusion of some molecules than of others. Such materials are known as semipermeable membranes. They are essential components of nearly all living things, and the development of new materials of this type is an important component of biomedical research.

The control of diffusion of molecules through a membrane can be accomplished by variations in the hydrophilicity of the polymer molecules that constitute the membrane. As in biological membranes, hydrophobic molecules are more likely to pass through the hydrophobic domains of a synthetic membrane than through the hydrophilic regions, and vice versa.

It will be fairly obvious from the earlier parts of this chapter that the synthesis methods used in polyphosphazene chemistry offer almost unique opportunities for modifying and fine-tuning the surface and bulk properties of a polymer. For example, variations in the ratios of hydrophilic and hydrophobic side groups can generate a whole range of "amphiphilic" properties that may be useful in membrane design and technology. The following example illustrates some of the opportunities that exist in this area of research.[90]

The methylamino group is hydrophilic. Trifluoroethoxy and phenoxy groups are hydrophobic. Thus, by varying the ratios of methylamino to trifluoroethoxy or phenoxy in structures such as Structures 3.53 or 3.54 over the range of 0–100% of each type of group, it is possible to vary the surface properties and the semipermeable membrane behavior.

The methylamino single-substituent polymer has a surface contact angle of 30°, whereas the trifluoroethoxy- and phenoxy-single-substituent polymers have values in the range of 100–107°. Although a rough relationship exists between

$$
\left[\begin{array}{c} NHCH_3 \\ | \\ -N=P- \\ | \\ OCH_2CF_3 \end{array} \right]_n
\qquad
\left[\begin{array}{c} NHCH_3 \\ | \\ -N=P- \\ | \\ O-\langle\bigcirc\rangle \end{array} \right]_n
$$

3.53   **3.54**

side group ratio and surface contact angle, the values can be distorted by the fact that hydrophilic side groups will be concentrated at the polymer surface if that surface was formed in contact with a hydrophilic material such as glass. Moreover, measured contact angles may change over time as side groups that were concentrated at the surface when the surface was formed become buried by molecular motions as the polymer is stored. The high chain mobility of the phosphazene skeleton tends to accelerate this process.

The semipermeability behavior of a polymer film can be measured by equipment that permits measurements of the diffusion of a small-molecule dye through the membrane. In the phosphazene systems investigated, the rate of diffusion varied with the side group ratio. A typical polymer with roughly 50% of the side groups as methylamino and 50% as trifluoroethoxy showed a faster transmission of dye molecules than did standard cellulose dialysis tubing.[90]

Membrane design and fabrication require more optimization than does the synthesis of the right type of polymer. For example, those phosphazene polymers that contained the highest ratios of methylamino groups were too brittle to be used as membranes (because of the high glass transition temperatures) and too soluble in aqueous media. However, the polymers could be made insoluble by radiation cross-linking (Scheme 3.15).

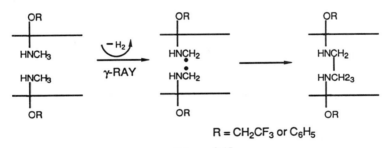

$$R = CH_2CF_3 \text{ or } C_6H_5$$

Scheme 3.15

3.10.6 Hydrogels Derived from Polyphosphazenes

A hydrogel is a three-dimensional matrix of lightly cross-linked polymer molecules that has absorbed a considerable amount of water. In practice, the polymer in its uncross-linked state should be soluble in water: the cross-links between chains prevent the formation of a true solution. Thus, hydrogels swell as they absorb water, but they do not dissolve. Moreover, the volume expansion is limited by the degree of cross-linking. The minimum number of cross-links required to form an insoluble matrix is approximately 1.5 per chain, and this yields a system with the maximum expansion possible without separation of the chains into a true solution. Thus, a hydrogel may be more than 95% water, yet it has structure, and, in that sense, it has much in common with living soft tissues.

Hydrogels have many potential uses in biomedicine, ranging from materials for the construction of soft tissue prostheses or tissuelike coatings, to linings for heart valves or artificial blood vessels, and matrices that imbibe drug molecules and release them by diffusion. Additional prospective applications include their use as substrates for enzyme or antigen immobilization and as materials that favor tissue ingrowth. Soft contact lenses and intraocular lenses are hydrogels. Hydrogels may be biostable or they may be bioerodable. Erosion can occur by hydrolysis of the chains or by cleavage of the crosslinks.

A number. of polyphosphazenes are soluble in water. Examples include polymers with methylamino, glucosyl, glyceryl, or alkyl ether side groups. Here, we will consider one example, a polymer that is also well suited to the "clean" method of radiation cross-linking.[91]

The polymer is poly[bis(methoxyethoxyethoxy)phosphazene] (Structure 3.37), also known as MEEP. Polymer 3.37 is quite sensitive to gamma-radiation-induced cross-linking, due to the presence of 11 carbon-hydrogen bonds in each side group. Thus, relatively low doses of gamma radiation (1.2 Mrad) convert it from a water-soluble polymer to a material that imbibes water to form a hydrogel (Scheme 3.16). The degree of water imbibition depends on the degree of cross-

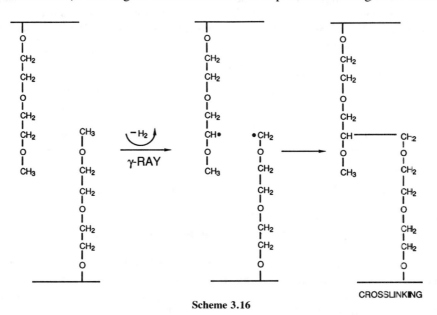

Scheme 3.16

linking, which, in turn, depends on the radiation dose. Hydrogels formed from this polymer are stable to water and appear to be interesting candidates for use as intraocular lenses, soft tissue prostheses, or as hydrophilic coatings for biomedical devices (see Figure 3.19).

3.10.7 Water-Soluble, Bioactive Polymers

In principle, water-soluble macromolecular drugs may be more effective than their small-molecule counterparts because of their restricted ability to escape

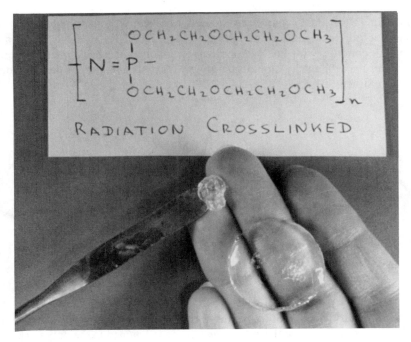

Figure 3.19   Lightly cross-linked poly[bis(methoxyethoxyethoxy)phosphazene] (small pellet of elastomer on the tip of the spatula) absorbs water to form a hydrogel (lower right).

through semipermeable membranes. Moreover, targeting groups can also be linked to the macromolecule to further increase the effectiveness of the drug. The macromolecular component of such drug-polymer conjugates should be biodegradable in order to prevent the eventual deposition of the polymer molecules at some site in the body, for example, in the spleen. The release of the small-molecule drug may precede, parallel, or follow hydrolytic breakdown of the polymer skeleton.

We have already seen how water solubility and hydrolytic degradability can be built into the carrier macromolecule by the use of specific side groups. Here, we will review an additional way in which drug molecules have been linked to polyphosphazenes—by coordination.

3.10.7.1 Polymer-bound platinum antitumor agents.

The platinum complex, $(NH_3)_2PtCl_2$, is a well-known antitumor agent. Because the drug is soluble in water, it is readily excreted through the kidneys and can cause severe kidney damage. Various procedures are employed clinically to minimize these side effects, but the linkage of the $PtCl_2$ component to a nonexcretable, water-soluble polymer offers an additional possibility for improving the effectiveness of the chemotherapy.

A polyphosphazene that is itself a base, and that can replace the ammonia ligands in $(NH_3)_2PtCl_2$, has been studied as a carrier macromolecule.[92] The polymer is poly[bis(methylamino)phosphazene], $[NP(NHCH_3)]_n$. It forms a coordination complex with $PtCl_2$. Coordination involves the backbone nitrogen atoms rather than the methylamino side groups. A representation of the structure is shown in Structure 3.55.

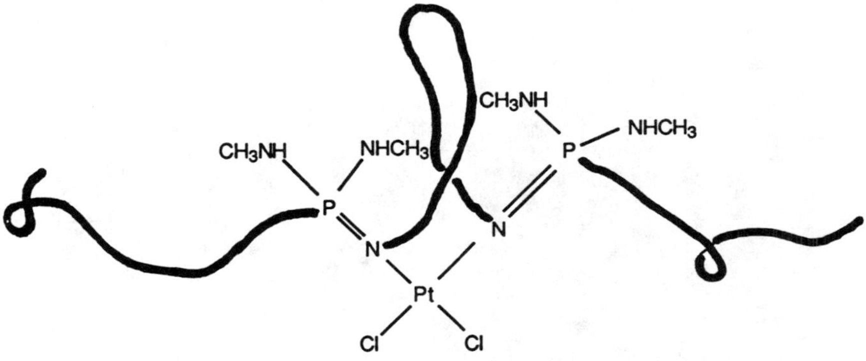

3.55

Preliminary tissue culture testing has been carried out using this polymer, and it appears that some of the antitumor activity is retained in the polymeric drug. This result is interesting and perhaps surprising since the active form of $(NH_3)_2PtCl_2$ is believed to be the small-molecule diaquo derivative, $(H_2O)_2PtCl_2$, which must penetrate the tumor cell membrane in order to affect DNA replication. Thus, more work is needed to understand the behavior of macromolecular drugs of this type, and the role played by dissociation or endocytosis (engulfment of the polymer by the cell) in their biological activity.

3.10.7.2 Other water-soluble systems.

It has been shown in earlier sections of this chapter that water solubility in polyphosphazenes can be generated not only by the presence of cosubstituent methylamino side groups, but also by glyceryl, glucosyl, or alkyl ether side groups. This provides a broad range of options for the tailoring of polymeric, water-soluble drugs. All of these options have not yet been explored in detail, but preliminary work has shown, for example, that water-soluble analogs of the anesthetic-bearing macromolecules discussed earlier are accessible. Water solubility is achieved if 50% of the side groups are methylamino units (Structure 3.56).

At another level, water-soluble polyphosphazenes are of interest as plasma extenders. In addition, specific polymers with pendent imidazolyl units have been studied as carrier macromolecules for heme and other iron porphyrins (Structures 3.57 and 3.58).[93,94] (In Structures 3.57 and 3.58 the ellipse and Fe symbol represent heme, hemin, or a synthetic heme analog.)

Finally, a new water-soluble polyphosphazene has recently been synthesized that has the structure shown in Structure 3.59.[95] This polymer has two attributes

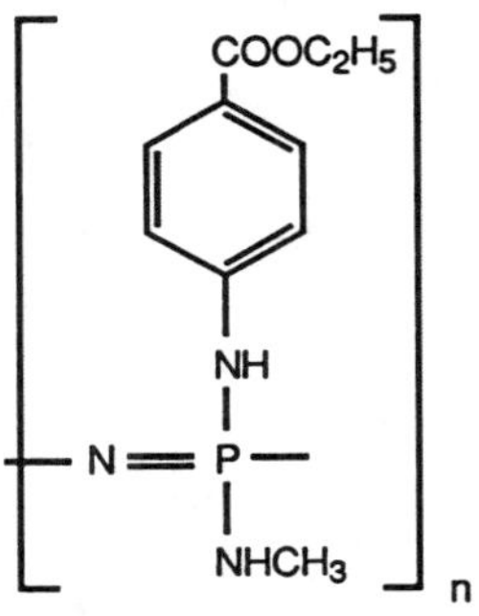

3.56

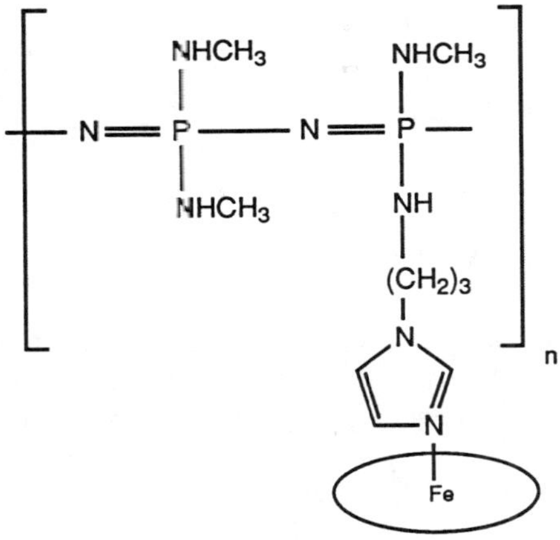

3.57

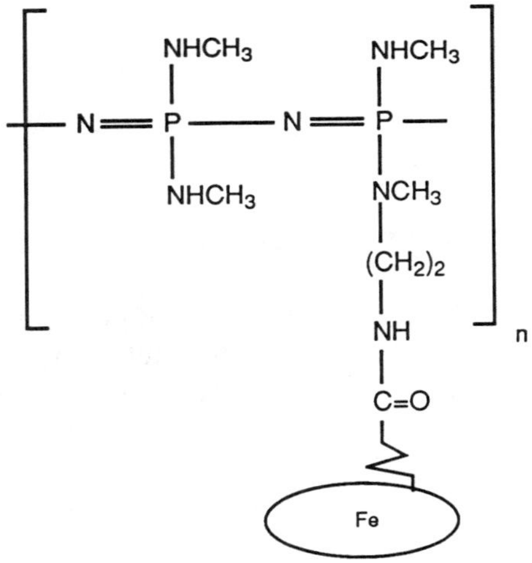

3.58

as a biomedical macromolecule. First, the pendent carboxylic acid groups are potential sites for condensation reactions with amines, alcohols, phenols, or other carboxylic acid units to generate amide, ester, or anhydride links to polypeptides or bioactive small molecules. Second, the polymer of Structure 3.59 forms ionic cross-links when brought into contact with di- or trivalent cations such as Ca^{++} or Al^{+++}. The cross-linking process converts the water-soluble polymer to a hydrogel, a process that can be reversed when the system is infused with a monovalent cation such as Na^+ or K^+. This property opens pathways for (1) removal of ions from living systems, (2) temporary and reversible gelation of a solution as a soft tissue substitute or for temporary sealing of a blood vessel, or (3) entrapment of drug molecules in the hydrogel and release of them by changes in pH or cation concentrations. Microencapsulated liver cells trapped in coatings of this polymer continue to metabolize and are candidates for use in artificial liver devices.[96]

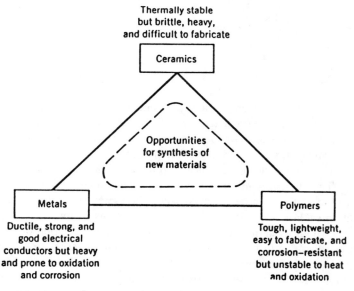

3.59

3.11 ORGANOMETALLIC POLYPHOSPHAZENES

Polymers form one corner of the three main materials areas (Figure 3.20). One of the main motivations for exploratory research work with inorganic polymers is to develop the area that lies between organic polymers, metals, and ceramics. The

Thermally stable but brittle, heavy, and difficult to fabricate

Ceramics

Opportunities for synthesis of new materials

Metals

Polymers

Ductile, strong, and good electrical conductors but heavy and prone to oxidation and corrosion

Tough, lightweight, easy to fabricate, and corrosion–resistant but unstable to heat and oxidation

Figure 3.20 The three classical areas of materials science. Advanced materials are now being discovered where these areas overlap.

development of polyphosphazene synthesis research has taken place with this in mind. For example, the use of a wide variety of organic side groups attached to a phosphazene backbone has permitted materials to be developed with "organic-like" properties. In this section, we consider polyphosphazenes that have metal atoms in the side group and that occupy the interface between polymers and metals.

Three approaches to the synthesis of metallophosphazenes have been explored, and these are summarized by the three repeating units shown as Structures 3.60, 3.61, and 3.62.

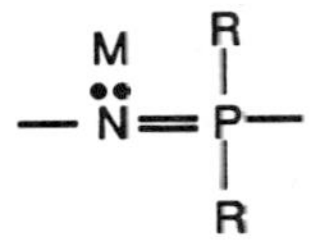

3.60 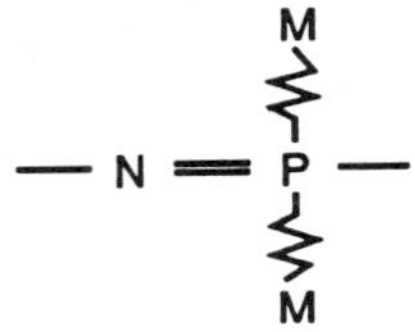   3.61 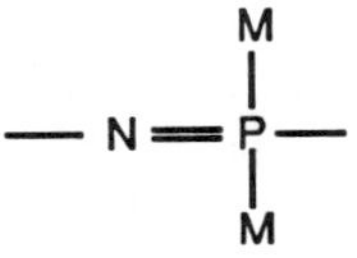   3.62

3.11.1 Coordination

Conceptually, the simplest approach to metallophosphazenes is the one shown in Structure 3.60, in which the lone pair electrons of the skeletal nitrogen atoms are used to coordinate metal atoms to the skeleton. Only one example will be given of this arrangement, although the principle could be applied widely. The example is the binding of $PtCl_2-$ to give the polymeric antitumor agent mentioned briefly in an earlier section. Specifically, the water-soluble, water-stable polyphosphazene shown in Structure 3.13 reacts with aqueous $PtCl_2$/18 crown-6 ether to give the polymer of Structure 3.55.[92] Only a low loading of $PtCl_2$ can be tolerated (approximately one platinum atom per 17 repeating units) because larger amounts tend to cross-link the chains and render the system insoluble. However, the evidence from model compound studies[97] suggests that it is the *skeletal* rather than the side group nitrogen atoms that form the coordination sites. This is presumably a consequence of the strong electron supply from the amino coordination character of the skeletal nitrogens.

3.11.2 Metals Linked to the Skeleton Through Organic or Inorganic Spacer Groups

A number of organic side groups have been utilized as "spacer" units to link transition metals to a polyphosphazene chain. Some of the spacer units are depicted in Structures 3.63–3.65.

Species of the type shown in Structure 3.63, in which a phosphine donor unit is linked to a skeletal phosphorus atom by a P—P bond are the least stable species, and the most difficult to prepare.[98] Specifically, they are synthesized by the reaction of P—Cl bonds in the polymer with lithium triethylborohydride,

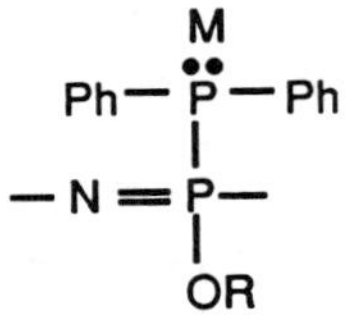

3.63 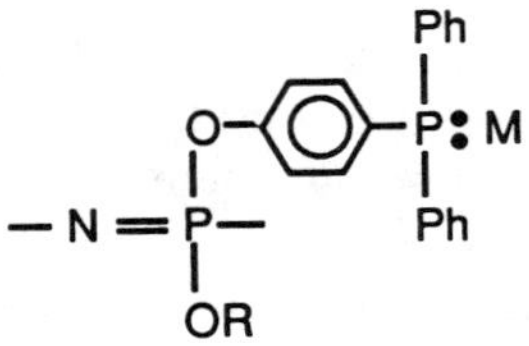   3.64 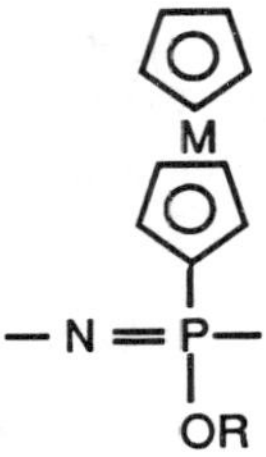   3.65

LiBEt$_3$H, which generates a P$^-$ site in place of the P—Cl bond. Treatment of the polymer with Ph$_2$PCl then yields side groups of the type shown in Structure 3.63. Reaction with Ru$_3$(CO)$_{12}$ results in the linkage of Ru$_3$(CO)$_{11}$ units (M) to the side groups.

The spacer units in Structure 3.64 are assembled from polyphosphazenes that bear p-bromophenoxy side groups (Structure 3.66) via a lithiation reaction and treatment with a diorganochlorophosphine. The chemistry is summarized in Scheme 3.17.[99] The polymer of Structure 3.67 coordinates to a variety of metallo species

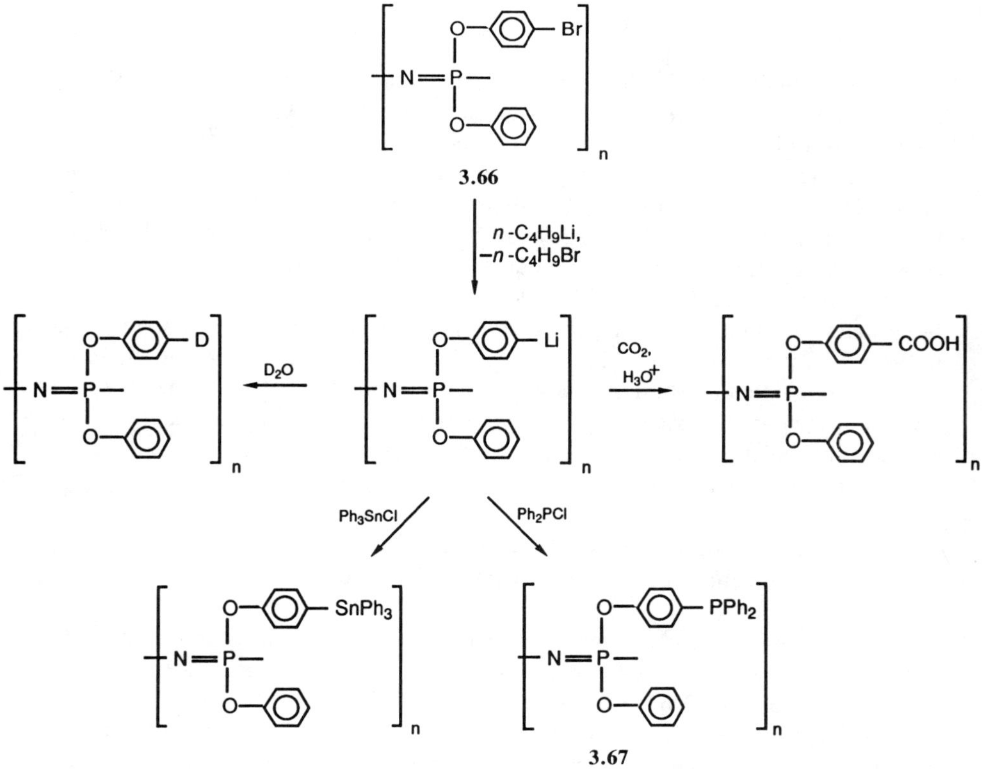

3.66

n-C$_4$H$_9$Li,
$-n$-C$_4$H$_9$Br

3.67

Scheme 3.17

(M),[100] including osmium cluster compounds and cobalt carbonyl hydroformylation catalysts. When used as a polymeric hydroformylation catalyst, this latter species showed how stable the polyphosphazene backbone is under the drastic conditions often needed for these types of reactions. The weakest bonds in the molecule proved to be those between the phosphine phosphorus atoms and the aromatic spacer groups.

The linkage of ferrocenyl and ruthenocenyl side units (Structure 3.65) to a polyphosphazene has been accomplished via the chemistry shown in Scheme 3.18.[101] The species of Structure 3.69 can also be prepared by the method dis-

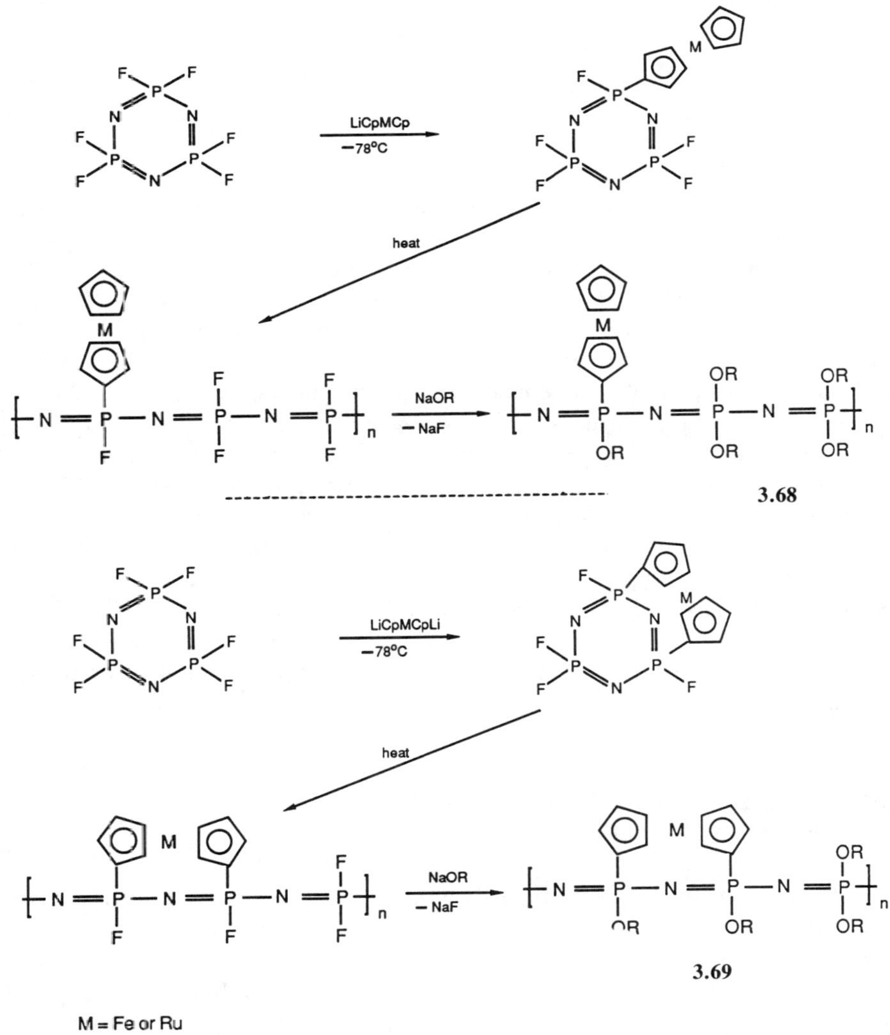

3.68

3.69

M = Fe or Ru

Scheme 3.18

cussed earlier from Structure 3.23. The polymers of Structures 3.68 and 3.69 are of interest for two reasons. First, they can be doped oxidatively with iodine to give polymers that are weak semiconductors. Second, when deposited on an electrode surface the polymers serve as electrode ''mediator'' coatings that assist electron transfer between a reagent in solution and the electrode.[102] Free ferrocene, for example, would not remain in the vicinity of the electrode surface. An electroinactive polymer is needed to immobilize the metallocene in the region of the electrode surface. Copolymerization of phosphazene trimers that bear ferrocenyl and ruthenocenyl side groups is possible, and the resultant polymers bear sites with different oxidation potentials. Such polymers are of particular interest in electrode mediator catalysis.

3.11.3 Polyphosphazenes with Metal-Phosphorus Bonds

Polymers that consist of metal atoms joined together in a linear array are quite rare. This is because such systems prefer to collapse to three-dimensional clusters—literally very small chunks of metal stabilized on the surface by ligands. However, many of the intriguing electrical properties proposed for linear metallo polymers might be realized if a nonmetallo polymer were used as a template and scaffold to stabilize a string of metal atoms and prevent cluster formation. This has been accomplished only in a very preliminary and tentative way, and mainly with small-molecule model systems rather than with high polymers.[103-105] Nevertheless, the results are instructive and perhaps predictive.

Consider the set of model compound reactions shown in Scheme 3.19. These are prototype reactions which, if applied at the high polymer level, might yield polymers of the type depicted (idealistically) in Structure 3.70 or 3.71.

Synthesis of species corresponding exactly to Structure 3.70 or 3.71 has not been accomplished, but polymers that bear metal atoms linked to phosphorus every five or so repeating units have been prepared by the chemistry shown in Scheme 3.20.[106] The aim of this type of work is to optimize the synthetic procedures to increase the loading of metal atoms to the point where metallic-like properties become evident.

3.12 LIQUID CRYSTALLINE AND HIGH REFRACTIVE INDEX POLYMERS

In polyphosphazene chemistry, the side groups usually define the properties. Nowhere is this more true than for those polymers which bear large, rigid aromatic side groups. Such side groups have a tendency to stack or undergo alignment in the solid state or even in the molten state to generate crystallinity or liquid crystallinity.[107-109] If two or more different side groups of this type are present, the tendency to generate order may be reduced, but the large number of pi electrons

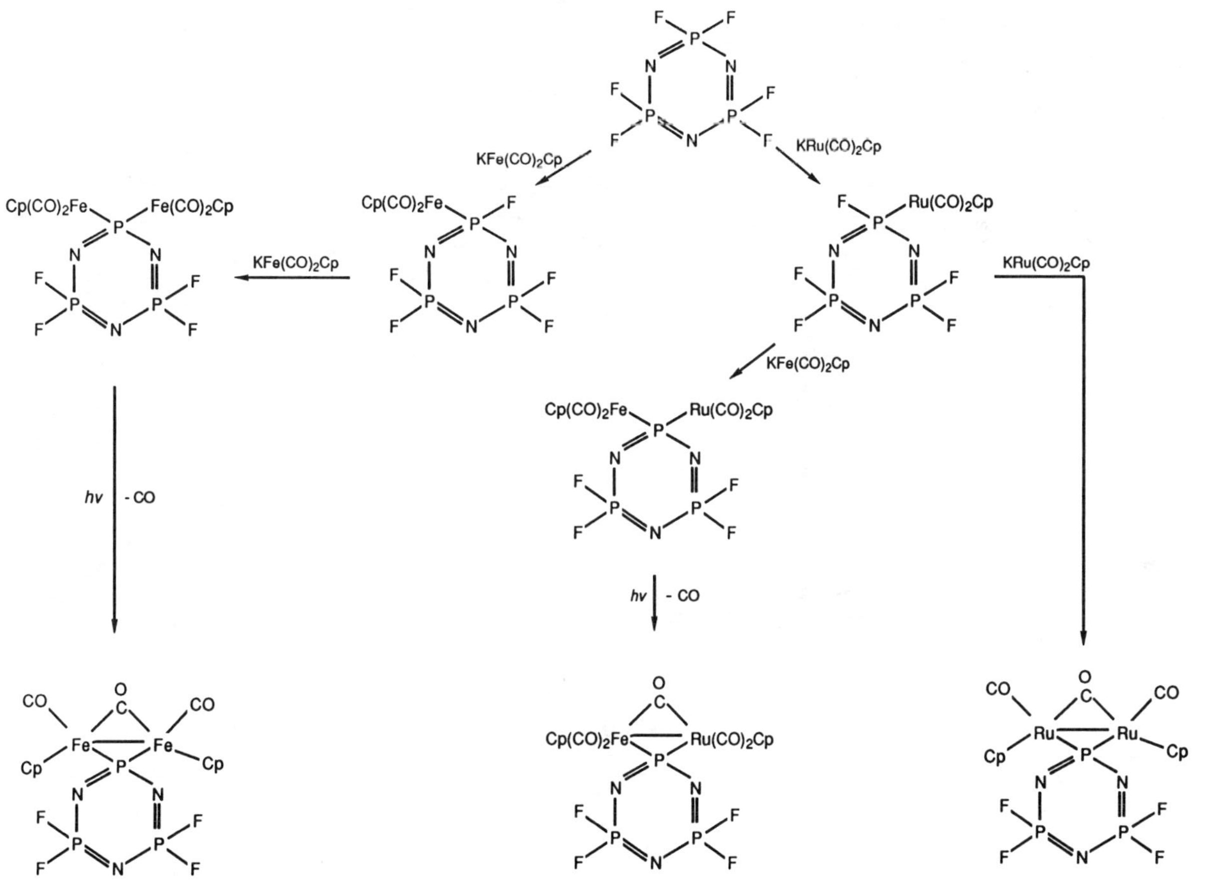

Scheme 3.19

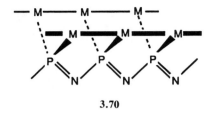

3.70

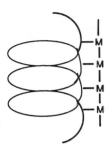

Polyphosphazene chain

3.71

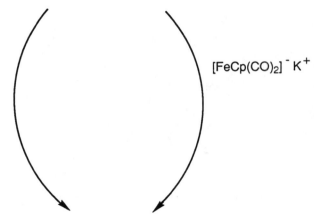

(1) LiBEt$_3$H

(2) FeCp(CO)$_2$I

[FeCp(CO)$_2$]$^-$ K$^+$

C$_p$ = Cyclopentadienyl, C$_5$H$_5$

Scheme 3.20

in the side groups will generate a high refractive index. Such polymeric glasses are of interest for their possible uses in lenses, optical waveguides, and perhaps eventually, as connectors for nonlinear optical devices in optical integrated circuits.

3.12.1 Liquid Crystalline Polyphosphazenes

First, we will consider the design of polyphosphazenes as side chain liquid crystalline materials. Side chain liquid crystalline polymers are a subclass of species described earlier as Structure 3.36. Liquid crystallinity occurs when the rigid side groups become organized, usually in the semiliquid state. The organization may be nematic (oriented but unlayered) or smectic (layered), as illustrated in Figure 3.21.

The essential components of such a system are:

1. A highly flexible polymer chain that responds to the alignment needs of the rigid "mesogenic" units rather than dictating the orientation of side groups. Polyphosphazene chains have this flexibility.
2. A flexible spacer group that decouples the thermal motions of the chain from the alignment of the mesogens. Oligoethyleneoxy spacer groups often serve this purpose.
3. The mesogen itself—biphenyl units, for instance, or aromatic azo groups.

The synthesis of one example of a liquid crystalline polyphosphazene is shown in Scheme 3.21.[110]

This orange-yellow polymer (Structure 3.72) changes from a glassy-microcrystalline material to the rubbery-microcrystalline state at the glass transition (T_g) at 79°C. Further heating raises the temperature to the microcrystallite melting transition (T_m) at 118°C. But even in the molten state, the polymer retains some

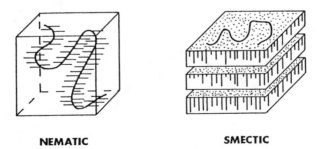

NEMATIC   **SMECTIC**

Figure 3.21 Side chain liquid crystallinity occurs when rigid side groups (mesogens) can form arrays or stacks. The appearance of this phenomenon requires a flexible main chain (such as a polyphosphazene or polysiloxane chain) and a flexible spacer unit that links the main chain to the mesogen and decouples the motions of the main chain from those of the mesogens. (Drawing by M. S. Connolly.)

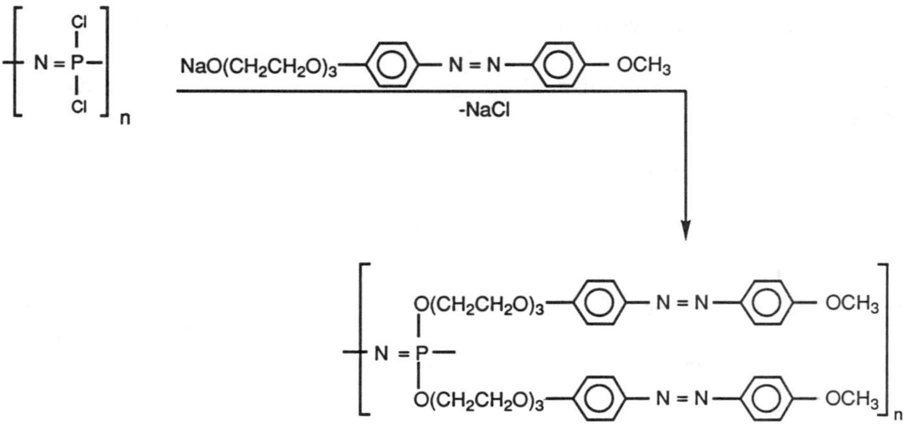

Scheme 3.21

3.72

order, as indicated by the pattern seen through a microscope fitted with crossed polarizers (Figure 3.22).

The texture of the pattern is indicative of a nematic-type liquid crystalline system. The birefringence obtained from the analogous cyclic trimer is also shown in Figure 3.22. Further heating of this polymer to 126°C breaks the liquid crystalline order, and the material now behaves as an isotropic liquid, showing only a dark field when viewed between crossed polarizers.

Other similar polymers with both aromatic azo mesogenic groups and "inert" cosubstituent groups (OCH_2CF_3) have also been found to form a liquid crystalline phase.[112] It seems clear that a wide range of related derivatives will behave similarly. However, the length of the spacer group between the mesogen and the

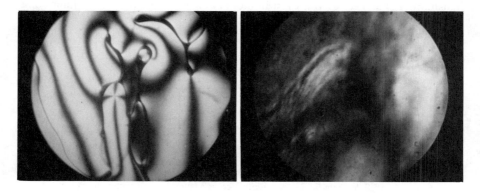

Figure 3.22 Photographs taken through a polarizing microscope of (a) a cyclic trimeric phosphazene bearing six p-$CH_3OC_6H_4N\!=\!NC_6H_4(OCH_2CH_2)_2O-$ side group units and (b) a related high polymer (Structure 3.72), with the samples heated above the normal melting point of 111–112°C for (a) and 118°C for (b).

chain is critical. If the spacer group in Structure 3.72 is shortened to only one ethyleneoxy unit, the liquid crystalline behavior disappears.[110]

3.12.2 Polyphosphazenes with Polyaromatic Side Units

The introduction of bulky side groups, such as those depicted in Structures 3.73–3.78 requires the use of special experimental conditions. If introduced via an aryloxide ion by the macromolecular substitutive route, these side groups impart and suffer considerable steric hindrance. Replacement of every chlorine atom along the phosphazene chain by organic side groups may be difficult.

Thus, forcing reaction conditions are usually needed, and these are provided by raising the temperature of the substitution reaction into the 120–150°C range. Because solvents such as tetrahydrofuran (which boil near 70°C) are normally used for these reactions, the substitution must be carried out in sealed, medium pressure reactors to prevent loss of the solvent. Using these reaction conditions, it has been possible to synthesize a broad range of polymers, including those shown in Structures 3.73–3.78.[113] Some of these polymers are microcrystalline. Others (usually

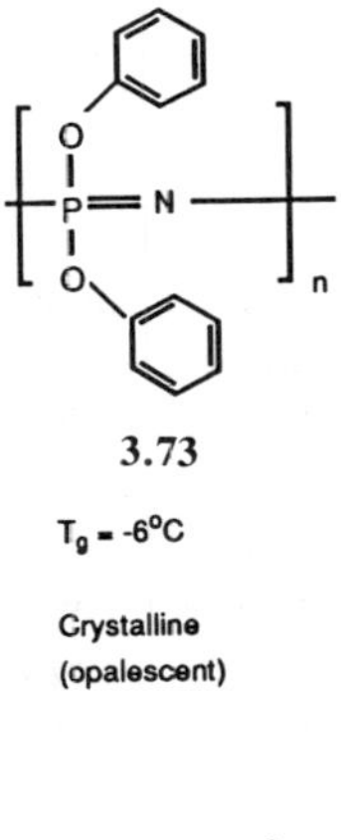

3.73

$T_g = -6°C$

Crystalline
(opalescent)

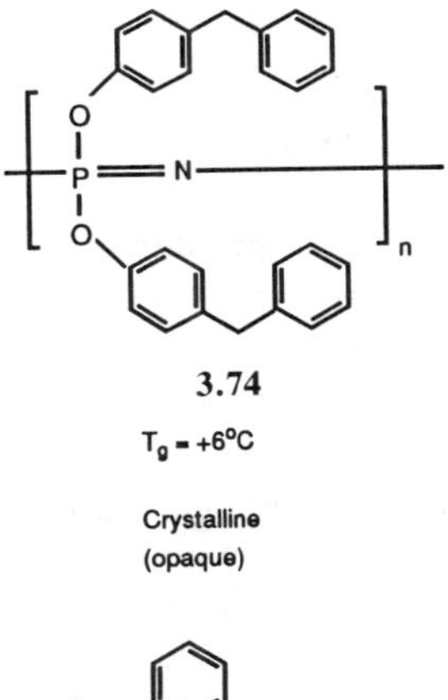

3.74

$T_g = +6°C$

Crystalline
(opaque)

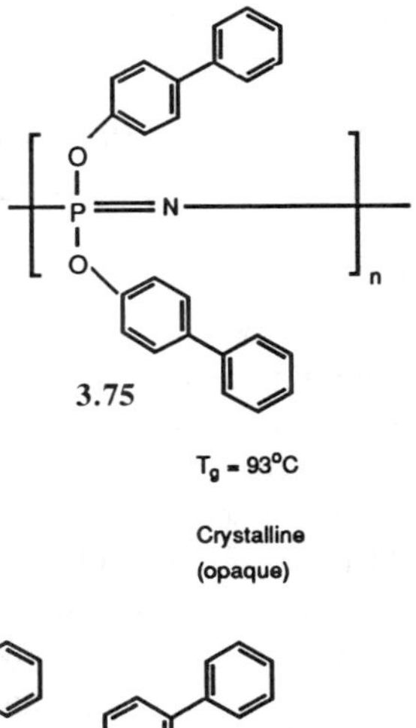

3.75

$T_g = 93°C$

Crystalline
(opaque)

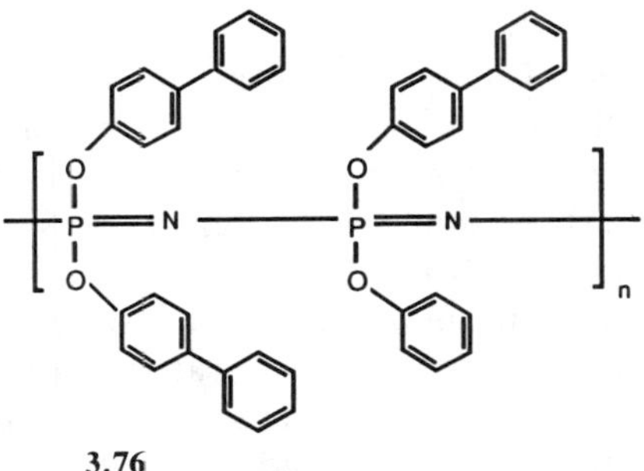

3.76

$T_g = 63°C$

Amorphous
(transparent)

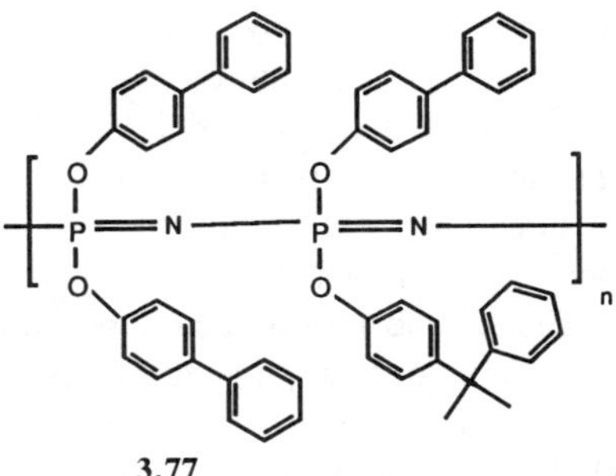

3.77

$T_g = 59°C$

Amorphous
(transparent)

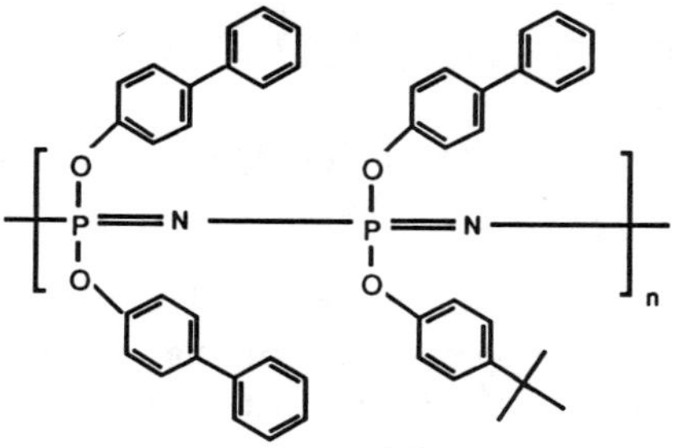

3.78

$T_g = 68°C$

Amorphous
(transparent)

those with two different side groups) are amorphous. All have high refractive indices because of the large number of pi-electrons in the side groups. The polymer of Structure 3.77 is an amorphous glass at temperatures up to 60°C and has a refractive index of 1.65. Hence, it and related polymers are good candidates for optical applications.

3.13 POLYCARBOPHOSPHAZENES
AND POLYTHIOPHOSPHAZENES

Long-chain polymers based on a skeleton of alternating phosphorus and nitrogen atoms may be considered as the ''parent'' systems for other macromolecules that contain phosphorus, nitrogen, and some other element in the skeletal system. The first examples of these new systems were provided by the poly(carbophosphazenes) which contain phosphorus, nitrogen, and carbon in the backbone structure.[114] They are prepared by the polymerization of a chlorocarbophosphazene cyclic compound (Structure 3.79), followed by the thermal polymerization of this species to the high polymer (Structure 3.80), and the subsequent replacement of the chlorine atoms in this macromolecule by organic side groups (Scheme 3.22).

An interesting feature of these polymers is that, for a given side group, the glass transition temperature is significantly higher than for the corresponding polyphosphazene. This implies that the presence of the skeletal carbon atoms brings about a decrease in the flexibility of the backbone, probably because a $C = N$ double bond has a higher barrier to torsion than a $P = N$ bond.

The second group of new polymers contain sulfur atoms in the skeleton in addition to the normal phosphorus and nitrogen atoms.[115] They too are synthesized by the ring-opening polymerization of a heterocyclic inorganic species (Structure 3.81), followed by macromolecular halogen replacement reactions carried out on the resultant polymer (Structure 3.82) (Scheme 3.23). The properties of the or-

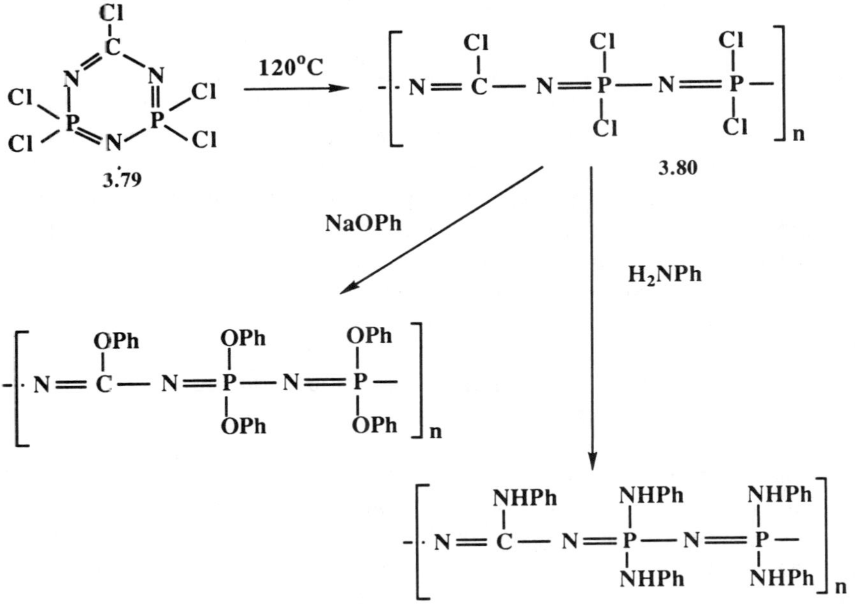

3.79 3.80

Scheme 3.22

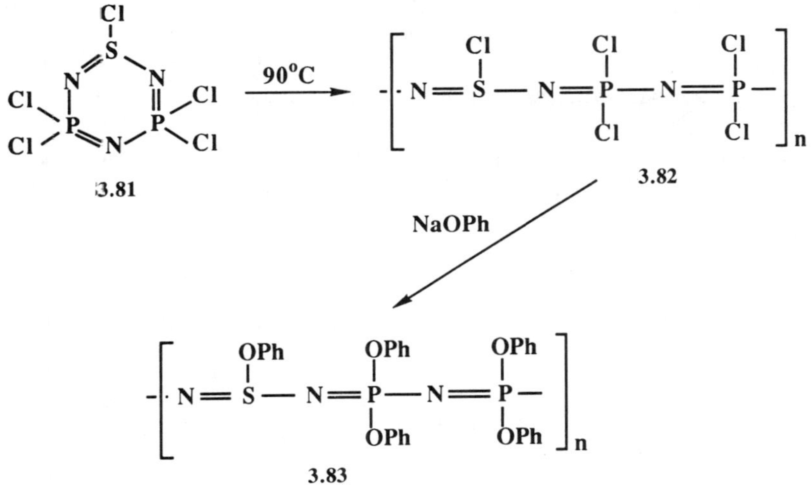

3.81 3.82 3.83

Scheme 3.23

gano-substituted polymers (Structure 3.83) are different again from those of poly(carbophosphazenes) and classical polyphosphazenes, and it is clear that phosphazene polymers with carbon or sulfur in the backbone are the forerunners of many new polymer systems that will be investigated in the coming years.

3.14 OTHER TYPES OF POLYPHOSPHAZENES

The foregoing sections have dealt with the largest class of polyphosphazenes—the linear or macrocyclic poly(organophosphazenes). But this chapter would not be complete without mention of three additional types of phosphazene polymers—those in which cyclophosphazene rings are linked to an *organic* polymer chain, cyclomatrix polymers formed by connecting phosphazene rings together by a three-dimensional network of organic short chains, and the inorganic ceramic-type phosphazenes formed by the pyrolysis of phosphazene rings or chains. These last two classes in particular provide an introduction to the subject of preceramic polymers to be discussed in Chapter 5.

3.14.1 Organic Polymers with Cyclophosphazene Side Groups

The phosphazene skeleton has some interesting properties quite apart from its behavior as a polymer chain. For one thing, the presence of phosphorus and nitrogen provide protection against burning of the material. Thus, organic polymers that contain *cyclic* small-molecule phosphazenes have a much lower flammability than do their counterparts in which the phosphazene is absent. For many years a fire-resistant rayon (regenerated cellulose) was manufactured that contains alkoxycyclophosphazenes physically trapped in the organic polymer fibers.[116]

An alternative approach is to synthesize organic polymers that have cyclophosphazene units bound *chemically* as side groups to the organic chain.[117-120] Examples are polymers such as Structure 3.85 prepared by polymerization of vinyl monomer (Structure 3.84). Polymers such as Structure 3.85 are flame retardant. Other polymers have been prepared by the copolymerization of vinyl- or allylcyclophosphazene compounds with classical vinyl monomers such as styrene.

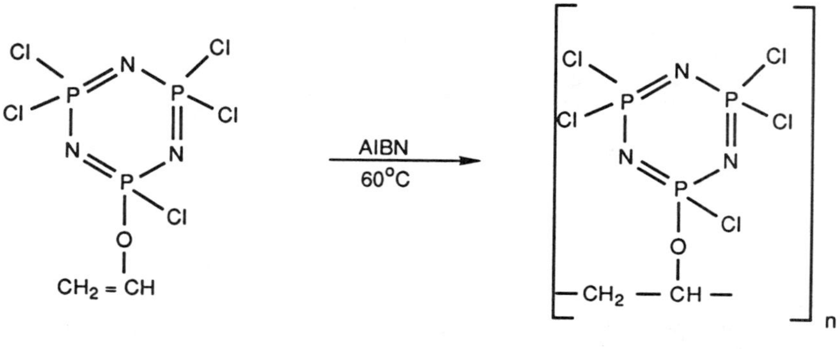

3.84   **3.85**

(Eq. 3.10)

3.14.2 Cyclolinear and Cyclomatrix Materials

The cyclic trimeric phosphazene ring is a stable unit with valuable technological properties, such as heat resistance and flame-retardant behavior. Thus, some
attempts have been made to take advantage of these properties by connecting cy-
clotriphosphazene rings together to form species of the type shown in Structure
3.86.

3.86

Two requirements must be met for the synthesis of polymers of this type.
First, the cyclophosphazene must have only two functional sites (usually P—Cl
bonds). Second, a difunctional linkage molecule is needed to form the chain. Aromatic or aliphatic diols or their sodium salts, or diamines, can serve this purpose.

If the cyclophosphazene precursor molecule has more than two reactive sites,
the linkage reaction will generate a three-dimensional structure rather than a linear
chain. Such materials are known as *cyclomatrix polymers*. These species form a
half-way stage between linear polymers and ceramics and occupy a critical position
in materials science, as illustrated earlier in Figure 3.20. The term "ultrastructure" covers a broad range of materials of this type.

The difference between a linear or macrocyclic polymer and an ultrastructure
is one of rigidity and thermal stability with both these properties increasing with
the density of cross-linking. Most ceramics such as silicates, silicon nitride, or
silicon carbide are totally inorganic materials held together by a tightly knit three-
dimensional array of covalent bonds. However, organic ultrastructures are also
known. One example is the melamine-formaldehyde resin in which *s*-triazine rings
are linked together by a three-dimensional framework of organic cross-links.

Similar cyclomatrix materials have been reported that are based on cyclotri-
phosphazene rather than an *s*-triazine rings. In these materials a difunctional nucleophile, such as a diamine or an aromatic diol or disodium diol, is used to link
together the cyclophosphazene rings, as illustrated in Scheme 3.24, with the final
structure approximating to the arrangement shown in Structure 3.87.[121]

Cyclophosphazenes with *p*-aminophenoxy side groups are readily converted
into ultrastructures by reaction with maleic anhydride.[122] These materials are reported to have a very high thermal stability, a property needed for use as high-
temperature coatings.

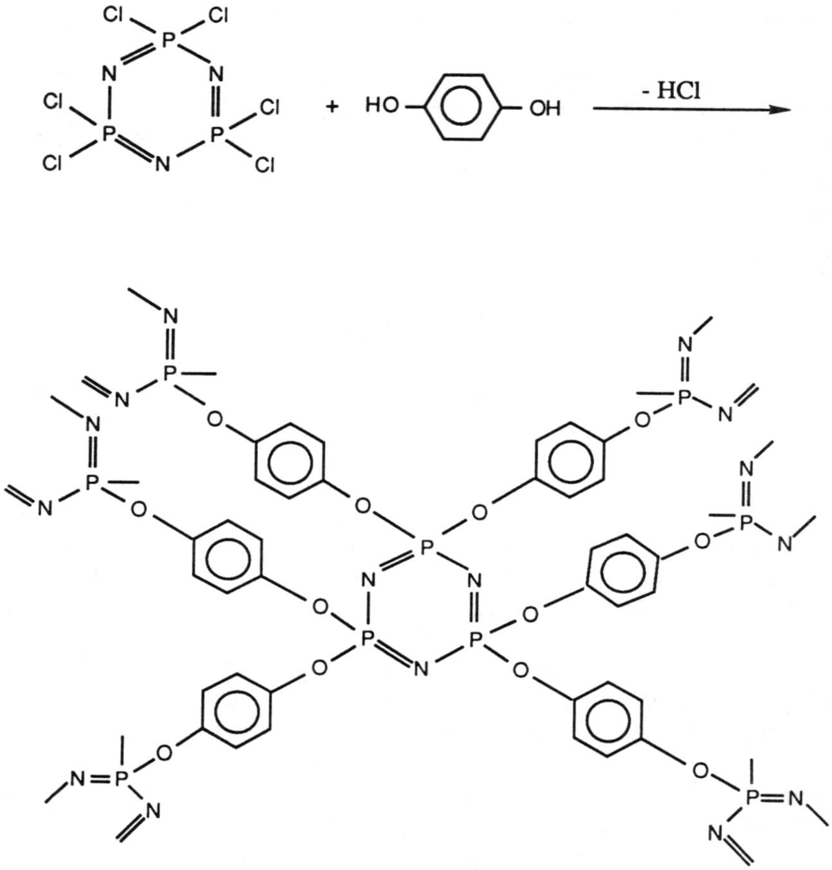

Scheme 3.24

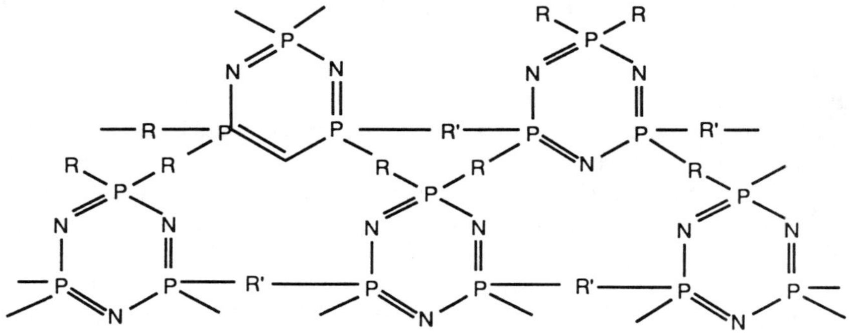

3.87

3.14.3 Inorganic Ceramics Derived from Phosphazenes

Ceramic-type materials that contain no organic linkage units can be prepared by the pyrolysis of cyclic or high polymeric aminophosphazenes. An example is shown in Scheme 3.25. Under appropriate conditions, pyrolysis products that cor-

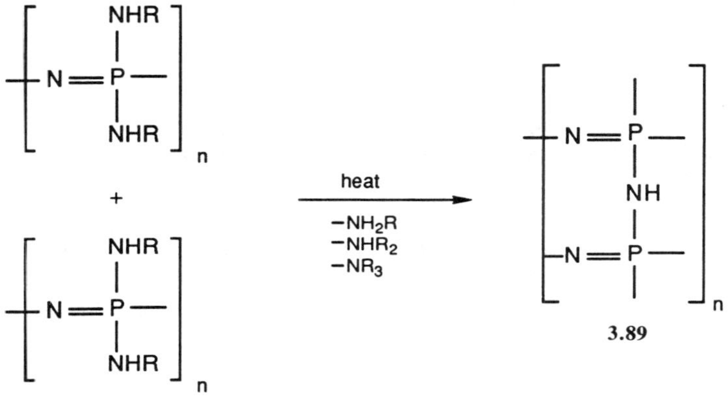

Scheme 3.25

respond to phosphorus nitride are formed.[123] The conversion of a formable polymer, such as Structure 3.88 into a ceramic (Structure 3.89) has many potential advantages for the controlled synthesis and fabrication of advanced ceramics. This principle is discussed in more detail in later chapters, especially for silicon-containing systems.

3.15 CONCLUSIONS

The polyphosphazene systems comprise some of the most diverse macromolecules yet discovered. Because of space limitations, many aspects of polyphosphazene chemistry have not been included in this survey. The reference list for this chapter covers less than one-tenth of the total number of publications and patents on this subject. The selection of topics for this chapter was made on the basis of fundamental chemistry, and many applied aspects have not been discussed. However, it will be clear that the fundamental chemistry has spawned a very broad range of potential technological developments. Many of these developments reflect the critical position occupied by polyphosphazene chemistry as a bridge between organic polymers, metals, and ceramics.

Compared to many areas of polymer chemistry, the field of polyphosphazenes is quite new. The first stable poly(organophosphazenes) were synthesized only in the 1960s at a time when the conventional wisdom held that all the major polymer systems had already been discovered. The sequence of synthetic discoveries in this area, outlined in Figure 3.23, and the increase in the number of papers

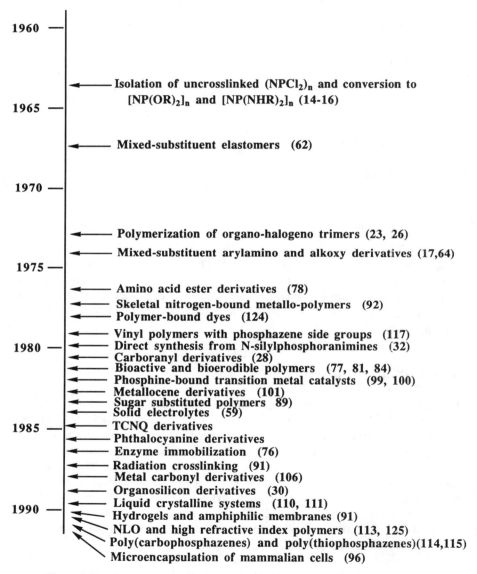

Figure 3.23 Approximate sequence of developments in the synthesis of polyphosphazenes since 1964. The numbers in parentheses refer to references in the main reference list for Chapter 3.

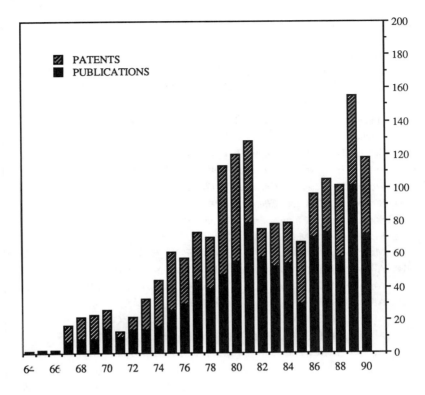

YEAR (from 1964)

Figure 3.24 Number of publications and patents on polyphosphazenes since the initial synthesis discoveries in 1964. The data for 1990 are for the first nine months only.

and patents (Figure 3.24), illustrate how the development of a new inorganic-based polymer system can occur.

In many respects, the polyphosphazenes are the prototype inorganic polymers, exemplifying the principles of ring-opening polymerization, macromolecular substitution reactions and their potential for molecular design, an enormous range of derivatives with the same backbone but different organic side groups, and a diversity of properties that covers and exceeds the range found in most organic polymer systems. Yet they represent just one skeletal elemental combination—that of phosphorus and nitrogen. Literally scores of other inorganic skeletal systems can be visualized, each of which may give rise to an equally diverse polymer chemistry. These possibilities are illustrated further in the chapters that follow, in which the emphasis is on inorganic macromolecules derived from the element silicon.

3.16 REFERENCES

1. Liebig, J. *Ann. Chem.* **1834,** *11,* 139.
2. Rose, H. *Ann. Chem.* **1834,** *11,* 131.
3. Gerhardt, C. *C. R. Acad. Sci.* **1846,** *22,* 858.
4. Laurent, A. *C. R. Acad. Sci.* **1850,** *31,* 356.
5. Gladstone, J. H.; Holmes, J. D. *J. Chem. Soc.,* **1864,** *17,* 225.
6. Wichelhaus, H. *Chem. Ber.* **1870,** *3,* 163.
7. Stokes, H. N. *Amer. Chem. J.* **1895,** *17,* 275.
8. Stokes, H. N. *Amer. Chem. J.* **1896,** *18,* 629, 780.
9. Stokes, H. N. *Amer. Chem. J.* **1897,** *19,* 782.
10. Stokes, H. N. *Amer. Chem. J.* **1898,** *20,* 740.
11. Staudinger, H. *Chem. Ber.* **1920,** *53,* 1073.
12. Meyer, K. H.; Mark, H. *Chem. Ber.* **1928,** *61,* 1939.
13. Meyer, K. H.; Lotmar, W.; Pankow, G. W. *Helv. Chim. Acta* **1936,** *19,* 930.
14. Allcock, H. R.; Kugel, R. L. *J. Am. Chem. Soc.* **1965,** *87,* 4216.
15. Allcock, H. R.; Kugel, R. L.; Valan, K. J. *Inorg. Chem.* **1966,** *5,* 1709.
16. Allcock, H. R.; Kugel, R. L. *Inorg. Chem.* **1966,** *5,* 1716.
17. Allcock, H. R.; Mack, D. P. *J. Chem. Soc., Chem. Commun.* **1970,** *11,* 685.
18. Allcock, H. R.; Chu, C. T.-W. *Macromolecules* **1979,** *12,* 551.
19. Allcock, H. R.; Patterson, D. B.; Evans, T. L. *J. Am. Chem. Soc.* **1977,** *99,* 6095.
20. Evans, T. L.; Allcock, H. R. *J. Macro. Sci.—Chem.* **1981,** *A16(1),* 409.
21. Allcock, H. R.; Mang, M. N.; McDonnell, G. S.; Parvez, M. *Macromolecules* **1987,** *20,* 2060.
22. Allcock, H. R. *Polymer* **1980,** *21,* 673.
23. Allcock, H. R.; Moore, G. Y. *Macromolecules* **1975,** *8,* 377.
24. Allcock, H. R.; Patterson, D. B. *Inorg. Chem.* **1977,** *16,* 197.
25. Manners, I.; Riding, G. H.; Dodge, J. A.; Allcock, H. R. *J. Am. Chem. Soc.* **1989,** *111,* 3067.
26. Prons, V. N.; Grinblat, M. P.; Klebanskii, A. L. *J. Gen. Chem. of USSR* **1971,** *41,* 475.
27. Ritchie, R. J.; Harris, P. J.; Allcock, H. R. *Macromolecules* **1979,** *12,* 1014.
28. Scopelianos, A. G.; O'Brien, J. P.; Allcock, H. R. *J. Chem. Soc., Chem. Commun.* **1980,** 198.
29. Allcock, H. R.; Lavin, K. D.; Riding, G. H. *Macromolecules* **1985,** *18,* 1340.
30. Allcock, H. R.; Brennan, D. J. *J. Organomet. Chem.* **1988,** *341,* 231.
31. Neilson, R. H.; Wisian-Neilson, P. *Chem. Rev.* **1988,** *88,* 541.
32. Wisian-Neilson, P.; Neilson, R. H. *J. Am. Chem. Soc.* **1980,** *102,* 2848.
33. Neilson, R. H.; Hani, R.; Wisian-Neilson, P.; Meister, J. J.; Roy, A. K.; Hagnauer, G. L. *Macromolecules* **1987,** *20,* 910.
34. Wisian-Neilson, P.; Ford, R. R.; Neilson, R. H.; Ray, A. K. *Macromolecules* **1986,** *19,* 2089.

35. Flindt, E. P.; Rose, H. Z. *Z. Anorg. Allg. Chem.* **1977,** *428,* 204.

36. Montague, R. A.; Matyjaszewski, K. *J. Am. Chem. Soc.* **1990,** *112,* 6721.

37. Hornbacker, E. D.; Li, H. M. U.S. Patent 4 198 381, 1980.

38. De Jaeger, R.; Helioui, M.; Puskavic, E. U.S. Patent 4 377 558, 1983.

39. Allcock, H. R.; Best, R. J. *Can. J. Chem.* **1964,** *42,* 447.

40. Allcock, H. R.; Gardner, J. E.; Smeltz, K. M. *Macromolecules* **1975,** *8,* 36.

41. Sennett, M. S.; Hagnauer, G. L.; Singler, R. E. *Polymer. Mater. Sci. Eng.* **1983,** *49,* 297. Fieldhouse, J. W.; Graves, D. F. U.S. Patent 4 226 840, 1980.

42. Seel, F.; Langer, J. **1956,** *68,* 461.

43. Seel, F.; Langer, J. *Z. Anorg. Allg. Chem.* **1958,** *295,* 316.

44. Allcock, H. R.; Kugel, R. L.; Stroh, E. G. *Inorg. Chem.* **1972,** *11,* 1120.

45. Allcock, H. R. *Acct. Chem. Res.* **1979,** *12,* 351.

46. McBee, E. T.; Allcock, H. R.; Caputo, R.; Kalmus, A.; Roberts, C. W. U.S. Government ASTIA Rep. AD 209,669; U.S. Government Printing Office: Washington, DC, 1959.

47. Ratz, R.; Schroeder, H.; Ulrich, H.; Kober, E.; Grundmann, C. *J. Am. Chem. Soc.* **1962,** *84,* 551.

48. Schmutz, J. L.; Allcock, H. R. *Inorg. Chem.* **1975,** *14,* 2433.

49. Allcock, H. R.; Lampe, F. W. *Contemporary Polymer Chemistry*, 2nd ed.; Prentice Hall: Englewood Cliffs, NJ, 1990.

50. Ibid., Chapter 18.

51. Allcock, H. R.; Allen, R. W.; Meister, J. J. *Macromolecules* **1976,** *9,* 950.

52. Allen, R. W.; Allcock, H. R. *Macromolecules* **1976,** *9,* 956.

53. Allcock, H. R.; Arcus, R. A.; Stroh, E. G. *Macromolecules* **1980,** *13,* 919.

54. Chatani, Y.; Yatsuyanagi, K. *Macromolecules* **1987,** *20,* 1042.

55. Dewar, M. J. S.; Lucken, E. A. C.; Whitehead, M. A. *J. Chem. Soc.* **1960,** 2423.

56. Allcock, H. R.; Lampe, F. W. *Contemporary Polymer Chemistry*, 2nd ed.; Prentice Hall: Englewood Cliffs, NJ, 1990; Chapter 18.

57. Allcock, H. R.; Austin, P. E.; Neenan, T. X.; Sisko, J. T.; Blonsky, P. M.; Shriver, D. F. *Macromolecules* **1986,** *19,* 1508.

58. Allcock, H. R.; Connolly, M. S.; Sisko, J. T.; Al-Shali, S. *Macromolecules* **1988,** *21,* 323.

59. Blonsky, P. M.; Shriver, D. F.; Austin, P. E.; Allcock, H. R. *J. Am. Chem. Soc.* **1984,** *106,* 6854.

60. Blonsky, P. M.; Shriver, D. F.; Austin, P. E.; Allcock, H. R. *Solid State Ionics* **1986,** *18, 19,* 258.

61. Bennett, J. L.; Dembek, A. A.; Allcock, H. R.; Heyen, B. J.; Shriver, D. F. *Chemistry of Materials* **1989,** *1,* 14.

62. Rose, S. H. *J. Polymer Sci.* **1968,** *B6,* 837.

63. Allen, G.; Lewis, C. J.; Todd, S. M. *Polymer* **1970,** *11,* 31.

64. Singler, R. E.; Schneider, N. S.; Hagnauer, G. L. *Polym. Eng. Sci.* **1975,** *15,* 321.

65. Tate, D. P. *J. Polymer Sci., Polymer Symp.* **1974,** *48,* 33.

66. Futamura, S.; Valaitis, J. K.; Lucas, K. R.; Fieldhouse, J. W.; Cheng, J. W.; Tate, D. P. *J. Polymer Sci., Polymer Phys.* **1980**, *18*, 767.

67. Penton, H. R. In *Inorganic and Organometallic Polymers*; Zeldin, M.; Wynne, K. J.; Allcock, H. R., Eds.; ACS Symposium Series; American Chemical Society: Washington, DC, **1988**; Vol. 360, p 277.

68. Allcock, H. R.; Brennan, D. J. *J. Organomet. Chem.* **1988**, *341*, 231.

69. Beres, J. J.; Schneider, N. S.; Desper, C. R.; Singler, R. E. *Macromolecules* **1979**, *12*, 566.

70. Allcock, H. R.; Mang, M. N.; Dembek, A. A.; Wynne, K. J. *Macromolecules* **1989**, *22*, 4179.

71. Kojima, M.; Magill, J. H. *Makromol. Chem.* **1985**, *186*, 649; *Polymer* **1985**, *26*, 1971.

72. Wade, C. W. R.; Gourlay, S.; Rice, R.; Hegyeli, A. In *Organometallic Polymers;* Carraher, C. E.; Sheats, J. E.; Pittman, C. U., Eds.; Academic: New York, 1978; p 289.

73. Allcock, H. R.; Cook, W. J.; Mack, D. P. *Inorg. Chem.* **1972**, *11*, 2584.

74. Fitzpatrick, R. J.; Allcock, H. R., unpublished results.

75. Neenan, T. X.; Allcock, H. R. *Biomaterials* **1982**, *3*, 2, 78.

76. Allcock, H. R.; Kwon, S. *Macromolecules* **1986**, *19*, 1502.

77. Allcock, H. R.; Hymer, W. C.; Austin, P. E. *Macromolecules* **1983**, *16*, 1401.

78. Allcock, H. R.; Fuller, T. J.; Mack, D. P.; Matsumura, K.; Smeltz, K. M. *Macromolecules* **1977**, *10*, 824.

79. Allcock, H. R.; Fuller, T. J.; Matsumura, K. *Inorg. Chem.* **1982**, *21*, 515.

80. Grolleman, C. W. J.; De Visser, A. C.; Volke, J. G. C.; Klein, C. P. A. T.; Van Der Goot, H.; Timmerman, H. *J. Controlled Release* **1986**, *3*, 143; **1986**, *4*, 119; **1986**, *4*, 133.

81. Allcock, H. R.; Fuller, T. J. *Macromolecules* **1980**, *13*, 1338.

82. Goedemoed, J. H.; De Groot, K. *Makromol. Chem., Macromol. Symp.* **1988**, *19*, 341.

83. Allcock, H. R.; Fuller, T. J. *J. Am. Chem. Soc.* **1981**, *103*, 2250.

84. Laurencin, C.; Koh, H. J.; Neenan, T. X.; Allcock, H. R.; Langer, R. S. *J. Biomed. Mater. Res.* **1987**, *21*, 1231.

85. Allcock, H. R.; Austin, P. E. *Macromolecules* **1981**, *14*, 1616.

86. Allcock, H. R.; Austin, P. E.; Neenan, T. X. *Macromolecules* **1982**, *15*, 689.

87. Allcock, H. R.; Neenan, T. X.; Kossa, W. C. *Macromolecules* **1982**, *15*, 693.

88. Allock, H. R.; Kwon, S. *Macromolecules* **1988**, *21*, 1980.

89. Allcock, H. R.; Scopelianos, A. G. *Macromolecules* **1983**, *16*, 715.

90. Allcock, H. R.; Gebura, M.; Kwon, S.; Neenan, T. X. *Biomaterials* **1988**, *19*, 500.

91. Allcock, H. R.; Kwon, S.; Riding, G. H.; Fitzpatrick, R. J.; Bennett, J. L. *Biomaterials* **1988**, *19*, 509.

92. Allcock, H. R.; Allen, R. W.; O'Brien, J. P. *J. Am. Chem. Soc.* **1977**, *99*, 3984.

93. Allcock, H. R.; Greigger, P. P.; Gardner, J. E.; Schmutz, J. L. *J. Am. Chem. Soc.* **1979**, *101*, 606.

94. Allcock, H. R.; Neenan, T. X.; Boso, B. *Inorg. Chem.* **1985**, *24*, 2656.

95. Allcock, H. R.; Kwon, S. *Macromolecules* **1989**, *22*, 75.

96. Cohen, S.; Bano, M. C.; Vissher, K. B.; Chow, M.; Allcock, H. R.; Langer, R. *J. Am. Chem. Soc.* **1990**, *112*, 7832.

97. Allen, R. W.; O'Brien, J. P.; Allcock, H. R. *J. Am. Chem. Soc.* **1977**, *99*, 3987.

98. Allcock, H. R.; Manners, I.; Mang, M. N.; Parvez, M. *Inorg. Chem.* **1990**, *29*, 522.

99. Allcock, H. R.; Fuller, T. J.; Evans, T. L. *Macromolecules* **1980**, *13*, 1325.

100. Allcock, H. R.; Lavin, K. D.; Tollefson, N. M.; Evans, T. L. *Organometallics* **1983**, *2*, 267.

101. Allcock, H. R.; Riding, G. H.; Lavin, K. D. *Macromolecules* **1985**, *18*, 1340; **1987**, *20*, 6.

102. Saraceno, R. A.; Riding, G. H.; Allcock, H. R.; Ewing, A. G. *J. Am. Chem. Soc.* **1988**, *110*, 7254.

103. Allcock, H. R.; Greigger, P. P.; Wagner, L. J.; Bernheim, M. Y. *Inorg. Chem.* **1981**, *21*, 716.

104. Allcock, H. R.; Wagner, L. J.; Levin, M. L. *J. Am. Chem. Soc.* **1983**, *105*, 1321.

105. Allcock, H. R.; Suszko, P. R.; Wagner, L. J.; Whittle, R. R.; Boso, B. *J. Am. Chem. Soc.* **1984**, *106*, 4966.

106. Allcock, H. R.; Mang, M. N.; McDonnell, G. S.; Parvez, M. *Macromolecules* **1987**, *20*, 2060.

107. Beres, J. J.; Schneider, N. S.; Desper, C. R.; Singler, R. E. *Macromolecules* **1979**, *12*, 566.

108. Desper, C. R.; Singler, R. E.; Schneider, N. S. In *IUPAC Proc. Macromol. Symp.*, Oxford, 1982, p. 682.

109. Schneider, N. S.; Desper, C. S.; Beres, J. J. In *Liquid Crystalline Order Polymers;* A. Blumstein, Ed.; Academic: New York, 1978, p 299.

110. Allcock, H. R.; Kim, C. *Macromolecules* **1989**, *22*, 2596.

111. Singler, R. E.; Willingham, R. A.; Lenz, R. W.; Furakawa, A.; Finkelmann, H. *Macromolecules* **1987**, *20*, 1727.

112. Singler, R. E.; Willingham, R. A.; Noel, C.; Friedrich, C.; Bosio, L.; Atkins, E. D. T.; Lenz, R. W. in Liquid-Crystalline Polymers. (ACS Symp. Series 435), 1990, Ch. 14.

113. Allcock, H. R.; Mang, M. N.; Dembek, A. A.; Wynne, K. J. *Macromolecules* **1989**, *22*, 4179.

114. Manners, I.; Renner, G.; Nuyken, O.; Allcock, H. R. *J. Am. Chem. Soc.* **1989**, *111*, 5478.

115. Dodge, J. A.; Manners, I.; Allcock, H. R.; Renner, G.; Nuyken, O. *J. Am. Chem. Soc.* **1990**, *112*, 1268.

116. Godfrey, L. E. A.; Schappel, J. W. *Ind. Eng. Chem. Prod. Res. Develop.* **1970**, *9*, 426.

117. Dupont, J. G.; Allen, C. W. *Macromolecules* **1979**, *12*, 169.

118. Allen, C. W.; Bright, R. P. *Macromolecules* **1986**, *19*, 571.

119. Allen, C. W. *J. Polym. Sci., Polym. Symp.* **1983**, *70*, 79.

120. Allen, C. W. in *Inorganic and Organometallic Polymers;* Zeldin, M.; Wynne, K. J.; Allcock, H. R., Eds.: ACS Symposium Series; American Chemical Society: Washington, DC., **1988,** *360*, 290.

121. Rice, R. G.; Ernest, M. V. French Patent 1 450 474, 1966.

122. Kumar, D.; Fohlen, G. M.; Parker, J. A. *J. Polymer Sci. (Chem.)* **1983,** *21*, 3155.

123. Allcock, H. R.; McDonnell, G. S.; Riding, G. H.; Manners, I., unpublished work.

124. Allcock, H. R.; Wright, S. D.; Kosydar, K. M. *Macromolecules* **1978,** *11*, 357.

125. Dembek, A. A.; Kim, C.; Allcock, H. R.; Devine, R. L. S.; Steier, W. H.; Spangler, W. *Chemistry of Materials* **1990,** *2*, 97.

4

Polysiloxanes and Related Polymers

4.1 INTRODUCTION

At the present time, polysiloxanes are unique among inorganic and semi-inorganic polymers. They have been the most studied by far, and are the most important with regard to commercial applications. Thus, it is not surprising that a large number of review articles exist describing the synthesis, properties, and applications of these materials.[1-36]

The Si—O backbone of this class of polymers endows it with a variety of intriguing properties. For example, the strength of these bonds gives the siloxane polymers considerable thermal stability, which is important for their use in high-temperature applications, for example, as heat-transfer agents and high-performance elastomers. The nature of the bonding and the chemical characteristics of the side groups give the chains a very low surface free energy and, therefore, highly unusual and desirable surface properties. Not surprising, poly(organophosphazenes) are widely used for example as mold-release agents, for waterproofing garments, and as biomedical materials.

Some unusual structural features of the chains give rise to physical properties that are also of considerable scientific interest. For example, the substituted Si atom and the unsubstituted O atom differ tremendously in size, giving the chain a very irregular cross section. This influences the way the chains pack in the bulk, amorphous state, which in turn, gives the chains extremely unusual equation-of-state properties (such as compressibilities). Also, the skeletal bond angles around the O atom are much larger than those around the Si, and this makes the planar all-*trans* form of the chain approximate a series of closed polygons. As a result, siloxane chains exhibit a number of interesting configurational characteristics. These structural features, and a number of properties and their associated applications, will be discussed in this chapter.

The major categories of homopolymers and copolymers to be discussed are[31] (1) linear siloxane polymers $[-SiRR'O-]$ (with various alkyl and aryl R,R' side groups), (2) sesquisiloxane polymers possibly having a ladder structure, (3) siloxane-silarylene polymers $[-Si(CH_3)_2OSi(CH_3)_2(C_6H_4)_m-]$ (where the skeletal phenylene units are either *meta* or *para*), (4) silalkylene polymers $[-Si(CH_3)_2(CH_2)_m-]$, and (5) random and block copolymers and blends of some of the foregoing. Topics of particular importance are the molecular structure, flexibility, transition temperatures, permeability, and other physical properties. Applications include high-performance fluids, elastomers, and coatings, surface modifiers, separation membranes, photoresists, soft contact lenses, body implants, and controlled-release systems. Also of interest is the conversion of silicon-containing materials to novel reinforcing fillers, to ceramics by the sol-gel technique, and to high-performance fibers by controlled pyrolyses.

4.2 HISTORY

The first reaction of relevance in this area was the conversion of elemental silicon to silicon tetrachloride $SiCl_4$ and to trichlorosilane $SiHCl_3$. These purely inorganic substances were then converted into organometallic species such as $RSiX_3$, by allowing them to react with diethyl zinc and related compounds.[2,3,15,23,28]

The entire area of organosilicon chemistry blossomed with Kipping's preparation of such compounds by the more convenient Grignard process. These silanes turned out to be of paramount importance since they hydrolyzed readily to form compounds containing Si—O bonds, both linear and cyclic.[12,21,23,28] The new materials were first called silicoketones or "silicones" by analogy with ketones in the organic area. Structural studies, however, showed that they did not contain the Si=O double bond. Thus, the silicone name is a misnomer, but it has persisted, at least in casual usage. However, the terms siloxanes and polysiloxanes are much to be preferred.

4.3 NOMENCLATURE

In most of the literature, the terminology consists simply of specifying the side groups and then the backbone. For example, the polymer having the repeat unit $[-Si(CH_3)_2O-]$ is called poly(dimethylsiloxane), and that having the repeat unit

$[-Si(CH_3)(C_6H_5)O-]$ is called poly(methylphenylsiloxane). Closely related polymers are the poly(silmethylenes) of repeat unit $[-SiRR'CH_2-]$ and the poly(siloxane-silphenylenes) of repeat unit $[-SiRR'OSiRR'C_6H_4-]$, in which the second oxygen atom in a doubled repeat unit is replaced by a phenylene group.

Because certain structures and structural segments appear over and over again in the siloxane area, several abbreviations are used in specialized areas of the literature.[23] The monofunctional unit $R_3SiO_{0.5}$ is designated "M," the difunctional unit R_2SiO "D," the trifunctional unit $RSiO_{1.5}$ "T," and the quadrifunctional SiO_2 "Q." For example, the dimer $(CH_3)_3SiOSi(CH_3)_3$ is termed "MM," the oligomer $(CH_3)_3Si[OSi(CH_3)_2]_{10}OSi(CH_3)_3$ is termed "MD$_{10}$M," and the cyclic trimer $[Si(CH_3)_2O]_3$ is called "D$_3$." Unprimed abbreviations are taken to mean the R substituents are methyl groups, since these are the most common and the most important. Primed abbreviations are used for other substituents, the most important of which is probably the phenyl group.

4.4 PREPARATION AND ANALYSIS

4.4.1 Preparation of Monomers

The elemental silicon on which the entire technology is based is typically obtained by reduction of the mineral silica with carbon at high temperatures:[15,23]

$$SiO_2 + 2\,C \longrightarrow Si + 2\,CO \qquad (Eq. 4.1)$$

The silicon is then converted directly to tetrachlorosilane by the reaction

$$Si + 2\,Cl_2 \longrightarrow SiCl_4 \qquad (Eq. 4.2)$$

As already mentioned, this can be used to form an organosilane by the Grignard reaction

$$SiCl_4 + 2\,RMgX \longrightarrow R_2SiCl_2 + 2\,MgClX \qquad (Eq. 4.3)$$

This relatively complicated reaction has been replaced by the so-called "direct process" or "Rochow process,"[12,15,21] that starts from elemental silicon. It is illustrated by the reaction

$$Si + 2\,RCl \longrightarrow R_2SiCl_2 \qquad (Eq. 4.4)$$

but the process also yields $RSiCl_3$ and R_3SiCl, which can be removed by distillation. Compounds of formula R_2SiCl_2 are extremely important, because they provide access to the preparation of a wide variety of substances having both organic and inorganic character.[37-40] Their hydrolysis gives dihydroxy structures that condense to give the basic $[-SiR_2O-]$ repeat unit. The nature of the product obtained depends critically on the reaction conditions.[37] Basic catalysts and higher temperatures favor higher molecular weight polymers that are linear. Acidic catalysts tend to produce cyclic small molecules or low molecular weight polymers.

The hydrolysis approach to polysiloxane synthesis has now been largely replaced by ring-opening polymerizations[15,28,37,41-45] of organosilicon cyclic trimers and tetramers, with the use of ionic initiation. These cyclic monomers are produced by the hydrolysis of dimethyldichlorosilane. Under the right conditions, at least 50 wt % of the products are cyclic oligomers. The desired cyclic species are separated from the mixture for use in ring-opening polymerizations such as those described next.

4.4.2 Ring-Opening Polymerizations

Cyclic siloxanes can undergo a ring-opening polymerization that is a chain-growth process. Free radicals are not useful as initiator species, because of the nature of the siloxane bond, but anionic and cationic initiators are very effective. The reaction is illustrated using the most common cyclic oligomers, the trimer hexamethylcyclotrisiloxane or the tetramer octamethylcyclotetrasiloxane,[41,42]

$$(SiR_2O)_{3,4} \longrightarrow [-SiR_2O-]_x \qquad \text{(Eq. 4.5)}$$

where R can be alkyl or aryl and x is the degree of polymerization. In principle the reaction is reversible, but in practice it is made essentially irreversible by the choice of monomer, initiator, and polymerization conditions. Because of this potential reversibility, however, it is important to remove all initiators after the reaction, and this is typically done by neutralization of the acidic or basic terminal chain residues. Alternatively, some initiators can be removed by volatilization or thermolysis. Frequently, end blocking is used to modify the ends of the chains by placing groups there that will increase thermal stability and prevent reequilibration.

Because it is frequently impossible to remove all traces of active species, some reorganization is almost inevitable. In these processes, siloxane bonds interchange so as to bring about variation in both molecular weight and in the relative amounts of cyclic and linear species. At equilibrium, a Gaussian distribution of molecular weights exists. The cyclic oligomers that occur most frequently are D_4 through D_6, and the amount obtained depends greatly on the "monomer" and on the polymerization conditions. They are typically present to the extent of 10–15 wt %. The lower molecular weight cyclics are generally removed from the polymer before it is used in a commercial application. Their presence is also of interest from a fundamental scientific point of view, in two respects. First, the extent to which they are formed can be used as a measure of chain flexibility.[46] Second, the various cyclic species produced can be purified using standard separation techniques, and then used to test theoretical predictions of the differences between otherwise identical cyclic and linear molecules.[47-51]

For anionic equilibrations, typical catalysts are alkali metal oxides and hydroxides, and bases in general. Initiation and propagation involve nucleophilic attack on the monomer, causing opening of the ring followed by chain extension. As is frequently the case in ionic polymerizations, the nature of the countercation, particularly its size, can have a large effect on the reaction. This polymerization is very different from most others.[37] Most polymerizations are energetically driven, by a decrease in enthalpy. The decrease in entropy that accompanies the linkage of many monomers into one chainlike structure is counteracted and overcome by the decrease in energy generated by the formation of new chemical bonds. In the siloxane case, the bonds linking the monomer units into the chain are similar in energy to those found in siloxane rings, and the net energy change is very small. However, there is an increase in entropy, presumably from increased internal molecular freedom of the siloxane segments in going from the cyclic structures to the linear chains. It is this increase in entropy that drives the polymerization reaction.

Cationically catalyzed polymerizations[28,52] have not received as much attention as the anionic variety. Typical cationic (acidic) catalysts in this case are Lewis acids. Yields and proportions of the various species are generally very similar to those obtained in anionic polymerizations, although the mechanism is very different. The reaction is thought to proceed through a tertiary oxonium ion formed by addition of a proton to one of the O atoms of the cyclic siloxane. Part of the mechanism may involve step growth, as well as the expected chain growth.

Polymerization of nonsymmetrical cyclic siloxanes gives stereochemically variable polymers $[-SiRR'O-]$ that are analogous to the totally organic vinyl and vinylidene polymers $[-CRR'CH_2-]$. In principle, it should be possible to prepare them in the same stereoregular forms (isotactic and syndiotactic) that have been achieved for some of their organic counterparts,[8,37] as mentioned in Chapter 2. Unfortunately, this has not yet been accomplished, and only the stereoirregular (atactic) form has been obtained. Unlike the two (regular) stereochemical forms, the atactic modification is inherently noncrystallizable.

In some cases, an end blocker such as $YR'SiR_2OSiR_2R'Y$ is used to form reactive $-OSiR_2R'Y$ chain ends.[53,54] Some of the uses of these materials are described later in this chapter.

Homopolymerizations of this type are discussed in detail elsewhere.[37,42]

4.4.3 Copolymerization

Polymerization of mixtures of monomers, such as $(SiR_2O)_m$ with $(SiR'_2O)_m$, can be used to obtain random copolymers. They are generally highly irregular in structure,[55] but in a chemical rather than a stereochemical sense. Correspondingly, they also show little if any crystallizability.

Copolymerizations of this type are also discussed in detail elsewhere.[37,42]

4.4.4 Structural Features

Several structural features make the siloxane backbone one of the most flexible in all of polymer science.[33] The reasons for this extraordinary flexibility can be seen from Figure 4.1. First, because of the nature of the bonding,[46,56,57] the Si—O skeletal bond has a length (1.64 Å) which is significantly larger than that (1.53 Å) of the C—C bond found in most organic polymers. As a result, steric

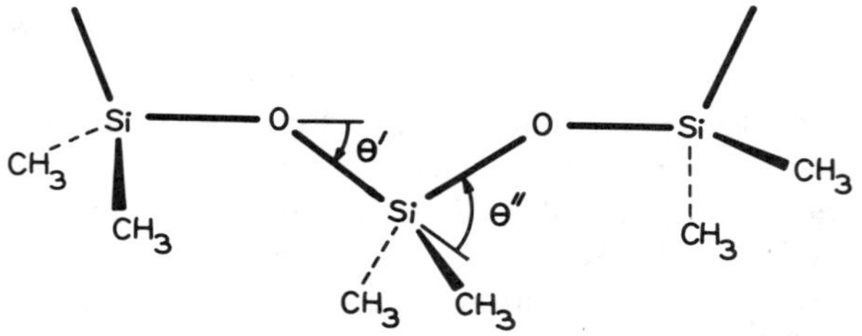

Figure 4.1 Sketch of the PDMS chain, showing some structural information relevant to its high flexibility. Reprinted with permission from P. J. Flory, *Statistical Mechanics of Chain Molecules*; Wiley-Interscience: New York, 1969. Copyright © 1969 John Wiley & Sons.

interferences or intramolecular congestion are diminished.[46] [This circumstance is true for inorganic and semi-inorganic polymers in general. Almost any single bond between a pair of inorganic atoms (Si—Si, Si—C, Si—N, P—N, etc.) is longer than the C—C bond]. Also, the oxygen skeletal atoms are unencumbered by side groups, and they are as small as an atom can be and still have the divalency needed to continue a chain structure. Finally the Si—O—Si bond angle $(180 - \theta')$ of $\sim 143°$ is much more open than the usual tetrahedral bonding $(\sim 110°)$,[46] and torsional rotations can occur without incurring a serious energy penalty. These structural features have the effect of increasing the dynamic flexibility of the chain.[46] They also increase its equilibrium flexibility, which is the ability of a chain to take on a compact shape when in the form of a random coil. (It is generally measured, inversely, by the mean-square end-to-end distance or radius of gyration of the chain in the absence of excluded-volume effects).

4.4.5 Elastomer Technology

Pure siloxane polymers are only rarely appropriate for use in technology. Numerous additives must generally be incorporated in order to improve their properties. A typical formulation contains the siloxane polymer, plus some or all of the following ingredients: reinforcing fillers, extending (nonreinforcing) fillers, processing aids, heat-aging additives, pigments, and curing agents (end-linking agents with associated catalysts or organic peroxides).[5,58]

The siloxane polymer usually has a rather high molecular weight and may have reactive ends for end-linking curing or vinyl side chains for peroxide curing.

The preferred reinforcing filler is high surface area silica, particularly that made by the fume process. This gives the greatest reinforcement and, because of its high purity, yields excellent electrical insulation properties. Silicas obtained from aqueous solutions impart only moderately good reinforcement and, because of the presence of water on the filler particles, can adversely affect the electrical properties of the elastomer. Carbon black provides some reinforcement, but can interfere with some types of peroxide cures. Also, its electrical conductivity can severely compromise the electrical properties of the material.[5,58] In some cases, silane coupling agents are used to improve the bonding between the reinforcing phase and the polymer.[59] These molecules typically have the structure X_3SiY, with X chosen to interact strongly with one phase and Y with the other. For example, if X is an alkoxy group, it can hydrolyze and react with OH groups on the surface of a filler particle.[15] Similarly, if Y is a vinyl group, it can be polymerized into the organic matrix being reinforced, proving enhanced filler-matrix bonding.[15]

One reason for using extending fillers is to reduce the cost of the compounded elastomer. They are exemplified by ground silica, kaolin, diatomaceous earth, and minerals such as calcium carbonate.

Coloring agents can be either organic or inorganic, but the former can adversely affect heat stability. Examples of suitable inorganic colorants are oxides and salts of iron, chromium, cobalt, titanium, and cadmium. Some not only provide color, but can also have some beneficial heat-aging effects.

Processing aids are particularly important in the case of elastomers that contain highly reinforcing silicas, since these fillers adsorb polymer chains so strongly to their surfaces that premature gelation can occur. These additives have a softening or plasticizing effect, thus delaying the occurrence of this complication.[5,58]

The nature of the curing (cross-linking) agents introduced depends on the particular chemical reaction chosen for generating the cross-links.[60-63] In the case of end-linking reactions, the end groups are generally either hydroxyl or vinyl units. For hydroxyl end groups, the end-linking agent may be tetraethoxysilane, which reacts by a condensation reaction, with a stannous salt used as a catalyst. For vinyl units, the end-linking agent can be an oligomeric siloxane that contains reactive $Si-H$ groups, with the H atoms adding to the double bonds in the polymeric siloxane. Platinum salts are catalysts for this type of addition curing reaction.[60-63]

Aliphatic or aromatic peroxide curing agents can be used. They may react with saturated alkyl groups or with vinyl side chains. Specific peroxides are chosen on the basis of their decomposition temperatures, and the reaction products they leave behind after the curing process is complete. Some peroxides used are bis(2,4-dichlorobenzoyl)peroxide, benzoyl peroxide, dicumyl peroxide, and di-*t*-butyl peroxide.[60-63]

The compounding process itself can be very complicated.[5,58] The amounts of some ingredients used are fixed by the stoichiometry of the reaction in which they participate. The end-functionalized polymer and the end-linking agent are, of course, in this category. The relative amounts of other ingredients are often chosen by experience or by trial and error. After all these amounts have been selected, they are mixed ("compounded") in conventional equipment, for example, a Banbury mixer. Although the resulting mixture can be fabricated immediately, it is common practice to permit the material to "recover" or "age" for a few days. If hardening occurs during this period, some remilling ("refreshening") will be required.

A variety of processing steps may then be carried out. Examples are compression molding, injection molding, transfer molding, extrusion, calendering, dispersion coating, and blowing into foams.[5,58]

Although some cures can take place at room temperature, most are carried out at elevated temperatures.[5,60-63] Conventional electric, gas, or forced-air ovens are used, but presumably microwave heating could be used as well.

4.4.6 Analysis and Testing

Infrared and ultraviolet spectroscopy are often used to determine the composition of siloxane copolymers, and of mixtures of siloxanes and silicates with other species.[16,23] These types of spectroscopy can also be used to monitor vinyl groups introduced to facilitate cross-linking, phenyl groups to suppress crystallization or to improve radiation resistance, or silanol end groups introduced during polymerization and used to determine number-average molecular weights, or for chemical reactions such as end-linking. Some important absorbances are those for $Si-O-Si$ groups at 1010 cm^{-1}, $Si(CH_3)_2$ groups at 800 cm^{-1}, $SiCH_3$ groups at 1260 cm^{-1}, and SiH groups at 2200 cm^{-1}. Not surprising, these and other methods are used for quality control in the commercial production of siloxane products.[16,23]

NMR is used for a variety of purposes, most of which parallel those used to characterize small-molecule systems.[16,23] In addition to ^1H and ^{13}C NMR, ^{29}Si NMR is frequently employed. These methods are used to characterize chemical composition, structural features, and conformational preferences. They are also used to characterize hybrid inorganic composites, and silica-type ceramics, in general.[64,65]

Specific functional groups are also analyzed by chemical methods. For example, the various chlorosilanes can be hydrolyzed and the resulting chloride ions determined by titration with silver salts. Similarly $Si-H$ groups can be determined by measurement of the amount of hydrogen gas evolved during their hydrolyses. As a final example, silanol groups can be monitored through measurement of the amount of methane gas evolved when they are reacted with methyl Grignard reagent.[16,23]

It is of course possible to determine the amount of silicon present in a sample by pyrolysis to silica, followed by atomic absorption methods.

Both mass spectrometry and gas chromatography are used to identify and determine the amounts of volatile siloxane-type materials.[16,23] The more important nonvolatile materials, including the polymers, can be characterized by liquid chromatography and by gel permeation chromatography (GPC), as was described in Chapter 2. When used analytically, these techniques give molecular-weight distributions. Used preparatively, they yield narrow-molecular-weight distribution fractions that are suitable for determining reliable structure-property relationships. Average molecular weights themselves can also be determined by a variety of techniques, including dilute solution viscometry, osmometry, ultracentrifugation, and light-scattering intensity measurements, as mentioned in Chapter 2. Other techniques for determining molecular-weight distributions, such as fractional precipitations, and gradient elutions are also used. Extractions with supercritical fluids seem particularly promising in this regard.[66,67]

Thermal properties are measured and evaluated by some of the methods also mentioned in Chapter 2. For identification of transition temperatures, measurements of heats of fusion, and so on, differential thermal analysis (DTA) and differential scanning calorimetry (DSC) are much used. Thermal stability is measured by thermogravimetric analysis (TGA), although this technique can give overly optimistic results unless used with great care.

Rheological measurements are of central importance in the processing of siloxane polymers. Typical studies include determination of the dependence of the bulk viscosity of the material on the average molecular weight, molecular-weight distribution, and rate of shear. Characterization of the effects of any branched chains or reinforcing fillers present is also of great importance.[15]

Most siloxane polymers are excellent insulators, and electrical properties relevant to this characteristic are also widely studied. Examples are resistivity, dielectric constant, dielectric losses, dielectric strength (resistance to electrical breakdown), and power factors.[15]

The use of siloxane polymers in applications such as separation membranes, drug release systems, and blood oxygenators requires extensive permeability studies. This, of course, also involves measurements of diffusivity and solubility.[15]

For some specialized applications, optical properties can be of crucial importance. Two examples are contact lenses and interlayers for glass windshields. Here, transparency measurements are of primary importance, but index of refraction (n) measurements are important for matching values of n for polymers and fillers.[15]

Applications in the biomedical area require extensive testing of biocompatibility. These include acute, dietary, and implant testing and monitoring toxicological effects such as carcinogenicity, mutagenicity, teratogenicity, and bacterial or fungal colonization.[15]

In the case of siloxane elastomers, the testing of mechanical properties is of particular importance. Elongation or tensile measurements are used almost to the exclusion of other types of mechanical tests, probably because of their simplicity. In this way, structural information is obtained about the networks, such as their degrees of cross-linking.[68] Measurements of the ultimate strength (modulus at rupture), and the maximum extensibility (elongation at rupture) are also important. Relating such properties to the chemical nature of the siloxane polymer, to the curing conditions, and to the nature and amounts of any reinforcing fillers is obviously a task of paramount importance in the area of elastomeric applications.[68]

Many applications of siloxane materials involve such a complicated array of properties that the ultimate evaluation has to involve a "use" test. Many surface applications, such as release coatings, are in this category. In this approach, the prospective material is tested directly under standard conditions chosen to mimic those under which the material would actually be employed.[15, 16, 23] The advantage of such a test is its direct connection with the desired application. The major disadvantage results from the fact that the underlying reasons why a material fails are not uncovered in a global test of this type.

4.5 GENERAL PROPERTIES

4.5.1 Conformations and Spatial Configurations

4.5.1.1 Symmetrically substituted polysiloxanes.
The first member of this series, poly(dimethylsiloxane) (PDMS), $[-Si(CH_3)_2O-]_x$, has been studied more extensively with regard to its configuration-dependent properties than any other chain molecule.[10,46] As can be seen from Figure 4.1, it is very similar in its structure to the polyphosphate chain to be discussed in Chapter 6 (Figure 6.6). As was already mentioned, the Si—O bond length in polysiloxanes is 1.64 Å, and bond angles at the Si and O atoms are 110° and 143°, respectively. This inequality of bond angles causes the all-*trans* form of the molecule to form a closed structure after approximately 11 repeat units. The torsional barrier for rotations about the skeletal bonds is very low, and this is one of the reasons for the very high dynamic flexibility and very low glass transition temperature of the PDMS chain.

Trans states are of lower energy than are *gauche* states in this chain.[10,46] This conformational preference may arise from favorable van der Waals interactions between pairs of CH_3 groups separated by four bonds in *trans* states. This is apparently more important than favorable coulombic interactions between oppositely charged Si and O atoms separated by three bonds, which are larger in *gauche* states because of the reduced distance. However, comparisons between experimental and theoretical values of various configuration-dependent properties yield a value for this energy difference that is significantly larger than that obtained from the usual semiempirical calculations of interactions between nonbonded atoms. Conformations $g^{\pm}g^{\mp}$ about O—Si—O skeletal bond pairs give rise to pentane-type interferences[10,46] between the bulky $Si(CH_3)_2$ groups and are therefore completely excluded. The same conformations about Si—O—Si bond pairs cause interferences between the smaller O atoms, and these can occur, but to only low incidence. Such conflicting arrangements, between groups separated by four bonds, can be visualized by rotations about pairs of consecutive skeletal bonds in Figure 4.1.

The equilibrium flexibility of PDMS can be characterized by its unperturbed dimensions, in particular its value of the characteristic ratio described in Chapter 2. Experimental values of this ratio are in the range 6.2–7.6, the precise value depending on the nature of the solvent used in the study.[10,46] The origin of this "specific solvent" effect is obscure but may involve highly specific interactions between solvent molecules and polymer segments in a way which changes the conformational preferences in the chain. The effect is significant apparently only in the case of polar polymers. The unperturbed dimensions appearing in the definition of the characteristic ratio also appear in the equations for the modulus of the chains when cross-linked into an elastomeric network. Not surprisingly, therefore, the specific solvent interactions can effect the modulus of swollen PDMS networks as well as the dimensions of isolated PDMS chains in solution.[68]

In any case, the characteristic ratio of PDMS is known to increase with increasing temperature. This is expected since the low-energy conformation is the closed polygon mentioned earlier,[10,46] and an increase in temperature provides thermal energy for switching from these compact, low-energy conformations to less compact higher-energy ones. Comparisons between the experimental and theoretical values of the characteristic ratio and its temperature coefficient gave values of the chain conformational energies which were then used to predict a number of other configuration-dependent properties. Dipole moments calculated in this way were found to be in excellent agreement with experiment in the range of small chain length; the agreement at longer chain length is less satisfactory, possibly because of the large specific solvent effect[69] already mentioned. Stress-optical coefficients have also been determined for the polymer, using networks both unswollen and swollen with a variety of solvents. Only very approximate qualitative agreement was obtained, presumably because of the vanishingly small optical anisotropy of the PDMS chain.[10]

Molecular mechanics and more sophisticated computational techniques are being applied extensively to siloxane conformational problems, to study both the chain backbone[70] and the side chains.[71]

It is interesting to note that the PDMS chain and polyphosphate chain have approximately the same characteristic ratio. Isolated gauche states, of relatively high spatial extension, are more prevalent in the polyphosphate chain, but pairs of *gauche* states ($g^{\pm}g^{\pm}$) of the same sign are less prevalent and the two effects largely offset one another.

Much experimental and theoretical work has been reported on the cyclization of dimethylsiloxane chains and on the study of the properties of these cyclics.[47-51] The cyclization process has been investigated for a very wide range of polymer chain lengths, but only the results for the behavior of long chains will be discussed here. (The interpretation of the results for shorter chains may be complicated by failure of the Gaussian distribution function employed for the end to distances, directional correlations between terminal bonds prior to cyclization, ring-strain contributions to the heat of the reaction, and the possible necessity of revising some of the statistical weight factors for the chains.) In the limit of large chain length, the agreement between theory and experiment is excellent, and thus is in strong support of the proposed model for PDMS. In addition, the cyclization studies cited have also provided useful information on excluded volume effects (their absence in the undiluted amorphous state and their magnitudes in solutions, particularly at high polymer concentrations), the critical chain length at which the Gaussian distribution becomes unacceptably inaccurate, the magnitude of specific solvent effects, and the validity of gel permeation chromatography theories pertaining to both linear and branched chain molecules.

As already mentioned, the cyclic species formed in some of these reactions are themselves of considerable interest. Their physical properties, and a comparison of them with those of the linear chains of the same degree of polymerization, have been investigated extensively.[47-51] Example of such comparisons are solution viscosity–molecular-weight relationships, bulk viscosities, densities, refractive indices, glass transition temperatures, ^{29}Si NMR chemical shifts, chain dimensions from neutron scattering, diffusion coefficients and their concentration dependence, thermal stability, second virial coefficients, radii of gyration, equilibrium shapes (from Monte Carlo simulations), static dielectric permittivities, particle-scattering functions, monolayer surface pressures, melting points, theta temperatures (at which the chains are unperturbed by excluded volume effects),[46] critical temperatures for phase separations, melt mobility (by excimer emission), and conformational dynamics (by ultrasonic relaxation measurements).

Other symmetrically substituted polysiloxanes have been investigated less thoroughly.[10] Poly(diethylsiloxane) $[-Si(C_2H_5)_2O-]_x$ has been reported to have a characteristic ratio of 7.7 ± 0.2, which is essentially the same value as that of PDMS. Its dipole moment is very difficult to measure because of the low polarity of the repeat unit, but it too is approximately the same as that of PDMS. These tentative results, if confirmed, suggest that this lengthening of the side chains must generate self-compensating effects. Furthermore, poly(di-n-propylsiloxane) $[-Si(C_3H_7)_2O-]_x$ has been reported to have a very large characteristic ratio, specifically, 13.0 ± 1.0. The high spatial extension in this chain could result from the fact that an articulated side chain such as $-CH_2CH_2CH_3$, could probably adopt more conformations in the case of *trans-gauche* states along the chain backbone than in the more restrictive *trans-trans* states shown in Figure 4.2. Although this "entropic destabilization" of compact *trans* states would increase the chain dimensions, so large an increase in the characteristic ratio over that for PDMS would not have been anticipated. This intuitive conclusion is supported by some rotational isomeric state calculations that do take into account the conformational variability of the siloxane side chains. It should be mentioned, however, that results of cyclization studies carried out on some stereochemically variable polysiloxanes (see the upcoming discussion) suggest that the characteristic ratio increases with increase in length or size of the side groups.

Cyclization studies have also been carried out on the chemical copolymers poly(ethylene, dimethylsiloxane) and poly(styrene, dimethylsiloxane).[47-51] Nu-

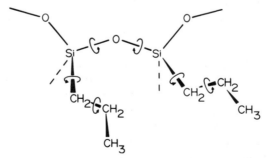

Figure 4.2 The poly(di-n-propylsiloxane) chain, showing the conformational variability of the propyl side chains. Reprinted with permission from J. E. Mark, *Macromolecules* **1978**, *11*, 627. Copyright 1978 American Chemical Society.

merous intramolecular interactions need to be taken into account in a chemical copolymer, and, consequently, the results on the copolymers have been given only a preliminary interpretation in terms of rotational isomeric state theory. Cyclization calculations have also been carried out for poly(dihydrogensiloxane) $[-SiH_2O-]_x$, but at present there are no experimental data available for comparison with theory.

Finally, melting point depression measurements have been conducted on several symmetrically substituted polysiloxanes, specifically the dimethyl, diethyl, di-n-propyl, and diphenyl polymers. Interpretation of such experimental results yields entropies of fusion. Although it is difficult to extract a reliable configurational entropy from this quantity, such results could help elucidate the configurational characteristics of the chains thus investigated.[10]

4.5.1.2 Stereochemically variable polysiloxanes.

In these unsymmetrically disubstituted chains, the substituents of one type can be on the same side of the all-*trans* chain, on opposite sides, or on either side in a random arrangement. These are the isotactic, syndiotactic, and atactic forms first mentioned in Chapter 2. One of the chains chosen to illustrate this stereochemical variability in Figure 2.19 was, in fact, poly(methylphenylsiloxane). As already mentioned, the relatively large Si—O bond length and Si—O—Si bond angle place apposed side groups at distances of separation (ca. 3.8 Å) at which there is a favorable energy of interaction. Conformational energy calculations on this polymer indicate that the attractions should be particularly strong in the case of a pair of phenyl groups on adjacent repeat units. Therefore, the chains should have a tendency to adopt conformations in which two phenyl groups are apposed on the same side of the chain.[10] For the syndiotactic polymer, this gives rise to a preference for gauche states, which confer relatively high spatial extension, but to a suppression of such states for the isotactic polymer. As a result, the characteristic ratio is predicted to be very small for the isotactic polymer and to increase monotonically and linearly with increases in the number of syndiotactic placements in the chain. These results are quite different from those calculated for monosubstituted $[-CHRCH_2-]_x$ vinyl or disubstituted $[-CRR'CH_2-]_x$ vinylidene chains, including the structurally analogous poly(α-methylstyrene) $[-C(CH_3)(C_6H_5)CH_2-]_x$.[10]

A characteristic ratio of 8.8 was reported for several samples of poly(methylphenylsiloxane),[10,72] at least some of which[73] were known to be essentially atactic. This experimental result, however, can be reproduced from the model only by assuming a relatively large fraction of syndiotactic placements; the temperature coefficient predicted for this degree of syndiotacticity is then also in good agreement with experiment. The assumption of significant syndiotacticity is in disagreement with NMR results[73] and with the results of cyclization studies, both of which suggest that poly(methylphenylsiloxane) is typically essentially atactic. The cyclization results, however, yield a prediction for the characteristic ratio which is significantly larger than the experimental value cited earlier. The two tentative conclusions regarding the stereochemical structure might be brought into closer agreement by improving the calculation of the interaction energy of two apposed phenyl groups to take into account the fact that they would be less exposed to favorable interactions with the solvent in such conformations. The effect is apparently quite important in polystyrene $[-CH(C_6H_5)CH_2-]_x$ but may be less so in poly(methylphenylsiloxane) because of the larger distance of separation between side groups in the siloxane polymers. Such revision could increase the number of isotactic placements which could be incorporated into the chain without decreasing its predicted value of the characteristic ratio to below its known experimental value. In any case, resolution of this point really requires reliable experimental values of the characteristic ratio, determined on samples of known stereochemical structure.

Cyclization measurements have also been conducted on other stereochemically variable polysiloxanes $[-Si(CH_3)RO-]_x$, where R was H, CH_2CH_3, $CH_2CH_2CH_3$, and $CH_2CH_2CF_3$. The basic conclusion from these investigations was that such polymers are generally atactic in structure and that increase in the length or size of the side chains tends to increase the characteristic ratio.[47-51]

Of course, it would be interesting to find catalysts that parallel the Ziegler-Natta catalysts used to prepare isotactic poly(α-olefin) polymers.[37] Poly-(methylphenylsiloxane) polymers, for example, having sufficient isotacticity (or syndiotacticity) to crystallize could be quite important.

4.5.1.3 Poly(dimethylsilmethylene).

This unusual polymer, $[-Si(CH_3)_2CH_2-]_x$, can be thought of either as a hydrocarbon analog of PDMS (in which O atoms are replaced by CH_2 groups) or as a silicon analog to polyisobutylene $[-C(CH_3)_2CH_2-]_x$ (in which Si atoms replace one of the two skeletal C atoms in the repeat unit).[10] The polymer is shown schematically in Figure 4.3. The Si—C bonds are 1.90 Å long and, in contrast to siloxane chains, the two types of skeletal bond angles are essentially identical and approximately tetrahedral. Most interesting is the fact that, since CH_2 and CH_3 groups have very similar interactions, this chain molecule should have some characteristics reminiscent of the idealized "freely rotating" chain.[46] This conclusion is supported by experimental evidence which indicates that the characteristic ratio of the polymer is relatively small and that both its unperturbed dimensions and dipole moments are essentially independent of temperature.[10]

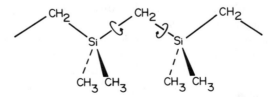

Figure 4.3 The poly(dimethylsilmethylene) chain. Reprinted with permission from J. E. Mark, *Macromolecules* **1978**, *11*, 627. Copyright 1978 American Chemical Society.

4.5.2 Flexibility of the Polymer Chains

4.5.2.1 Equilibrium flexibility. The equilibrium type of flexibility mentioned earlier has a profound effect on the melting point T_m of a polymer, as was mentioned in Chapter 2. High flexibility in this equilibrium sense means high conformational randomness, and thus high entropy of fusion and low melting point. This entropy can be reduced by a stretching process, in what is called "strain-induced crystallization," as described in Figure 2.40. The crystallites thus generated can be very important since they may provide considerable reinforcement for a network. Most polysiloxane elastomers have melting points that are unfortunately too low to benefit from this effect, however.

Examples of ways in which a polymer may be made less flexible, or more rigid, are shown in Figure 4.4.[74] They involve the combination of two chains into a ladder structure, insertion of rigid units such as p-phenylene groups into the chain backbone, or the addition of bulky side groups. The sesquisiloxane polymer shown in the upper portion of this figure would be in this category.[75,76] Attempts have been made to prepare it using the reaction shown in Figure 4.5,[30] but the structure shown may not have been obtained. This stiffness decreases the entropy of fusion and thus increases the melting point T_m. For example, if the chains are combined

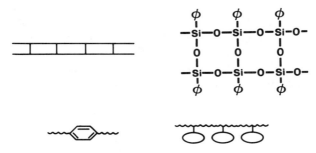

Figure 4.4 Some ways for making a polymer more rigid. Reprinted with permission from J. E. Mark, *Physical Chemistry of Polymers*, ACS Audio Course C-89, American Chemical Society, Washington, DC, 1986. Copyright 1986 American Chemical Society.

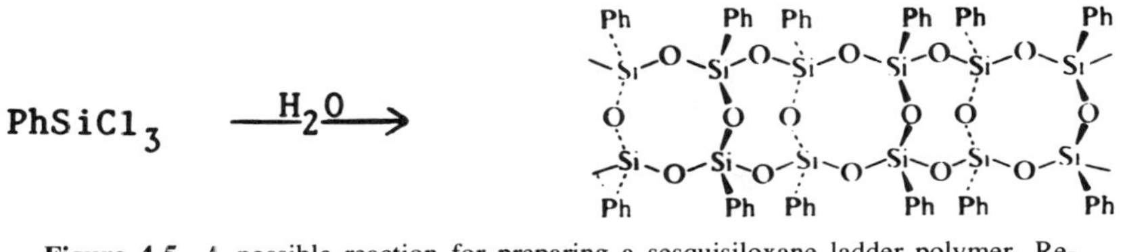

Figure 4.5 A possible reaction for preparing a sesquisiloxane ladder polymer. Reprinted with permission from H. R. Allcock and F. W. Lampe, *Contemporary Polymer Chemistry*, 2nd ed. Prentice Hall: Englewood Cliffs, NJ, 1990. Copyright 1990 Prentice Hall.

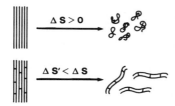

Figure 4.6 Sketch explaining the increase in melting point with increase in chain rigidity. Reprinted with permission from J. E. Mark, *Physical Chemistry of Polymers*, ACS Audio Course C-89, American Chemical Society, Washington, DC, 1986. Copyright 1986 American Chemical Society.

into a ladder structure, they cannot disorder themselves as much as when they are separate, as is shown in Figure 4.6.[74] The same argument holds for the other two methods for increasing T_m. This decreased equilibrium flexibility is paralleled by a decreased dynamic flexibility, and thus by an increased glass transition temperature T_g. Another advantage of the ladder structure is its resistance to degradative chain scission. It will not be degraded into two shorter ladder structures except in the unlikely event that two single-chain scissions occur nearly directly across from one another.[77]

Insertion of a silphenylene group $[-Si(CH_3)_2C_6H_4-]$ into the backbone of the PDMS repeat unit yields the siloxane *meta*- and *para*-silphenylene polymers shown in Figure 4.7.[68,78,79] The T_g of the former polymer is increased to $-48°C$ [relative to $-125°C$ for poly(dimethylsiloxane)], but no crystallinity has been detected to date.[80] Since the repeat unit is symmetric, it should be possible to induce crystallinity by stretching, as described in Figure 2.40. The explanation here is the same as that given in Figure 4.6 except that the chains are prevented from completely disordering themselves by being held elongated by the stretching force, rather than by the structural features of the chains. As expected, the *p*-silphenylene group has a larger stiffening effect, increasing the T_g to $-18°C$ and giving rise to crystallinity with a T_m of 148°C. The resulting polymer is thus a *thermoplastic siloxane*. Apparently, ortho (*o*) silphenylene units have not been introduced in this

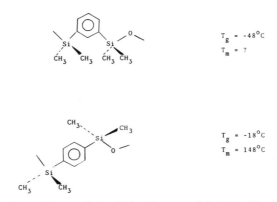

Figure 4.7 *Meta*- and *para*-silphenylene polymers and their transition temperatures. Reprinted with permission from J. E. Mark and B. Erman, *Rubberlike Elasticity: A Molecular Primer*; Wiley-Interscience: New York, 1988. Copyright © 1988 John Wiley & Sons.

way. They are probably much harder to incorporate because of steric problems and, even if they could be incorporated, would not be expected to have as large a stiffening effect on the chain.

Silarylene polymers that contain more than one phenylene group in the repeat unit could be of considerable interest because of the various *meta*, *para* combinations that could presumably be synthesized.

Figure 4.4 also has a sketch showing the use of bulky side groups to make a chain stiffer. This can be illustrated by replacing one of the methyl groups in the poly(dimethylsiloxane) repeat unit by a phenyl group. The resulting polymer, poly(methylphenylsiloxane), has a glass transition temperature of $-28°C$,[73] which is significantly higher than the value, $-125°C$, shown by poly(dimethylsiloxane).[49]

It is intriguing that even some flexible siloxane polymers form mesomorphic (liquid-crystalline) phases. Some illustrative data are given in Table 4.1.[33, 81-87] Both poly(diethylsiloxane) and poly(di-*n*-propylsiloxane) show two crystalline modifications as well as a mesomorphic phase. (The other major class of semi-inorganic polymers, the polyphosphazenes, are also relatively flexible, and show similarly interesting behavior.[10, 88])

Some polysiloxanes form liquid-crystalline phases because of the presence of relatively stiff side chains.[89, 90] They have been widely studied, particularly with regard to the effect of deformation of the elastomeric polysiloxane phase on the mesomorphic behavior exhibited by the side chains.

Although the polysiloxanes are much more flexible than their organic counterparts, the polysilanes seem to fall between these two extremes. In this sense, it is instructive to compare a polysilane with its hydrocarbon analog in terms of chain flexibility. For example, relevant conformational energy calculations have recently been carried out on polysilane itself $[-SiH_2-]_x$.[91, 92] Some conformational energies, shown as a function of two consecutive skeletal rotation angles ϕ, were in fact presented in Figure 2.23.[86] The results suggest that the lowest energy conformation should be a sequence of *gauche* states ($\phi = \pm120°$) of the same sign.[91, 92] This is in contrast to polyethylene $[-CH_2-]_x$, which has a preference for *trans* states ($\phi = 0°$). Such preferences generally dictate the regular conformation cho-

TABLE 4.1   Liquid-Crystalline Siloxane Polymers[a]

Polymer	$\sim T$, °C	Transition		
$[-Si(CH_3)_2O-]$	-40	Cryst	$\longrightarrow$	isotropic
$[-Si(C_2H_5)_2O-]$	-60	Cryst	$\longrightarrow$	cryst'
	0	Cryst'	$\longrightarrow$	mesomorphic
	40	Mesomorphic	$\longrightarrow$	isotropic
$[-Si(n\text{-}C_3H_7)O-]$	-55	Cryst	$\longrightarrow$	cryst'
	60	Cryst'	$\longrightarrow$	mesomorphic
	205	Mesomorphic	$\longrightarrow$	isotropic

[a]Reproduced by permission from J. M. Zeigler and F. W. G. Fearon, eds., *Silicon-Based Polymer Science. A Comprehensive Resource*; Advances in Chemistry Series, 1990, *224*. Copyright 1990 American Chemical Society.

sen by a polymer chain when it crystallizes. Polyethylene crystallizes in the all-*trans* planar zigzag conformation,[46] and it would be interesting to determine whether polysilane crystallizes in the predicted helical form generated by placing all of its skeletal bonds in *gauche* states of the same sign.

The calculations also predict that polysilane should have a higher equilibrium flexibility than polyethylene,[92] and solution characterization techniques could be used to test this expectation. Dynamic flexibility can also be estimated from such energy maps, by determining the barriers between energy minima. Relevant experimental results could be obtained by a variety of dynamic techniques.[93]

4.5.2.2 Dynamic flexibility.

Dynamic flexibility refers to a molecule's ability to *change* spatial arrangements by rotations around its skeletal bonds. The more flexible a chain is in this sense, the more it can be cooled before the chains lose their flexibility and mobility and the bulk material becomes glassy. Chains with high dynamic flexibility thus generally have very low glass transition temperatures T_g.[94] Since exposing a polymer to a temperature below its T_g generally causes it to become brittle, low values of T_g can be advantageous, particularly in the case of fluids and elastomers.

Structural changes that increase a chain's equilibrium rigidity generally also increase its dynamic rigidity and thus increase T_g. Conversely, the very high flexibility of PDMS is the origin of its low T_m ($-40°C$)[95] as well as its very low T_g, $-125°C$.[95] The general effect of increased rigidity is therefore to increase a polymer's "softening temperature," which is T_m if the polymer is crystalline and T_g (typically $\sim 2/3\, T_m$ in °K) if it is not.[77]

4.5.3 Permeability

Siloxane polymers have much higher permeability to gases than most other elastomeric materials. For this reason, they have long been of interest for gas separation membranes, the goal being to vary the basic siloxane structure to improve selectivity without decreasing permeability. The repeat units of some of the polymers which have been investigated[96-100] include $[-Si(CH_3)RO-]$, $[-Si(CH_3)XO-]$, $[-Si(C_6H_5)RO-]$, $[-Si(CH_3)_2(CH_2)_m-]$, $[-Si(CH_3)_2(CH_2)_mSi(CH_3)_2O-]$, and $[-Si(CH_3)_2(C_6H_4)_mSi(CH_3)_2O-]$, where R is typically an *n*-alkyl group and X is an *n*-propyl group made polar by substitution of atoms such as Cl or N. Unfortunately, structural changes that increase the gas-diffusion selectivity are generally found to decrease the permeability, and vice versa. In contrast to the polysiloxanes, the polysilanes $[-SiRR'-]$ are thought to have greater variations in permeability.[101] Differences in permeability between the polysilanes and polysiloxanes may be due to the absence of skeletal oxygen atoms, or to the fact that polysilanes tend to be glassy rather than elastomeric at room temperature. An important comparison would therefore be between a polysilane and polysiloxane having the same side groups, at a temperature high enough for both polymers to be elastomeric, or low enough for both to be glassy.

TABLE 4.2   Some Gas Permeability Information[a]

Polymer	Gas	$10^8 P^b$	(P_{O_2}/P_{N_2})
$[-Si(CH_3)_2O-]$	O_2	6.0	1.9
	N_2	3.1	
$[-C(Si(CH_3)_3)=C(CH_3)-]$	O_2	72.	1.7
	N_2	42.	

[a]Reproduced by permission from J. M. Zeigler and F. W. G. Fearon, eds., *Silicon-Based Polymer Science. A Comprehensive Resource*; Advances in Chemistry Series, 1990, *224*. Copyright 1990 American Chemical Society.
[b]Units of $cm^3(STP)cm/(cm^2\ s\ cm\ Hg)$.

Also of interest is the recent observation that the presence of a trimethylsilyl group $[-Si(CH_3)_3]$ as a side chain in an acetylene-derived repeat unit increases the permeability of the polymer to a value above that of PDMS! The specific polymer is poly[1-(trimethylsilyl)-1-propyne],[102-107] and some comparisons between it and PDMS are given in Table 4.2.[33] Remarkably, its permeability coefficient P is about an order of magnitude higher than that of PDMS without much decrease in selectivity (as measured by the ratio of the P values for oxygen and nitrogen). The greatly increased values of P are apparently due to the unusually high solubility of gases in this polymer.[103] Studies of the effects of substituting the trimethylsilyl group onto other polymer backbones are in progress.

Another type of membrane designed as an artificial skin coating for burns also exploits the high permeability of siloxane polymers.[17, 108] The inner layer of the membrane consists primarily of protein and serves as a template for the regenerative growth of new tissue. The outer layer is a sheet of silicone polymer which not only provides mechanical support, but also permits outward escape of excess moisture while preventing ingress of harmful bacteria.

Soft contact lenses as prepared from PDMS provide a final example, as is shown in Figure 4.8.[33] The oxygen required by the eye for its metabolic processes must be obtained primarily by inward diffusion from the air rather than through blood vessels. PDMS is ideal for such lenses[17] because of its high oxygen permeability, but it is too hydrophobic to be adequately wetted by the fluid covering the

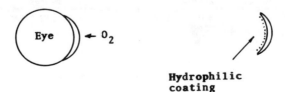

Hydrophilic coating

Figure 4.8 PDMS soft contact lens. An example of the use of grafting to change only the surface properties of a polymeric material. Reprinted with permission from J. M. Zeigler and F. W. G. Fearon, eds. *Silicon-Based Polymer Science: A Comprehensive Resource*; Advances in Chemistry Series, 1990, *224*. Copyright 1990 American Chemical Society.

eye. This prevents the lens from "feeling right," and can also cause a serious adhesion of the lens to the eye itself. One way to remedy this is to graft a thin layer of a hydrophilic polymer to the inner surface of the lens. Because of the thinness of the coating, the high gas permeability of the PDMS is essentially unaffected.

The contact lens application just cited illustrates the use of one of the most striking properties of PDMS, its superb transparency. In fact, it retains considerable transparency even when filled (reinforced) with rather large amounts of silica! This unusual optical property can be experienced firsthand, by examining the sheet of filled PDMS included with this book. It was generously provided by the Dow Corning Corporation of Midland, MI.

4.5.4 Stability, Safety Aspects, and Environmental Impact

Siloxane polymers possess a number of properties that seem almost contradictory. One example is the tremendous stability and durability shown by polysiloxanes in a wide variety of applications.[23, 109] One reason for this stability is the fact that the chain is already in a high oxidation state, and reduction takes place only at very high temperatures. Most scientists are familiar with the now-commonplace constant-temperature bath containing "silicone oil" which operates at high temperatures for years without any evidence of thermal degradation. Body implants made of polysiloxanes show little evidence of degradation, hydrolytic or otherwise, after decades of useful service, due to their resistance to hydrolysis and oxidation. (The inertness of the siloxanes should not be much of a surprise, if one thinks of them as simple hydrocarbon modifications of the silicates we commonly refer to as "glass.") In spite of this, polysiloxanes do not present severe environmental problems for reasons discussed shortly. For example, in the case of a spill, or rupture of an electrical device such as a transformer, the polymers released degrade completely and relatively rapidly under normal environmental conditions.[23, 109]

Such degradation can occur in water, in air, and particularly in the soil. This occurs because the polymers come into contact with one or more reactive species which can cause their degradation. One of these is the nitrate ion present in natural waterways. It is a source of atomic oxygen, and from it, hydroxyl radicals which initiate the degradation process. Another reagent is ozone, split by UV light into oxygen atoms, followed again by the production of hydroxyl radicals. It is interesting to note that UV light itself, has very little effect on the siloxane structure. Only the very shortest wavelengths present in sunlight have any influence and, in this case, cause generation of methyl radicals from the side groups. Polysiloxanes are resistant to all types of radiation, particularly if they contain aromatic groups, for example, the phenyl groups in poly(methylphenylsiloxane).

Even when methyl radicals are replaced by silanol units, the surface of the material does not remain hydrophilic (water wettable) very long. Either the silanol groups condense with other silanol units to restore the siloxane structure, or unmodified chain segments migrate to the surface. In any case, a "self-repair" mechanism underlies the "recoverability" of siloxane surfaces, and this is an important part of their durability.

Clay minerals present in many soils have high surface areas with strongly acidic groups on their surfaces. These materials can react with siloxane chains and reorganize them into much smaller molecules. In fact, water readily reacts with the Si—O bond in the presence of catalytic amounts of either acids or bases. Some of these small molecules are volatile enough to evaporate into the atmosphere. Others become capped with silanol (—SiOH) groups, and this is frequently sufficient to make them water soluble, and thus environmentally degradable. At later stages in the process, even the hydrocarbon groups are affected. Although the organosiloxane structure is completely unknown in nature, the introduction of these small molecules into the biosphere is thought to be entirely harmless. Furthermore, some of the degradation processes lead ultimately to silica, water, and carbon dioxide or inorganic carbonates.[23, 109]

It will be clear that the degradative reactions experienced by siloxane chains generally generate silanol groups on the molecules, usually at their ends. Pairs of such silanol groups can condense with one another, forming new siloxane linkages. As was mentioned, this condensation reaction provides an interesting "healing" mechanism for the siloxane backbone.

4.5.5 Some Additional Unusual Properties of Poly(dimethylsiloxane)

Some of the unusual physical properties exhibited by PDMS are summarized in Table 4.3.[2, 33, 110–112] Atypically low values are exhibited for the characteristic pressure[103, 104] (a corrected internal pressure, which is widely used in the study of liquids), the bulk viscosity η, and the temperature coefficient of η.[2] Also, entropies of dilution and excess volumes on mixing PDMS with solvents are much lower than can be accounted for by theory.[110, 111] Finally, as has already been mentioned, PDMS has a surprisingly high permeability.

TABLE 4.3   Some Unusual Properties of PDMS[a]

Property	Experimental Result
Characteristic pressure	Unusually low
Bulk viscosity η	Unusually low
Temperature coefficient of η	Unusually low
Entropies of dilution	Significantly lower than theory
Excess volumes on mixing	Significantly lower than theory
Permeability	Unusually large

[a]Reproduced by permission from J. M. Zeigler and F. W. G. Fearon, eds., *Silicon-Based Polymer Science. A Comprehensive Resource*; Advances in Chemistry Series, 1990, *224*. Copyright 1990 American Chemical Society.

Another striking feature of siloxane polymers is their unusual surface properties.[113-117] Fluorosiloxane polymers[118] have been studied most recently in this regard,[115-116] but the properties to be described are characteristic of a number of different members of the polysiloxane family. Specifically, their surface properties permit them to serve in a variety of seemingly contradictory roles. For example, siloxanes can be both antifoaming agents and foam stabilizers, both paper-release coatings and pressure-sensitive adhesives, both water-repellents and dewatering agents, and both emulsifiers and deemulsifiers.[113,114] This paradox is explained by the differing ways in which the siloxane chains interact with the other species present. For example, in foam technology it is critically important whether the siloxane dissolves in the liquid phase or stays at the liquid-gas interface. Similar questions arise in other applications, and specific properties are generated by an appropriate choice of side group; addition of special polar groups, ionic groups, or reactive functional groups; or copolymerization with completely different classes of comonomer.

Also of interest is the recent demonstration that poly(methylphenylsiloxane) can undergo a reversible droplet-monolayer transition,[119] which could be relevant to the spreading of polymers on surfaces.

In the most general terms, the unusual surface properties of polysiloxanes are due to their low surface energies and surface tensions.[113,114] These characteristics are then understandable in terms of two important features of the chains themselves. The first is the very low intermolecular forces between the side chains, which are the methyl groups in PDMS, the commonest of the polymers having these unusual surface properties. The second is the remarkable flexibility of the siloxane backbone, which permits the chains to easily arrange and rearrange themselves so as to place the methyl groups at their surfaces and interfaces. A particularly interesting example of this is the already cited ease with which a damaged polysiloxane surface quickly regenerates the surface characteristics of the original material.

Although the molecular origin of the unusual properties of siloxane polymers is still not known definitively, a number of suggestions have been put forward. One involves low intermolecular interactions, as mentioned in the preceding paragraph. Others focus on differences between the nonpolar alkyl groups and the polar Si—O backbone[114] or the very high rotational and oscillatory freedom of the methyl side groups in PDMS, the most important of the polysiloxanes.[21] Still another focuses on the chain's irregular cross section (very large at the substituted Si atoms and very small at the unsubstituted O atoms[111]).

4.6 REACTIVE HOMOPOLYMERS

4.6.1 Types of Reactions

In the typical ring-opening polymerization mentioned in Section 4.4.2, reactive hydroxyl groups are automatically formed at the ends of the chains.[11,42] Substitution reactions carried out on these chain ends can then be used to convert

TABLE 4.4   Reactive Siloxane
Polymers[a]

$$XSi(CH_3)_2O[-Si(CH_3)_2O-]_xSi(CH_3)_2X$$

X	Reactant
OH	Alkoxysilanes $[Si(OC_2H_5)_4]$
H	Unsaturated groups
$CH=CH_2$	Active H atoms
	Free radicals

[a]Reproduced by permission from J. M. Zeigler and F. W. G. Fearon, eds., *Silicon-Based Polymer Science. A Comprehensive Resource*; Advances in Chemistry Series, 1990, *224*. Copyright 1990 American Chemical Society.

them into other reactive functional groups. These functionalized polymers can undergo a variety of subsequent reactions, some of which are listed in Table 4.4.[33]

For example, hydroxyl-terminated chains can undergo condensation reactions with alkoxysilanes (orthosilicates).[11,120] A difunctional alkoxysilane leads to chain extension, and a tri- or tetrafunctional one to network formation. Correspondingly, addition reactions with di- or triisocyanates provide other possibilities. Similarly, vinyl-terminated chains can react with molecules having active hydrogen atoms.[11,120]

A pair of vinyl or other unsaturated groups can also be linked by their direct reactions with free radicals. Similar end groups can be placed on siloxane chains by the use of an end blocker during polymerization,[53,54] as mentioned earlier. Reactive groups such as vinyl units can, of course, be introduced as side chains by random copolymerizations involving, for example, methylvinylsiloxane trimers or tetramers.[11]

4.6.2 Block Copolymers

One of the most important uses of end-functionalized polymers is the preparation of block copolymers.[53,54] The reactions are identical to the chain extensions already mentioned, except that the sequences being joined are chemically different. In the case of the $-OSiR_2R'Y$ chain ends mentioned, R' is typically $(CH_2)_{3-5}$ and Y can be NH_2, OH, COOH, $CH=CH_2$, and so on. The siloxane sequences containing these ends have been joined to other polymeric sequences such as carbonates, ureas, urethanes, amides, and imides.

4.6.3 Elastomeric Networks

The networks formed by reacting functionally terminated siloxane chains with an end linker of functionality three or greater have been used extensively to study the molecular aspects of rubberlike elasticity.[68,120-128] They are preferred for this purpose since there are relatively few complications from side reactions during their preparation. They are "model" networks in that a great deal is known about their structure by virtue of the very specific chemical reactions used for their synthesis. For example, if a stoichiometric balance exists between chain ends and functional groups on the end linker, the critically important molecular weight M_c between cross-links is equal to the molecular weight of the chains prior to their end linking. Also, the functionality of the cross-links (number of chains emanating from one of them) is simply the functionality of the end-linking agent. Finally, the molecular weight distribution of the network chains is the same as that of the starting polymer, and there should be few if any dangling-chain irregularities.

Since these networks have a known degree of cross-linking (as measured inversely by M_c), they can be used to test the molecular theories of rubberlike elasticity, particularly with regard to the possible effects of interchain entanglements.[68, 120-128] Intentionally imperfect networks can also be prepared, as is described in Figure 4.9.[120] Such networks have known numbers and lengths of dangling chains (those attached to the network at only one end), and thus the effects of these irregularities on elastomeric properties can be determined. In the first method, the stoichiometry is unbalanced so that there is an excess of chain ends over functional groups on the end-linking molecules. The limitation in this method is the fact that the dangling chains have to have the same average length as the elastically effective chains (those attached to the network at both ends). The second method avoids this limitation by the use of monofunctionally terminated chains of whatever length is desired.

These reactions can also be used to form networks that interpene-

(a) Excess difunctional chains

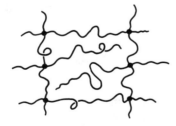

(b) Monofunctional chains

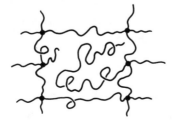

Figure 4.9 Two end-linking techniques for preparing networks having known numbers and lengths of dangling chains. Reprinted with permission from J. E. Mark, *Acc. Chem. Res.* **1985,** *18,* 202. Copyright 1985 American Chemical Society.

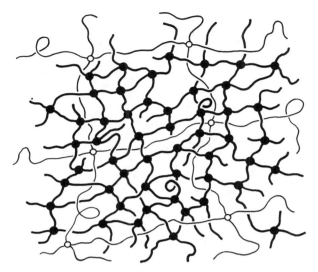

Figure 4.10   Sketch of two interpenetrating networks. Reprinted with permission from J. E. Mark, *Acc. Chem. Res.* **1985**, *18*, 202. Copyright 1985 American Chemical Society.

trate,[17, 120, 129–135] as illustrated in Figure 4.10.[120] For example, one network could be formed by a condensation end-linking of hydroxyl-terminated short chains and the other by a simultaneous but independent addition end-linking of vinyl-terminated long chains. Interpenetrating networks are of interest because they can be unusually tough, and could have unusual dynamic mechanical properties.

One of the most interesting classes of model networks is the bimodal type. These networks consist of very short chains intimately end-linked with the much longer chains that are representative of elastomeric materials.[120–122, 136–141] Such a network is shown in Figure 4.11,[138] where the short chains are arbitrarily drawn thicker than the long ones. These materials have unusually good elastomeric properties, specifically large values of both the ultimate strength and maximum extensibility. Possibly the short chains contribute primarily to the strength, in that their very limited extensibilities should give very high values for the modulus. The long chains might then contribute primarily by increasing the maximum extensibility by somehow delaying the spread of the rupture nuclei necessary for the catastrophic failure of the material.

An additional bonus exists if the chains in the bimodal network readily undergo strain-induced crystallization. It has been observed that the extent to which the bimodal networks are superior to their unimodal counterparts is larger at lower temperatures. This indicates that the bimodal character of the chain length facilitates strain-induced crystallization. Perhaps the short chains are particularly well oriented by an imposed elongation, and in these orientations can readily act as nucleation sites for the crystallization process.

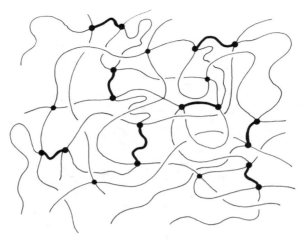

Figure 4.11 Sketch of a bimodal network. Reprinted with permission from J. E. Mark, *Makromol. Chem. Suppl.* **1979**, *2*, 87. Copyright 1979 Hüthig and Wepf Verlag.

Monte Carlo simulations based on rotational isomeric state model,[142, 143] as well as analytical calculations,[144] have helped elucidate the molecular origins of this interesting type of reinforcement.

Also, not only do short chains improve the ultimate properties of elastomers, but long chains improve the impact resistance of the much more heavily cross-linked thermosets.[140]

4.6.4 Trapping of Cyclic Oligomers

If relatively large PDMS cyclic oligomers are present in reaction systems where linear PDMS chains are end linked, some of the cyclic species will be permanently trapped by one or more network chains threading through them.[145] This is illustrated by cyclics B, C, and D in Figure 4.12.[146] One interesting result is the observation that the percent cyclic trapped does not depend on the amount of time elapsed between mixing the two components and then end-linking the linear chains. This is certainly consistent with the very high mobility of siloxane chains, as described in preceding sections. In any case, interpretation of the fraction trapped as a function of ring size, using rotational isomeric theory and Monte Carlo simulations, has provided useful information about the spatial configurations of cyclic molecules, including the effective hole size they present in the undiluted, amorphous state.[146]

This technique can also be used to form a network that has no cross-links whatsoever. Mixing linear chains with large amounts of cyclics and then *di*functionally end-linking them can give sufficient cyclic interlinking to yield a "chain mail" or "Olympic" network,[147-149] as is illustrated in Figure 4.13.[148] Such materials could have unusual equilibrium mechanical properties. For exam-

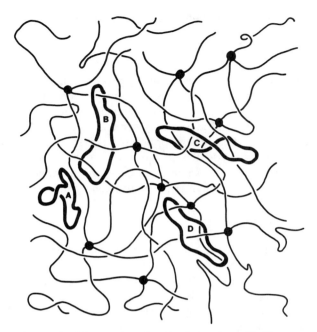

Figure 4.12 Sketch of the trapping of cyclics during the end-linking preparation of a network. Reprinted with permission from L. C. DeBolt and J. E. Mark, *Macromolecules* **1987,** *20,* 2369. Copyright 1987 American Chemical Society.

ple, they might deform, at least in part, by a mechanism very different from that for more conventional elastomers (the entropy-decreasing stretching out of the network chains). Their dynamic mechanical properties could well also be intriguingly different.

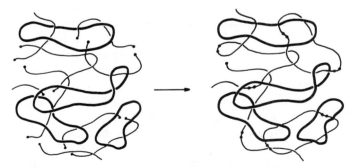

Figure 4.13 Preparation of a "chain mail" or "Olympic" network consisting entirely of interlooped cyclics. Reprinted with permission from L. Garrido, J. E. Mark, S. J. Clarson, and J. A. Semlyen, *Polym. Commun.* **1985,** *26,* 53. Copyright © 1985 Butterworth-Heinemann Ltd.

4.7 COPOLYMERS AND INTERPENETRATING NETWORKS

4.7.1 Random Copolymers

These materials may be prepared by the copolymerization of a mixture of cyclic oligomers as opposed to the homopolymerization of a single type of cyclic.[11,42,55,150,151] Although the resulting polymer can be quite blocky, taking the reaction to equilibrium can give a polymer that is essentially random in its chemical sequencing.[54,151] One reason for doing copolymerizations is to introduce functional species, such as hydrogen or vinyl side groups, along the chain backbone to facilitate cross-linking. Another reason is the introduction of sufficient chain irregularity to make the polymer inherently noncrystallizable.

4.7.2 Block Copolymers

As already mentioned, the sequential coupling of functionally terminated chains of different chemical structure can be used to make block copolymers,[37,152,153] including those in which one or more of the blocks is a polysiloxane.[54,154,155] If the blocks are relatively long, separation into a two-phase system almost invariably occurs. Frequently, one type of block will be in a continuous phase and the other will be dispersed in it in domains having an average size the order of a few hundred angstroms. Such materials can have unique mechanical properties not available from either species when present simply in homopolymeric form. Sometimes similar properties can be obtained by the simple blending of two or more polymers.[156]

4.7.3 Interpenetrating Networks

In this type of material, two networks are formed, either simultaneously or sequentially, in such a way as to interpenetrate one another. They thus "communicate" with one another through interchain interactions rather than through covalent bonds. A particularly simple example would be the already-mentioned simultaneous formation of two PDMS networks, one by a condensation end-linking reaction and the other by an addition end-linking reaction, with the two types of chains mixed at the molecular level.[132]

A more complex example, with additional novel properties, is the preparation of interpenetrating networks (IPNs) between PDMS and an organic thermoplastic polymer such as a nylon, polyurethane, or polyester. This process, called Rimplast® by its developers,[17] involves a chemical reaction, between a vinyl-functional polysiloxane blended into the thermoplastic, and a hydride-functional PDMS blended into another portion of the same thermoplastic. Small amounts of platinum catalyst are also present in both types of components. The amounts of polysiloxane present is typically around 10%. Pellets of both components are placed in an extruder or other high-temperature processor, where they melt into a uniform mass, at approximately 300°C. The reaction between the two complementary types of PDMS results in a network evenly distributed throughout the thermoplastic in a type of semi-IPN. The advantage of the resulting composite material is the fact that it has the good properties of both the PDMS (good lubricity, abrasion resistance, and dielectric properties) and the thermoplastic (good mechanical strength and molding characteristics).

4.8 APPLICATIONS

4.8.1 Medical

Numerous medical applications have been developed for siloxane polymers.[11,17,21] Prostheses, artificial organs, facial reconstruction, and tubing and catheters, for example, take advantage of the inertness, stability, and pliability of the polysiloxanes. Artificial skin, contact lenses, and drug delivery systems utilize their high permeability as well.

Figure 4.14 shows the range of diameters of "silastic" medical-grade siloxane tubing available for medical applications. The smallest tubing has an internal diameter of only 0.012 in. (0.031 cm) and an outer diameter of only 0.025 in. (0.064 cm)! Such materials are tested extensively with regard to sensitization of skin, tissue cell culture, and implant studies.

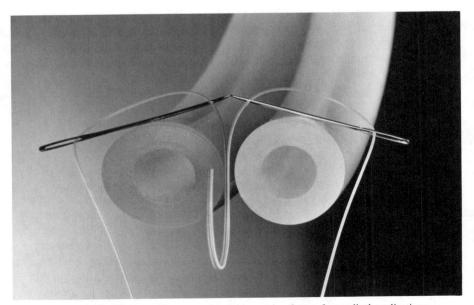

Figure 4.14 Siloxane polymer used in tubing and catheters for medical applications.
© Dow Corning Corporation 1991. Reproduced by permission.

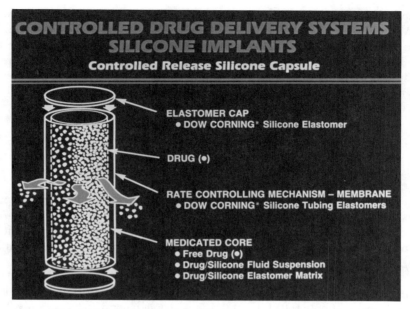

Figure 4.15 A controlled release drug-delivery system which utilizes both siloxane elastomers and fluids. © Dow Corning Corporation 1991. Reproduced by permission.

The use of polysiloxanes in drug delivery systems is illustrated in Figure 4.15. The desired goal is to have the drug released at a relatively constant rate (zero-order kinetics), and this ideal behavior is approached by placing the drug inside a siloxane elastomeric capsule, which is then implanted in an appropriate location in the body. The drug within the capsule can be either in the free state, in a fluid suspension, or mixed or dissolved into an elastomeric matrix. Release rates for drugs which are much more hydrophilic than the polysiloxanes, for example, melatonin and sulfanilamide, are frequently relatively slow. They can be increased, however, by incorporating solvents or fillers in the capsule.

4.8.2 Nonmedical

Typical nonmedical applications include high-performance elastomers, membranes, electrical insulators, water repellents, antifoaming agents, mold-release agents, adhesives, protective coatings, release control agents for agricultural chemicals, encapsulation media, mold-making materials, layers in high-tech laminates, and hydraulic, heat-transfer, and dielectric fluids.[11,21] They are based on the same properties of polysiloxanes just mentioned and also on their ability to modify surfaces and interfaces (for example, as water repellents, antifoaming agents, and mold-release agents). Two particularly recent examples are their uses as polymer electrolyte complexes[157] and in microlithographic applications.[158]

Figure 4.16   An electronic circuit board being given a protective polysiloxane coating.
© Dow Corning Corporation 1991. Reproduced by permission.

Figure 4.16 shows a typical nonmedical application of a siloxane polymer. In this case, a liquid, UV-curable polysiloxane is being used to protect a printed circuit board. The polymer being used was specifically chosen for its low viscosity, moisture resistance, and easy repairability. Also, some materials of this type can be treated so that thin spots in the coating are readily apparent. Although dip coating is illustrated in the figure, flow coating and spray coating can also be used.

One of the most impressive applications of polysiloxanes, particularly in the arts, is their use in making molds of intricate surfaces. This is illustrated in Figure 4.17. In this case, the surface to be copied was vertical and nonmovable, so a ''nonslumping'' end-linkable paste had to be used rather than a liquid. After the surface was coated, it was cured to give a remarkably faithful reproduction which was easily removed from the original surface.

Impregnating masonry and mineral-based works of art such as statues with polysiloxanes can protect them from corrosive agents in the atmosphere and from acid rain. In the final example, a polysiloxane appears as an interlayer in the types of plastic and glass laminates used for increased safety in windshields and canopies in aircraft. The flexibility and thermal stability of these polymers are great advantages in the case of high-performance aircraft, which can experience wide varia-

Figure 4.17 A polysiloxane room temperature vulcanizate (RTV) being used to make a mold of an intricately carved surface. © Dow Corning Corporation 1991. Reproduced by permission.

tions in temperature. In this application, a polysiloxane is chosen to give the highest transparency and good adhesion to the plastic or glass outer layers. The excellent transparency of such composites is readily seen in the samples displayed in Figure 4.18.

Figure 4.18 High-transparency polysiloxane elastomer used as a stress relieving interlayer in the type of safety glass or plastic used in windshields and canopies. © Dow Corning Corporation 1991. Reproduced by permission.

4.9 SILICA-TYPE MATERIALS (ULTRASTRUCTURES)

4.9.1 Sol-Gel Ceramics

A relatively new area that involves silicon-containing materials is the synthesis of ultrastructure materials, for example, materials in which structure can be controlled at the level of around 100 Å. An example of such a synthesis is the hydrolysis of alkoxysilanes or silicates to give silica.[159-174] The process is complicated, involving polymerization and branching, but a typical overall reaction may be written

$$Si(OR)_4 + 2\ H_2O \longrightarrow SiO_2 + 4\ ROH \qquad (Eq.\ 4.6)$$

where the $Si(OR)_4$ organometallic species is typically tetraethoxysilane (tetraethylorthosilicate) (TEOS). In this application, the organometallic compound is hydrolyzed, condensed to polymeric chains, the chains become more and more branched, and finally a highly swollen gel is formed. It is first dried at moderately low temperatures to remove volatile species, and then is fired into a porous ceramic object. It is then densified, and machined into the final ceramic part. Not surprisingly, the production of ceramics by this novel route has generated a great deal of interest. Its advantages, over the usual "heat-and-beat" (e.g., sintering) approach to ceramics, is (1) the higher purity of the starting materials, (2) the relatively low temperatures required, (3) the possibility of controlling the "ultrastructure" of the ceramic (to reduce the microscopic flaws that lead to brittleness), (4) the ease with which ceramic coatings can be formed, and (5) the ease with which ceramic alloys can be prepared, for example, by hydrolyzing mixtures (solutions) of silicates and titanates.

4.9.2 In-Situ Preparations of Multiphase Systems

4.9.2.1 Ceramic-reinforced polymers.
The same hydrolyses can be carried out within a polymeric matrix to generate particles of the ceramic material, typically with an average diameter of a few hundred angstroms.[120, 123, 175-179] Considerable reinforcement of elastomers, including those prepared from PDMS, can be achieved in this way. This method for introducing reinforcing particles could have a number of advantages over the conventional approach in which preexisting filler particles are blended into the uncross-linked elastomer before its vulcanization. This time-honored technique is difficult to control because the filler particles are generally badly agglomerated and the polymer is of sufficiently high molecular weight to make the viscosity of the mixture exceedingly high. Thus, the blending technique is energy intensive and time consuming, and frequently not entirely successful.

In contrast, because of the nature of the in-situ precipitation, the particles are well dispersed and are essentially unagglomerated (according to electron microscopy). The mechanism for their growth seems to involve simple homogeneous nucleation, and since the particles are separated by polymer, they do not have the opportunity to coalesce. A typical transmission electron micrograph of such a filled material is shown in Figure 4.19.[175] The particles are relatively monodisperse, with most of them having diameters in the range of 200–300 Å.

The remainder of this section provides some additional details on the use of this technique.[178]

First, it should be mentioned that the sol-gel method is quite general, in that a variety of organometallic materials can be hydrolyzed, photolyzed, or thermolyzed to give reinforcing, ceramic-type particles. For example, titanates can be hydrolyzed to titania, aluminates hydrolyzed to alumina, and metal carbonyls photolyzed or thermolyzed to metals or metal oxides. (Some of the metal or metal oxide particles are particularly interesting since they can be manipulated with an external magnetic field during the curing process.) Also, the method can be used in a variety of polymers (organic as well as inorganic, nonelastomeric as well as elastomeric.) Even purely hydrocarbon polymers can be reinforced in this way, provided the organometallic material is chosen so as to have significant miscibility with the polymer. The technique is also general in that it is much used in an entirely different area, the sol-gel approach to the generation of ceramic objects, or monoliths, as mentioned above.[159–174]

A variety of catalysts work well in the typical hydrolysis reactions used,

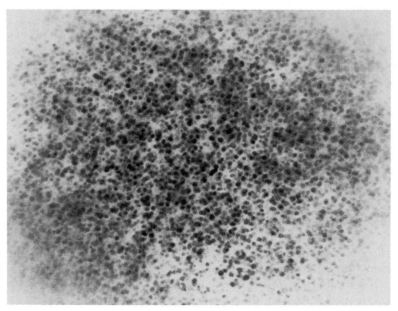

Figure 4.19 Electron micrograph of a PDMS network containing in-situ precipitated silica particles. Reprinted with permission from Y. P. Ning, M. Y. Tang, C. Y. Jiang, J. E. Mark, and W. C. Roth, *J. Appl. Polym. Sci.* **1984,** *29,* 3209. Copyright © 1984 John Wiley & Sons.

including acids, bases, and salts. Basic catalysts give precipitated phases that are generally well-defined particles, whereas the acidic catalysts give more poorly defined, diffuse particles.[175] In some cases, particles are not formed at all, and bicontinuous (interpenetrating) phases result.

These reactions can be carried out in three ways.[178] In the first, the polymer is cross-linked and then swelled with the organometallic reagent, which is then hydrolyzed in situ. In the second, hydroxyl-terminated chains are blended with enough of the organometallic compound (TEOS) to both end link them and generate silica by the hydrolysis reaction. Thus, curing and filling take place simultaneously, in a one-step procedure. In the final technique, TEOS is blended into a polymer that has end groups (e.g., vinyl units) that are unreactive under hydrolysis conditions. The silica is then formed in the usual manner (eq. 4.6), and the mixture is dried. The resultant slurry of polymer and silica is stable and can be cross-linked at a later time using any of the standard techniques, including vinyl-silane coupling.

Interesting ''aging'' effects are frequently observed in these systems. If the precipitated particles are left in contact with the hydrolysis catalyst and water they appear to reorganize, so that their surfaces become better defined and their sizes become more uniform.[177] The process seems quite analogous to the ''Ostwald ripening'' much studied by colloid chemists.

PDMS networks filled with a variety of types of silica, introduced in various ways, have been compared with regard to their stress-strain behavior in elongation. The materials prepared by this new in-situ technique were found to have considerably improved properties.[178]

A variety of techniques have been used to characterize these in-situ filled elastomers.[178] Density measurements, for example, yield information about the nature of the particles. Specifically, the density of the ceramic-type particles is significantly less than that of silica itself, and this suggests that the particles presumably contain some unhydrolyzed alkoxy groups or voids, or both.

The low-temperature properties of some of these peculiarly filled materials have also been studied, by the differential scanning calorimetry technique mentioned in Chapter 2. Of particular interest is the way in which reinforcing particles affect the crystallization of a polysiloxane, both in the undeformed state and at high elongations.

As already mentioned, electron microscopy (both transmission and scanning) has been used to characterize the precipitated particles. The information obtained in this way includes (1) the nature of the precipitated phase (particulate or nonparticulate), (2) the average particle size, if particulate, (3) the distribution of particle sizes, (4) the degree to which the particles are well defined, and (5) the degree of agglomeration of the particles.

A number of x-ray and neutron scattering studies have been carried out on these filled elastomers.[178,179] Although the results are generally consistent with those obtained by electron microscopy, some intriguing differences exist. Of particular interest is the observation that some fillers which appear to be particulate in electron microscopy appear to consist of a continuously interpenetrating phase by scattering measurements. Additional experiments of this type will certainly be forthcoming.

It is also possible to obtain reinforcement of a siloxane elastomer by polymerizing a monomer such as styrene to yield hard glassy domains within the elastomer. In PDMS, low concentrations of styrene give low-molecular-weight polymer that acts more like a plasticizer than a reinforcing filler. At higher styrene concentrations, roughly spherical polystyrene (PS) particles are formed, and good reinforcement is obtained. The particles thus generated are relatively easy to extract from the elastomeric matrix. This means that little effective bonding exists between the two phases. It is possible, however, to get excellent bonding onto the filler particles. One way is to include some $R'Si(OC_2H_5)_3$ in the hydrolysis, where R' is an unsaturated group. The R' groups on the particle surfaces then participate in the polymerization, thereby bonding the elastomer chains to the reinforcing particles. Alternatively, the $R'Si(OC_2H_5)_3$ can be used as one of the end-linking agents, to place unsaturated groups at the cross-links. Their participation in the polymerization would then tie the PS domains to the elastomer's network structure.

The PS domains have the disadvantage of having a relatively low glass transition temperature ($T_g = 100°C$) and in being totally amorphous. Above their T_g they would soften and presumably lose their reinforcing ability. For this reason, similar studies have been carried out using poly(diphenylsiloxane) as the reinforcing phase.[180] This polymer is crystalline, and measurements on copolymers containing diphenylsiloxane blocks indicate it has a melting point (and thus a softening temperature) of about $550°C$.[154, 155]

It is possible to convert the essentially spherical PS particles just described into rodlike ellipsoidal particles.[178] First, the PS-PDMS composite is heated to a temperature well above the T_g of PS. It is then stretched uniaxially, and is cooled while in the stretched state. The particles are thereby deformed into prolate ellipsoids and retain this shape when cooled. When the deforming force is removed, the elastomer retracts, but only part of the way back to its original dimensions. The particles themselves have been characterized using both scanning and transmission electron microscopy. This gives values for their axial ratios and provides a measure of the extent to which their axes were aligned in the direction of stretching. In these anisotropic materials, elongation moduli in the direction of the stretching were found to be significantly larger than those of the untreated PS-PDMS elastomer, whereas in the perpendicular direction they were significantly lower. This was to be expected from the anisotropic nature of these systems.

4.9.2.2 Polymer-modified ceramics.

If the hydrolyses in silane-polymer systems are carried out using relatively large amounts of the silane, then the silica generated can become the continuous phase, with the elastomeric polysiloxane dispersed in it.[181-191] Again, a variety of ceramic components and polymeric components have been studied. The resultant composite is a polymer-modified glass or ceramic, frequently of very good transparency. Although its thermal sta-

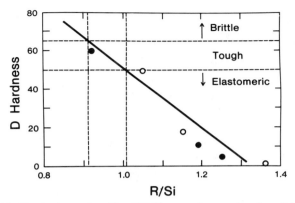

Figure 4.20   The hardness of a silica-PDMS composite as a function of the relative numbers of alkyl groups and silicon atoms. Reprinted with permission from J. E. Mark and C. C. Sun, *Polym. Bull.* **1987,** *18,* 259. Copyright 1987 Springer-Verlag.

bility will be inferior to that of the ceramic component, there are many applications for ceramic-type materials where this is not a serious problem.

As might be expected, the properties of these materials depend greatly on the relative amounts of the two phases. Properties of particular interest are modulus, impact resistance, ultimate strength, maximum extensibility, viscoelastic responses, and transparency. Their hardness, for example, can be varied by control of the molar ratio of alkyl R groups to Si atoms, as illustrated in Figure 4.20.[183] Low values of R/Si yield a brittle ceramic, and high values give a relatively hard elastomer. The most interesting range, R/Si $\cong$ 1, can yield a relatively tough ceramic of reduced brittleness.

A final topic worth mentioning is the use of inorganic or semi-inorganic polymers as precursors for ceramics.[159-162, 167] In this application, a polymer is fabricated, typically into a fiber, which is then cross-linked to hold its shape during subsequent processing steps. It is then pyrolyzed into a pure ceramic material, for example, to silicon carbide. Because such materials are extraordinarily strong, they are used in high-technology applications, for example, as fibers in superstrong composites.

4.10 REFERENCES

1. Allcock, H. R. *Heteroatom Ring Systems and Polymers.* Academic: New York, 1967.

2. Noll, W. *Chemistry and Technology of the Silicones.* Academic: New York, 1968.

3. Meals, R. In *Kirk-Othmer Encyclopedia of Chemical Technology;* John Wiley: New York, 1969.

4. Borisov, S. N.; Voronkov, M. G.; Lukevits, E. Ya. *Organosilicon Heteropolymers and Heterocompounds;* Plenum: New York, 1970.

5. Bobear, W. J. In *Rubber Technology*, M. Morton, Ed.; Van Nostrand Reinhold: New York, 1973.

6. Allcock, H. R. *Sci. Am.* **1974**, *230(3)*, 66.

7. Zeldin, M. *Polym. News* **1976**, *3*, 65.

8. Elias, H. G. *Macromolecules;* Plenum: New York, 1977; Vol. 2.

9. Noshay, A.; McGrath, J. E. *Block Copolymers: Overview and Critical Survey;* Academic: New York, 1977.

10. Mark, J. E. *Macromolecules* **1978**, *11*, 627.

11. Warrick, E. L.; Pierce, O. R.; Polmanteer, K. E.; Saam, J. C. *Rubber Chem. Technol.* **1979**, *52*, 437.

12. Rochow, E. G. *CHEMTECH* **1980**, *10*, 532.

13. Peters, E. N. In *Encyclopedia of Chemical Technology;* Wiley-Interscience: New York, 1981; Vol. 13.

14. Carraher, C. E.; Sheats, J. E.; Pittman, C. U. *Advances in Organometallic and Inorganic Polymers;* Marcel Dekker, Inc.: New York, 1982.

15. Stark, F. O.; Falender, J. R.; Wright, A. P. In *Comprehensive Organometallic Chemistry*, Wilkinson, G., Ed.; Pergamon: Oxford, 1982; Vol. 2.

16. Smith, A. L. *Analysis of Silicones;* Krieger: Malabar, FL, 1983.

17. Arkles, B. *CHEMTECH* **1983**, *13*, 542.

18. Allcock, H. R. *Chem. Eng. News*, March 18, 1985, p 22.

19. Archer, R. D. In *Encyclopedia of Materials Science and Engineering*, Bever, M. B., Ed.; Pergamon: Oxford, 1986, Vol. 3.

20. Zeldin, M. In *Encyclopedia of Materials Science and Engineering*, Bever, M. B., Ed.; Pergamon: Oxford, 1986.

21. Rochow, E. G. *Silicon and Silicones;* Springer-Verlag: Berlin, 1987.

22. Rheingold, A. In *Encyclopedia of Polymer Science and Engineering*, 2nd ed.; Wiley-Interscience: New York, 1987.

23. Hardman, B.; Torkelson, A. In *Encyclopedia of Polymer Science and Engineering*, 2nd ed.; Wiley-Interscience: New York, 1987.

24. ACS Symposium Series; Zeldin, M.; Wynne, K. J.; Allcock, H. R. *Inorganic and Organometallic Polymers;* American Chemical Society: Washington, DC, 1988; Vol. 360.

25. Pittman, C. U.; Sheats, J.; Carraher, C. E.; Zeldin, M.; Currell, B. *Preprints. ACS Division of Polymeric Materials Science and Engineering* **1989**, *61*, 91.

26. *The Chemistry of Organic Silicon Compounds;* Patai, S.; Rappoport, Z., Eds.; John Wiley: New York, 1989.

27. Lazar, M.; Bleha, T.; Rychly, J. *Chemical Reactions of Natural and Synthetic Polymers;* John Wiley: New York, 1989.

28. Kendrick, T. C.; Parbhoo, B.; White, J. W. In *The Chemistry of Organic Silicon Compounds;* Patai, S.; Rappoport, Z., Eds.; John Wiley: New York, 1989.

29. Carraher, C. E. *Polym. News* **1990**, *14*, 204.

30. Allcock, H. R.; Lampe, F. W. *Contemporary Polymer Chemistry*, 2nd ed.; Prentice Hall: Englewood Cliffs, NJ, 1990.

31. *Silicon-Based Polymer Science. A Comprehensive Resource;* Zeigler, J. M.; Fearon, F. W. G., Eds.; Advances in Chemistry Series; American Chemical Society: Washington, DC, 1990; Vol. 224.

32. Barton, T. J.; Boudjouk, P. In *Silicon-Based Polymer Science. A Comprehensive Resource;* Zeigler, J. M.; Fearon, F. W. G., Eds.; Advances in Chemistry Series; American Chemical Society: Washington, DC, 1990; Vol. 224.

33. Mark, J. E. In *Silicon-Based Polymer Science. A Comprehensive Resource;* Zeigler, J. M.; Fearon, F. W. G., Eds.; Advances in Chemistry Series; American Chemical Society: Washington, DC, 1990; Vol. 224.

34. Weyenberg, D. R.; Lane, T. H. In *Silicon-Based Polymer Science. A Comprehensive Resource;* Zeigler, J. M.; Fearon, F. W. G., Eds.; Advances in Chemistry Series; American Chemical Society: Washington, DC, 1990; Vol. 224.

35. *Inorganic and Metal-Containing Polymeric Materials;* Sheats, J.; Carraher, Jr., C. E.; Zeldin, M.; Currell, B.; Pittman, Jr., C. U., Eds.; Plenum: New York, 1991.

36. *Siloxane Polymers;* Clarson, S. J.; Semlyen, J. A., Eds.; Prentice Hall: Englewood Cliffs, NJ, 1991.

37. Odian, G. *Principles of Polymerization,* 2nd ed.; Wiley-Interscience: New York, 1981.

38. West, R.; Barton, T. J. *J. Chem. Educ.* **1980,** *57,* 165, 334.

39. Bock, H. *Angew. Chem. Int. Ed. Engl.* **1989,** *28,* 1627.

40. Goodwin, G. B.; Kenney, M. E. In *Silicon-Based Polymer Science. A Comprehensive Resource;* Zeigler, J. M.; Fearon, F. W. G., Eds.; Advances in Chemistry Series; American Chemical Society: Washington, DC, 1990; Vol. 224.

41. *Ring-Opening Polymerization;* Frisch, K. C.; Reegen, S. L., Eds.; Marcel Dekker: New York, 1969.

42. McGrath, J. E.; Riffle, J. S.; Banthia, A. K.; Yilgor, I.; Wilkes, G. L. In *Initiation of Polymerization;* Bailey, Jr., F. E., Ed.; American Chemical Society: Washington, DC, 1983.

43. McGrath, J. E. In *Ring-Opening Polymerization;* ACS Symposium Series; McGrath, J. E., Ed.; American Chemical Society: Washington, DC, 1985; Vol. 286.

44. Saam, J. C. In *Silicon-Based Polymer Science. A Comprehensive Resource;* Zeigler, J. M.; Fearon, F. W. G., Eds.; Advances in Chemistry Series. American Chemical Society: Washington, DC, 1990; Vol. 224.

45. Chojnowski, J. In *Siloxane Polymers;* Clarson, S. J.; Semlyen, J. A., Eds.; Prentice Hall: Englewood Cliffs, NJ, 1991.

46. Flory, P. J. *Statistical Mechanics of Chain Molecules;* Wiley-Interscience: New York, 1969.

47. Semlyen, J. A. *Pure Appl Chem.* **1981,** *53,* 1797.

48. *Cyclic Polymers;* Semlyen, J. A., Ed.; Elsevier: London, 1986.

49. Clarson, S. J.; Semlyen, J. A.; Dodgson, K. *Preprints. Div. of Polym. Chem., Inc., Am. Chem Soc.* **1990,** *31(1),* 563.

50. Semlyen, J. A. In *Siloxane Polymers;* Clarson, S. J.; Semlyen, J. A., Eds.; Prentice Hall: Englewood Cliffs, NJ, 1991.

51. Kuo, C. M.; Clarson, S. J. *Preprints. Div. of Polym. Chem., Inc., Am. Chem Soc.* **1990,** *31(1),* 550.

52. Wilczek, L.; Rubinsztajn, S.; Chojnowski, J. *Macromol. Chemie* **1986,** *187,* 39.

53. Yilgor, I.; Riffle, J. S.; McGrath, J. E. In *Reactive Oligomers;* Harris, F. W.; Spinelli, H. J., Eds.; ACS Symposium Series; American Chemical Society: Washington, DC, 1985; Vol. 282.

54. McGrath, J. E.; Sormani, P. M.; Elsbernd, C. S.; Kilic, S. *Macromol. Chem., Macromol. Symp.* **1986,** *6,* 67.

55. Siemelis, M.; Lee, M.; Saam, J. C. *Preprints. Div. of Polym. Chem., Inc., Am. Chem Soc.* **1990,** *31(1),* 38.

56. Oberhammer, H.; Boggs, J. E. *J. Am. Chem. Soc.* **1980,** *102,* 7241.

57. Lukevics, E.; Pudova, O.; Sturkovich, R. *Molecular Structure of Organosilicon Compounds;* John Wiley: New York, 1989.

58. Studebaker, M. L.; Beatty, J. R. In *Science and Technology of Rubber;* Eirich, F. R., Ed.; Academic Press: New York, 1978.

59. Pluddemann, E. D. *Silane Coupling Agents;* Plenum Press: New York, 1982.

60. Coran, A. Y. In *Science and Technology of Rubber;* Eirich, F. R., Ed.; Academic Press: New York, 1978.

61. Brydson, J. A. *Rubber Chemistry;* Appl. Sci.: London, 1978, Chapter 15.

62. Coran, A. Y. In *Encyclopedia of Polymer Science and Engineering,* 2nd ed.; Wiley-Interscience: New York, 1987.

63. Thomas, D. R. In *Siloxane Polymers;* Clarson, S. J.; Semlyen, J. A., Eds.; Prentice Hall: Englewood Cliffs, NJ, 1991.

64. Williams, E. A. *Preprints. Div. of Polym. Chem., Inc., Am. Chem Soc.* **1990,** *31(1),* 119.

65. Garrido, L.; Ackerman, J. L.; Chang, C.; Mark, J. E. *Preprints. Div. of Polym. Chem., Inc., Am. Chem Soc.* **1990,** *31(1),* 147.

66. Elsbernd, C. S.; Spinu, M.; Krukonis, V. J.; Gallagher, P. M.; Mohanty, D. K.; McGrath, J. E. In *Silicon-Based Polymer Science. A Comprehensive Resource;* Zeigler, J. M.; Fearon, F. W. G., Eds.; Advances in Chemistry Series; American Chemical Society: Washington, DC, 1990; Vol. 224.

67. Elsbernd, C. S.; DeSimone, J. M.; Hellstern, A. M.; Smith, S. D.; Gallagher, P. M.; Krukonis, V. J.; McGrath, J. E. *Preprints, Div. of Polym. Chem., Inc., Am. Chem Soc.* **1990,** *31(1),* 673.

68. Mark, J. E.; Erman, B. *Rubberlike Elasticity: A Molecular Primer;* Wiley-Interscience: New York, 1988.

69. Kucukyavuz, Z.; Al-Ghezawi, N. N. *Eur. Polym. J.* **1990,** *26,* 653.

70. Gregorias, S.; Lane, T. H. In *Silicon-Based Polymer Science. A Comprehensive Resource;* Zeigler, J. M.; Fearon, F. W. G., Eds.; Advances in Chemistry Series; American Chemical Society: Washington, DC, 1990; Vol. 224.

71. Grigoras, S. *Preprints, Div. of Polym. Chem., Inc., Am. Chem Soc.* **1990,** *31(1),* 697.

72. Clarson, S. J.; Dodgson, K.; Semlyen, J. A. *Polymer* **1987,** *28,* 189.

73. Clarson, S. J.; Semlyen, J. A.; Dodgson, K. *Polymer,* in press.

74. Mark, J. E. In ACS Short Course Manual *Polymer Chemistry*, Mark, J. E.; Odian, G., Eds.; American Chemical Society: Washington, DC, 1984.

75. Brown, Jr., J. F. *J. Polym. Sci., Part C* **1963**, *1*, 83.

76. Helminiak, T. E.; Berry, G. C. *J. Polym. Sci., Polym. Symp.* **1978**, *65*, 107.

77. Billmeyer, Jr., F. W. *Textbook of Polymer Science*, 3rd ed.; Wiley-Interscience: New York, 1984.

78. Dunnavant, W. R. *Inorg. Macro. Rev.* **1971**, *1*, 165.

79. Hani, R.; Lenz, R. W. In *Silicon-Based Polymer Science. A Comprehensive Resource;* Zeigler, J. M.; Fearon, F. W. G., Eds.; Advances in Chemistry Series; American Chemical Society: Washington, DC, 1990; Vol. 224.

80. Lee, C. L., Dow Corning Corporation, Midland, MI, personal communications.

81. Beatty, C. L.; Karasz, F. E. *J. Polym. Sci., Polym. Phys. Ed.* **1975**, *13*, 971.

82. Godovsky, Yu. K.; Makarova, N. N.; Papkov, V. S.; Kuzmin, N. N. *Makromol. Chem.* **1985**, *6*, 443.

83. Godovsky, Y. K.; Papkov, V. S. *Makromol. Chem., Makromol. Symp.* **1986**, *4*, 71.

84. Friedrich, J.; Rabolt, J. F. *Macromolecules* **1987**, *20*, 1975.

85. Godovsky, Y. K.; Volegova, I. A.; Valetskaya, L. A.; Rebrov, A. V.; Novitskaya, L. A.; Rotenburg, S. I. *Polym. Sci., U.S.S.R.* **1988**, *30*, 329.

86. Moller, M.; Siffrin, S.; Kogler, G.; Oelfin, D. *Makromol. Chem., Symp.* **1990**, *34*, 171.

87. Miller, K. J.; Grebowicz, J.; Wesson, J. P.; Wunderlich, B. *Macromolecules* **1990**, *23*, 849.

88. Schneider, N. S.; Desper, C. R.; Beres, J. J. In *Liquid-Crystalline Order in Polymers;* Blumstein, A., Ed.; Academic: New York, 1978.

89. Hammerschmidt, K.; Finkelmann, H. *Makromol. Chem.* **1989**, *190*, 1089.

90. Zentel, R. *Angew. Chem., Int. Ed. Engl., Adv. Mater.* **1989**, *28*, 1407.

91. Damewood, Jr., J. R.; West, R. *Macromolecules* **1985**, *18*, 159.

92. Welsh, W. J.; DeBolt, L.; Mark, J. E. *Macromolecules* **1986**, *19*, 2978.

93. Ferry, J. D. *Viscoelastic Properties of Polymers*, 3rd ed.; John Wiley: New York, 1980.

94. Hiemenz, P. C. *Polymer Chemistry: The Basic Concepts;* Marcel Dekker: New York, 1984.

95. *Polymer Handbook*, 3rd ed.; Brandrup, J.; Immergut, E. H., Eds.; Wiley-Interscience: New York, 1989.

96. Reports from Dow Corning Corporation, Syracuse University, and the University of Cincinnati, under Contract No. 5082-260-0666 from the Gas Research Institute, Chicago, IL.

97. Bhide, B. D.; Stern, S. A. *J. Appl. Polym. Sci.* **1989**, *38*, 2131.

98. Gordon, S. M.; Koros, W. J. *J. Polym. Sci., Polym. Phys. Ed.* **1990**, *28*, 795.

99. Stern, S. A. In *Siloxane Polymers;* Clarson, S. J.; Semlyen, J. A., Eds.; Prentice Hall: Englewood Cliffs, NJ, 1991.

100. Kamiya, Y.; Naito, Y.; Hirose, T.; Mizoguchi, K. *J. Polym. Sci., Polym. Phys. Ed.* **1990**, *28*, 1297.

101. West, R. W., University of Wisconsin, personal communication.
102. Masuda, T.; Isobe, E.; Higashimura, T.; Takada, K. *J. Am. Chem. Soc.* **1983,** *105,* 7473.
103. Ichiraku, Y.; Stern, S. A.; Nakagawa, T. *J. Membrane Sci.* **1987,** *34,* 5.
104. Masuda, T.; Higashimura, T. *Adv. Polym. Sci.* **1987,** *81,* 121.
105. Kita, H.; Sakamoto, T.; Tanaka, K.; Okamoto, K.-I. *Polym. Bulletin* **1988,** *20,* 349.
106. Masuda, T.; Higashimura, T. In *Silicon-Based Polymer Science. A Comprehensive Resource;* Zeigler, J. M.; Fearon, F. W. G., Eds.; Advances in Chemistry Series; American Chemical Society: Washington, DC, 1990; Vol. 224.
107. Baker, G. L.; Klausner, C. F.; Gozdz, A. S.; Shelburne III, J. A.; Bowmer, T. N. In *Silicon-Based Polymer Science. A Comprehensive Resource;* Zeigler, J. M.; Fearon, F. W. G., Eds.; Advances in Chemistry Series; American Chemical Society: Washington, DC, 1990; Vol. 224.
108. Yannas, I. V.; Burke, J. F. *J. Biomed. Mater. Res.* **1980,** *14,* 65.
109. Lentz, C. W. *Ind. Res. & Dev.,* April 1980, p 139.
110. Shih, H.; Flory, P. J. *Macromolecules* **1972,** *5,* 758.
111. Flory, P. J.; Shih, H. *Macromolecules* **1972,** *5,* 761.
112. McLure, I. A. In *Siloxane Polymers;* Clarson, S. J.; Semlyen, J. A., Eds.; Prentice Hall: Englewood Cliffs, NJ, 1991.
113. Owen, M. J. *CHEMTECH* **1981,** *11,* 288.
114. Owen, M. J. In *Silicon-Based Polymer Science. A Comprehensive Resource;* Zeigler, J. M.; Fearon, F. W. G., Eds.; Advances in Chemistry Series; American Chemical Society: Washington, DC, 1990; Vol. 224.
115. Owen, M. J. *Preprints. Div. of Polym. Chem., Inc., Am. Chem Soc.* **1990,** *31(1),* 332.
116. Kobayshi, H.; Owen, M. J. *Preprints. Div. of Polym. Chem., Inc., Am. Chem Soc.* **1990,** *31(1),* 334.
117. Owen, M. J. In *Siloxane Polymers;* Clarson, S. J.; Semlyen, J. A., Eds.; Prentice Hall: Englewood Cliffs, NJ, 1991.
118. Cornelius, D. J.; Monroe, C. M. *Polym. Eng. Sci.* **1985,** *25,* 467.
119. Granick, S.; Kuzmenka, D. J.; Clarson, S. J.; Semlyen, J. A. *Langmuir* **1989,** *5,* 144.
120. Mark, J. E. *Acc. Chem. Res.* **1985,** *18,* 202, and references cited therein.
121. Gottlieb, M.; Macosko, C. W.; Benjamin, G. S.; Meyers, K. O.; Merrill, E. W. *Macromolecules* **1981,** *14,* 1039.
122. Mark, J. E. *Polym. J.* **1985,** *17,* 265, and references cited therein.
123. Mark, J. E. *Br. Polym. J.* **1985,** *17,* 144, and references cited therein.
124. Saam, J. C. In *Encyclopedia of Materials Science and Engineering;* Bever, M. B., Ed.; Pergamon: Oxford, 1986.
125. Opperman, W.; Rennar, N. *Prog. Coll. Polym. Sci.* **1987,** *75,* 49.
126. Erman, B.; Mark, J. E. *Annu. Rev. Phys. Chem.* **1989,** *40,* 351.
127. Mark, J. E. In *Frontiers of Macromolecular Science;* Saegusa, T.; Higashimura, T.; Abe, A., Eds.; Blackwell: London, 1989.

128. Clarson, S. J.; Mark, J. E. In *Siloxane Polymers;* Clarson, S. J.; Semlyen, J. A., Eds.; Prentice Hall: Englewood Cliffs, NJ, 1991.

129. Frisch, H. L.; Frisch, K. C.; Klempner, D. *CHEMTECH* **1977,** *7,* 188.

130. Sperling, L. H. *Interpenetrating Polymer Networks and Related Materials;* Plenum: New York, 1981.

131. Frisch, K. C.; Klempner, D.; Frisch, H. L. *Polym. Eng. Sci.* **1982,** *22,* 1143.

132. Mark, J. E.; Ning, Y.-P. *Polym. Eng. Sci.* **1985,** *25,* 824.

133. Sperling, L. H. *CHEMTECH* **1988,** *18,* 104.

134. Arkles, B.; Crosby, J. In *Silicon-Based Polymer Science. A Comprehensive Resource;* Zeigler, J. M.; Fearon, F. W. G., Eds.; Advances in Chemistry Series; American Chemical Society: Washington, DC, 1990; Vol. 224.

135. Frisch, H. L. In *Siloxane Polymers;* Clarson, S. J.; Semlyen, J. A., Eds.; Prentice Hall: Englewood Cliffs, NJ, 1991.

136. Jeram, E. M.; Striker, R. A. U.S. Patent 3 957 713, 1976.

137. Lee, C. L.; Maxson, M. T.; Stebleton, L. F. U.S. Patent 4 162 243, 1979.

138. Mark, J. E. *Makromol. Chem. Suppl.* **1979,** *2,* 87.

139. Galiatsatos, V.; Mark, J. E. In *Silicon-Based Polymer Science. A Comprehensive Resource;* Zeigler, J. M.; Fearon, F. W. G., Eds.; Advances in Chemistry Series; American Chemical Society: Washington, DC, 1990; Vol. 224.

140. Tang, M. Y.; Letton, A.; Mark, J. E. *Colloid Polym. Sci.* **1984,** *262,* 990.

141. Clarson, S. J.; Galiatsatos, V.; Mark, J. E. *Macromolecules* **1990,** *23,* 1504.

142. Mark, J. E.; Curro, J. G. *J. Chem. Phys.* **1983,** *79,* 5705.

143. Curro, J. G.; Mark, J. E. *J. Chem. Phys.* **1984,** *80,* 4521.

144. Erman, B.; Mark, J. E. *J. Chem. Phys.* **1988,** *89,* 3314, and relevant references cited therein.

145. Clarson, S.; Mark, J. E.; Semlyen, J. A. *Polym. Commun.* **1986,** *27,* 244.

146. DeBolt, L. C.; Mark, J. E. *Macromolecules* **1987,** *20,* 2369.

147. de Gennes, P. G. *Scaling Concepts in Polymer Physics;* Cornell University: Ithaca, NY, 1979.

148. Garrido, L.; Mark, J. E.; Clarson, S. J.; Semlyen, J. A. *Polym. Commun.* **1985,** *26,* 53.

149. Rigbi, Z.; Mark, J. E. *J. Polym. Sci., Polym. Phys. Ed.* **1986,** *24,* 443.

150. Babu, G. N.; Christopher, S. S.; Newmark, R. A. *Macromolecules* **1987,** *20,* 2654.

151. Ziemelis, M. J.; Saam, J. C. *Macromolecules* **1989,** *22,* 2111.

152. Keohan, F. L.; Hallgren, J. E. In *Silicon-Based Polymer Science. A Comprehensive Resource;* Zeigler, J. M.; Fearon, F. W. G., Eds.; Advances in Chemistry Series; American Chemical Society: Washington, DC, 1990; Vol. 224.

153. Lupinski, J. H.; Policastro, P. P. *Polym. News* **1990,** *15,* 71.

154. Ibemesi, J.; Gvozdic, N.; Keumin, M.; Lynch, M. J.; Meier, D. J. *Polym. Prepr., Am. Chem. Soc., Div. Polym. Chem.* **1985,** *26(2),* 18.

155. Ibemesi, J.; Gvozdic, N.; Kuemin, M.; Tarshiani, Y.; Meier, D. J. In *Polymer-Based Molecular Composites;* Schaefer, D. W.; Mark, J. E., Eds.; Materials Research Society Symposium Volume, Pittsburgh, PA, 1990.

156. Manson, J. A.; Sperling, L. H. *Polymer Blends and Composites;* Plenum: New York, 1976.

157. Smid, J.; Fish, D.; Khan, I. M.; Wu, E.; Zhou, G. In *Silicon-Based Polymer Science. A Comprehensive Resource;* Zeigler, J. M.; Fearon, F. W. G., Eds.; Advances in Chemistry Series; American Chemical Society: Washington, DC, 1990; Vol. 224.

158. Reichmanis, E.; Novembre, A. E.; Tarascon, R. G.; Shugard, A.; Thompson, L. F. In *Silicon-Based Polymer Science. A Comprehensive Resource;* Zeigler, J. M.; Fearon, F. W. G., Eds.; Advances in Chemistry Series; American Chemical Society: Washington, DC, 1990; Vol. 224.

159. Hench, L. L.; Ulrich, D. R. *Ultrastructure Processing of Ceramics, Glasses, and Composites;* Wiley-Interscience: New York, 1984.

160. *Better Ceramics Through Chemistry*, Materials Research Society Symposium Proceedings; Brinker, C. J.; Clark, D. E.; Ulrich, D. R., Eds.; North-Holland: New York, 1984; Vol. 32.

161. Klein, L. C. *Ann. Rev. Mater. Sci.* **1985**, *15*, 227.

162. *Science of Ceramic Chemical Processing;* Hench, L. L.; Ulrich, D. R., Eds.; Wiley-Interscience: New York, 1986.

163. Roy, R. *Science* **1987**, *238*, 1664.

164. Brinker, C. J.; Bunker, B. C.; Tallant, D. R.; Ward, K. J.; Kirkpatric, R. J. In *Inorganic and Organometallic Polymers;* Zeldin, M.; Wynne, K. J.; Allcock, H. R., Eds.; ACS Symposium Series; American Chemical Society: Washington, DC, 1988; Vol. 360.

165. Sakka, S.; Kamiya, K.; Yoko, Y. In *Inorganic and Organometallic Polymers;* Zeldin, M.; Wynne, K. J.; Allcock, H. R., Eds.; ACS Symposium Series; American Chemical Society: Washington, DC, 1988; Vol. 360.

166. Ulrich, D. R. *CHEMTECH* **1988,** *18*, 242.

167. *Ultrastructure Processing of Advanced Ceramics;* MacKenzie, J. D.; Ulrich, D. R., Eds.; John Wiley: New York, 1988.

168. Hench, L. L.; West, J. K. *Chem. Rev.* **1990,** *90*, 33.

169. *Polymer-Based Molecular Composites;* Schaefer, D. W.; Mark, J. E., Eds.; Materials Research Society Symposium Volume 171, Pittsburgh, PA, 1990.

170. Brinker, C. J.; Scherer, G. W. *The Physics and Chemistry of Sol-Gel Processing;* Academic: New York, 1990.

171. Keefer, K. D. In *Silicon-Based Polymer Science. A Comprehensive Resource;* Zeigler, J. M.; Fearon, F. W. G., Eds.; Advances in Chemistry Series; American Chemical Society: Washington, DC, 1990; Vol. 224.

172. Doughty, D. H.; Assink, R. A.; Kay, B. D. In *Silicon-Based Polymer Science. A Comprehensive Resource;* Zeigler, J. M.; Fearon, F. W. G., Eds.; Advances in Chemistry Series; American Chemical Society: Washington, DC, 1990; Vol. 224.

173. Holmes, R. R. *Chem. Rev.* **1990,** *90*, 17.

174. Ulrich, D. R. *Chem. & Eng. News*, January 1, 1990, p 28.

175. Ning, Y. P.; Tang, M. Y.; Jiang, C. Y.; Mark, J. E.; Roth, W. C. *J. Appl. Polym. Sci.* **1984,** *29*, 3209.

176. Mark, J. E. *CHEMTECH* **1989,** *19*, 230.

177. Xu, P.; Wang, S.; Mark, J. E. In *Better Ceramics Through Chemistry IV;* Brinker, C. J.; Clark, D. E.; Ulrich, D. R.; Zelinkski, B. J. J., Eds.; Materials Research Society Symposium Volume 180, Pittsburgh, PA, 1991.

178. Mark, J. E.; Schaefer, D. W. In *Polymer-Based Molecular Composites;* Schaefer, D. W.; Mark, J. E., Eds.; Materials Research Society Symposium Volume 171, Pittsburgh, PA, 1990.

179. Schaefer, D. W.; Mark, J. E.; McCarthy, D.; Jian, L.; Sun, C.-C.; Farago, B. In *Polymer-Based Molecular Composites;* Schaefer, D. W.; Mark, J. E., Eds.; Materials Research Society Symposium Volume 171, Pittsburgh, PA, 1990.

180. Wang, S.; Mark, J. E. *J. Mat. Sci.* **1990,** *25,* 65.

181. Schmidt, H. *Mater. Res. Soc. Symp. Proc.* **1984,** *32,* 327.

182. Huang, H.-H.; Orler, B.; Wilkes, G. L. *Polym. Bull.* **1985,** *14,* 557.

183. Mark, J. E.; Sun, C. C. *Polym. Bull.* **1987,** *18,* 259.

184. Schmidt, H. K. In *Inorganic and Organometallic Polymers;* ACS Symposium Series; Zeldin, M.; Wynne, K. J.; Allcock, H. R., Eds.; American Chemical Society: Washington, DC, 1988; Vol. 360.

185. Wilkes, G. L.; Huang, H.-H.; Glaser, R. H. In *Silicon-Based Polymer Science. A Comprehensive Resource;* Zeigler, J. M.; Fearon, F. W. G., Eds.; Advances in Chemistry Series; American Chemical Society: Washington, DC, 1990; Vol. 224.

186. Chujo, Y.; Ihara, E.; Kure, S.; Suzuki, K.; Saegusa, T. *Preprints. Div. of Polym. Chem., Inc., Am. Chem. Soc.* **1990,** *31(1),* 59.

187. Wilkes, G. L.; Brennan, A. B.; Huang, H.-H.; Rodrigues, D.; Wang, B. In *Polymer-Based Molecular Composites;* Schaefer, D. W.; Mark, J. E., Eds.; Materials Research Society Symposium Volume 171, Pittsburgh, PA, 1990.

188. Brennan, A. B.; Huang, H.-H.; Wilkes, G. L. *Preprints. Div. of Polym. Chem., Inc., Am. Chem Soc.* **1990,** *31(1),* 105.

189. Spinu, M.; Arnold, C.; McGrath, J. E. *Preprints. Div. of Polym. Chem., Inc., Am. Chem. Soc.* **1990,** *31(1),* 125.

190. Wang, B.; Huang, H.-H.; Wilkes, G. L. *Preprints. Div. of Polym. Chem., Inc., Am. Chem. Soc.* **1990,** *31(1),* 146.

191. Rodrigues, D. E.; Wilkes, G. L. *Preprints. Div. of Polym. Chem., Inc., Am. Chem. Soc.* **1990,** *31(1),* 227.

5

Polysilanes and Related Polymers

5.1 INTRODUCTION

In polysilane polymers, the polymer backbone is made up entirely of silicon atoms. Therefore these materials differ from other important inorganic polymers, the siloxanes and phosphazenes, in which the polymer chain is heteroatomic. Structurally they are more closely related to homoatomic organic polymers such as the polyolefins. However, because the units in the main chain are all silicon atoms, the polysilanes exhibit quite unusual properties.

The cumulated silicon-silicon bonds in the polymer chain allow extensive *electron delocalization* to take place. As a consequence, the electronic and photochemical behavior of the polysilanes is very different from that of most other inorganic or organic polymers, in which electron delocalization is much less important. Many of the technical uses, as well as the remarkable properties, of polysilanes result from this unusual mobility of the sigma electrons.

Linear polysilane polymers, properly called poly(silylene)s, can be obtained as homopolymers or copolymers. Continuation of the polysilane chain consumes two of the four valences of each silicon atom; the other two are taken up by pendent groups, which may be the same (Structure 5.1) or different (5.2). Copolymers (5.3), which contain two or more kinds of silicon atoms, can be made up from units like those in Structure 5.1 or 5.2. A typical example is the copolymer of

$$
\begin{array}{cccc}
\underset{\underset{R}{|}}{\overset{\overset{R}{|}}{-\!(\!Si\!)_n\!-}} & \underset{\underset{R'}{|}}{\overset{\overset{R}{|}}{-\!(\!Si\!)_n\!-}} & \underset{\underset{R^2}{|}}{\overset{\overset{R^1}{|}}{-\!(\!Si\!)_n\!}}\,\underset{\underset{R^4}{|}}{\overset{\overset{R^3}{|}}{(\!Si\!)_m\!-}} & \underset{\underset{Me}{|}}{\overset{\overset{Me}{|}}{-\!(\!Si\!)_n\!}}\,\underset{\underset{Me}{|}}{\overset{\overset{Ph}{|}}{(\!Si\!)_m\!-}}
\end{array}
$$

$$
\begin{array}{cccc}
\textbf{5.1} & \textbf{5.2} & \textbf{5.3} & \textbf{5.4}
\end{array}
$$

Me_2Si and $PhMeSi$ units, poly(dimethylsilylene-co-phenylmethylsilylene) (Structure 5.4), which bears the popular name "polysilastyrene."

The pendent groups are typically organic units and can include alkyl, aryl, substituted aryl, hydrogen, Me_3Si, ferrocenyl, and so on. An unlimited number of different polymers are possible, and more than a hundred compositions have been described in the literature. The properties of the polysilanes, like those of the polyphosphazenes, depend greatly on the nature of the substituent groups. Polysilanes cover the entire range of properties from highly crystalline and insoluble, through partially crystalline, flexible solids, to glassy amorphous materials and rubbery elastomers.

Until recently, it was thought that polysilanes would be either intractable or unstable. In his famous book, Eugene G. Rochow, the father of the silicone industry, dismissed polysilane polymers in this way:

> It follows that, even though suitable procedures were to be found for the alkylation of longer silicon chains, the products would be subject to oxidation, to thermal dissociation, and to hydrolysis in the presence of alkalies. It seems unlikely that any combination of substituents could stabilize these chains sufficiently to allow their practical application to polymeric materials.[1]

Yet we know now that polysilane polymers are stable to heat up to almost 300°C, are inert to oxygen at ordinary temperatures, and are only mildly susceptible to hydrolysis. The principal weakness of polysilanes, as materials, is not any of these reasons; it is that they become degraded when exposed to ultraviolet light.

Undoubtedly Rochow thought of the polysilanes as structural polymers like the polysiloxanes, and never envisioned that polysilanes might have other sorts of uses. But today, polysilanes are being investigated for numerous technological applications. They are now sold commercially as precursors to silicon carbide ceramics. Polysilanes show considerable promise as ultraviolet acting photoresists for microelectronics, and as photoconductors in electrophotography; they are active as free radical photoinitiators for organic reactions and show marked nonlinear optical properties which may make them useful in laser and other optical technology. These practical uses for polysilanes will be discussed later in this chapter.

5.2 HISTORY

Oligomeric phenylpolysilanes, including the perphenylated rings $(Ph_2Si)_n$, $n = 4$ to 6, were studied more than 70 years ago by F. S. Kipping and his students,[2] but high polymers were almost unknown until the late 1970s, and the belief that silicon had limited capability for catenation persisted until quite recently.

The first clear description of a polysilane polymer appears in a classic paper by C. A. Burkhard of the General Electric Co. Laboratories published in 1949.[3] Burkhard combined 1 pound of sodium metal with 700 mL of dimethyldichlorosilane and heated the mixture in a steel autoclave to make poly(dimethylsilylene), $(Me_2Si)_n$ (eq. 5.1). This polymer is a highly crystalline material, which decomposes above 250°C without melting and is essentially insoluble in all organic solvents. Although we know now that $(Me_2Si)_n$ is atypical among polysilanes, Burkhard's description left no doubt that $(Me_2Si)_n$ is quite intractable, and this perhaps contributed to the neglect suffered by the polysilanes over the next three decades.

$$Me_2SiCl_2 + 2Na \longrightarrow (Me_2Si)_n + 2NaCl \qquad \text{(Eq. 5.1)}$$

Even though polysilane high polymers were largely ignored between 1949 and 1975, fundamental studies on linear and cyclic oligosilanes were being carried out in several laboratories. The linear permethylpolysilanes, $Me(SiMe_2)_nMe$ ($n = 2$ to 12), surprisingly were found to have strong ultraviolet absorption bands, which became more intense and shifted to longer wavelength as the number of Si atoms in the chain increased.[4] This behavior is reminiscent of that found for conjugated polyenes and was the first indication that electrons in the silicon-silicon bonds of polysilanes might be delocalized.

The cyclic peralkylsilane oligomers, $(R_2Si)_n$ with $n = 4$ to 6, manifested especially strong electron delocalization.[5] These rings are structurally analogous to those of the cycloalkanes, since the silicon atoms form four sigma bonds. However, the electronic properties of the cyclosilanes more nearly resemble those of aromatic hydrocarbons such as benzene. One example of such behavior is their reduction to anion radicals. Aromatic hydrocarbons such as naphthalene can be reduced, electrolytically or with alkali metals, to deeply colored anion radicals in which an unpaired electron occupies the lowest unoccupied molecular orbital (LUMO) of the hydrocarbon (eq. 5.2).

$$\text{(Eq. 5.2)}$$

Colorless   **Green**

Since they contain an unpaired electron, anion radicals can be studied by electron spin resonance spectroscopy. The cyclosilanes, $(RR'Si)_n$ where $n = 4$ to 6, similarly undergo reduction to strongly colored (eq. 5.3) anion radicals whose electron spin resonance spectra indicate that the unpaired electron is completely delocalized over the ring.

$$(Me_2Si)_5 \xrightarrow[-60°]{e^-, THF} (Me_2Si)_5^{\overline{\bullet}}$$

$$\textbf{Colorless} \qquad\qquad\qquad \textbf{Blue}$$

(Eq. 5.3)

Cyclosilanes are synthesized by alkali metal reduction of diorganodichlorosilanes, usually in ethereal solvents. One outgrowth of the study of these compounds was the finding that under nonequilibrium conditions, large rings may be formed, containing up to 40 silicon atoms.[6] Although their molecular weights are not very high, these compounds may be considered as early examples of polysilane polymers.

Interest in polysilanes was aroused in 1975, when Yajima found that the permethyl polymer $(Me_2Si)_n$, or its cyclic oligomer $(Me_2Si)_6$, could be transformed into silicon carbide by heating to high temperatures.[7] Shortly after this, in the course of an attempt at Wisconsin to make cyclic oligosilanes containing phenyl and methyl groups, Me_2SiCl_2 and $PhMeSiCl_2$ were cocondensed with sodium metal (eq. 5.4).

$$Me_2SiCl_2 + PhMeSiCl_2 \xrightarrow[110°]{Na, C_7H_8} \underset{\underset{Me}{|}}{\overset{\overset{Me}{|}}{{-}{\left(Si\right)_n}}}\underset{\underset{Me}{|}}{\overset{\overset{Ph}{|}}{{\left(Si\right)_m}{-}}}$$

(Eq. 5.4)

Instead of the desired ring compounds the product was mainly a linear polymer (Structure 5.4), but unlike the permethyl polymer described earlier, this copolymer was meltable and quite soluble in organic solvents.[8,9] Evidently the introduction of phenyl groups along the chain greatly reduces the crystallinity so that the polymer becomes thermoplastic and soluble. This copolymer also proved useful as a precursor to silicon carbide. Soluble polysilane polymers were discovered independently at Union Carbide Corporation[10] and Sandia Laboratories[11] at almost the same time.

The growth of interest in poly(silylene)s is indicated in Figure 5.1, which shows the number of literature publications on polysilanes plotted versus time. Note that no publications appeared between Burkhard's 1949 report and 1978, but during the past 10 years scientific activity in this field has increased dramatically. The interest in polysilanes stems partly from their novel constitution and behavior and partly from their potential utility. The possibility that soluble polysilanes might be precursors to silicon carbide was evident when they were first synthesized, but the other technological uses were not so obvious and arose only in the course of research.

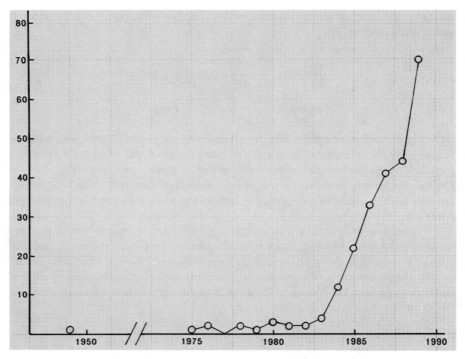

Figure 5.1 Number of papers published on polysilane polymers per year, plotted against time, showing the dramatic growth in the literature since 1984.

5.3 SYNTHESIS

5.3.1 Condensation of Dichlorosilanes with Alkali Metals

Polysilanes are usually made by dehalogenation of diorganodichlorosilanes with sodium metal in an inert diluent, following essentially the original method of Burkhard.[3] Single dichlorosilanes yield homopolymers, while mixtures of dichlorosilanes lead to copolymers.

$$RR'SiCl_2 \xrightarrow[> 100°C]{Na,\ solv.} (RR'Si)_n \qquad (Eq.\ 5.5)$$

$$R^1R^2SiCl_2 + R^3R^4SiCl_2 \xrightarrow[> 100°C]{Na,\ solv.} (R^1R^2Si)_n(R^3R^4Si)_n \qquad (Eq.\ 5.6)$$

The best results are obtained using finely divided sodium, above its melting point, 98°C.[12] The sodium may be added as a preformed sodium dispersion, or dispersed by vigorous stirring in the diluent. Toluene is convenient as a diluent because its boiling point of 111°C allows the reaction to be carried out at reflux temperature. An efficient condenser is useful because the heat of reaction is substantial. Other solvents such as alkanes (decane) or ethers (diglyme) may be used, either alone or mixed with toluene or xylene. Often the chlorosilane(s) are added slowly to the stirred sodium and diluent, but the reaction can also be carried by "inverse addition" in which preformed sodium dispersion is added to a stirred solution of the chlorosilane(s). Potassium metal, or sodium-potassium alloy, may be used in place of sodium with many chlorosilanes.*

Since cyclic oligomers are the exclusive products at equilibrium, high polymers are formed only when the reaction is kinetically controlled. The yields of high polymer obtained in the condensation depend upon the substituent groups on silicon and the exact reaction conditions, and can be quite low. The oligomeric fraction, consisting mostly of cyclic polysilanes which cannot be converted to polymer, often makes up more than 50% of the products. Some typical yields from the literature are given in Table 5.1. Unfortunately no general best procedure can be stated, because the reaction conditions must be optimized for each particular set of substituents. A further problem is that in laboratory syntheses, yields and molecular weights of the polymers are not easily reproducible. Probably this is because certain reaction variables, such as the purity of the chlorosilanes, the state of subdivision of the sodium, and the addition rate, are rather difficult to control.

The reactions leading to polymer formation from chlorosilanes and sodium appear to be complex and are not well understood. When sufficient sodium is present many chains (before workup) are terminated with silyl anions, consistent with an anionic condensation mechanism. Polysilane polymers prepared by sodium condensation typically show a bimodal molecular weight distribution (Figure 5.2), suggesting that at least two mechanisms may be involved in chain extension. An important observation is that some polymer with very high molecular weight is formed early in the reaction.[13] Although the polymerization reaction is necessarily a condensation, its kinetics therefore somewhat resemble those of chain reactions such as the polymerization of olefins, in which chains leading to high molecular weight polymer may occur immediately. (The formation of most condensation polymers proceeds stepwise, with the molecular weight building up only slowly.) Under ultrasound irradiation, polymerization is greatly accelerated, so that it can be carried out at much lower temperatures (60°C); the resulting polymer is of high molecular weight and monomodal (Figure 5.3).[14]

A careful kinetic study of the sodium condensation of methyl n-hexyldichlorosilane has been carried out by Worsfold.[13, 15] In the absence of catalysts, an

---

*Before isolating the polymer, a reactive monochlorosilane, R_3SiCl, is sometimes added to react with and terminate anionically ended chains. Excess sodium is then discharged by adding a small amount of an alcohol, followed by water. The solution is then washed with water to remove salts, and the organic layer is separated and filtered. Addition of a poor solvent such as ethanol or 2-propanol precipitates the polymer, leaving most of the oligomeric by-product in solution. The separated polymer is heated under vacuum to remove solvent.

TABLE 5.1　Examples of Polysilane Polymers

Polymer	Yield, %	$M_w \times 10^{-3}$	n/m
$(n\text{-PrSiMe})_n$	32	64, 13	
$(n\text{-BuSiMe})_n$	34	110, 6	
$(n\text{-HexSiMe})_n$	11	520, 20	
$(\text{cy-HexSiMe})_n$	20	1200, 19	
$(n\text{-}C_{12}H_{25}SiMe)_n$	8	1350, 9	
$(n\text{-Pr}_2Si)_n$	39	1160, 27	
$(n\text{-Bu}_2Si)_n$	34	110, 6	
$(n\text{-Hex}_2Si)_n$	6	2000, 1.2	
$(\text{PhSiMe})_n$	55	190, 8	
$(p\text{-MeOC}_6H_4SiMe)_n$	12	13	
$(p\text{-biphenylSiMe})_n$	40	80	
$(\text{PhCH}_2CH_2SiMe)_n$	35	290, 4.4	
$[(p\text{-}n\text{-BuC}_6H_4)_2Si]_n$	6.2	230, 2.1	
$(\text{PhSiMe}_2SiMe)_n$	9	13	
Copolymers			
$(\text{Me}_2Si)_n\ (n\text{-HexSiMe})_n$	57	170, 10	1.52
$(\text{Me}_2Si)_n\ (\text{PhSiMe})_m$	60	900, 70	1.51
$(\text{Me}_2Si)_n\ (\text{cyHexSiMe})_n$	63	150, 8	1.49
$(\text{PhC}_2H_4SiMe)_n\ (\text{PhSiMe})_n$	18	400, 3	1.0
$(\text{Me}_2Si)_n\ (\text{Ph}_2Si)_m$	70	350, 2	1.13
$(n\text{-HexSiMe})_n\ (\text{Ph}_2Si)_m$	51	360, 3	0.83
$(n\text{-Hex}_2Si)\ (\text{Ph}_2SiMe)$	17	270	0.67

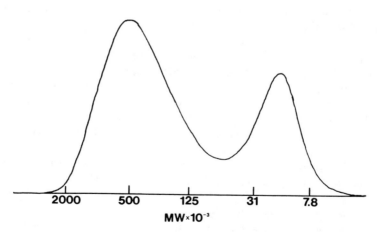

Figure 5.2　Gel permeation chromatograph for $(n\text{-HexSiMe})_n$, showing bimodal molecular weight distribution. Reprinted with permission from R. West, *J. Organometal. Chem.* **1986,** *300,* 327.

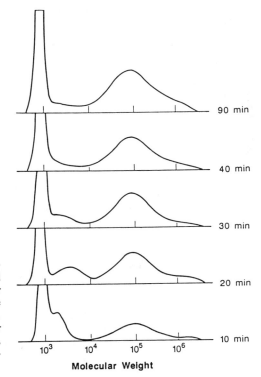

Figure 5.3 GPC curves showing polymerization of $(n\text{-HexSiMe})_n$ under ultrasound irradiation, to give a monomodal molecular weight distribution of polymer. The large peak near 10^{-3} is due to cyclic oligomers. Reprinted with permission from S. Gauthier and D. J. Worsfold, *Macromolecules* **1989,** *22,* 2213. Copyright 1989 American Chemical Society.

induction period is observed (Figure 5.4). Addition of small amounts of a crown ether (15-crown-5) eliminates the induction period, but does not otherwise change the rate of disappearance of the dichlorosilane during the early part of the condensation. In the presence of crown ether, the yield of high polymer is, however, increased, and the molecular weight distribution is strongly affected, becoming nearly monomodal (Figure 5.5). The surface area of the sodium also affects the product distribution, but not the maximum rate of disappearance of the monomer.

A suggested mechanism consistent with this evidence is shown in Scheme 5.1.[13] In this model the initiation step, thought to be very slow, is the reaction of $RR'SiCl_2$ with sodium to produce the ion pair $RR'SiCl^-,Na^+$ (eq. 5.7). The propagation step (eq. 5.8), which is fast but nevertheless rate determining, is the reaction of anion-terminated chains with dichlorosilane to add one silicon unit and produce a chlorine-ended chain. The latter is thought to be reduced rapidly to the anionic form by reaction with sodium. The polysilane chains will alternately be terminated with silyl anions or Si—Cl. This is consistent with evidence from chemical derivatization that both may be present. The phase-transfer catalyst, when added, may clean the sodium surface by solubilizing absorbed sodium salts and

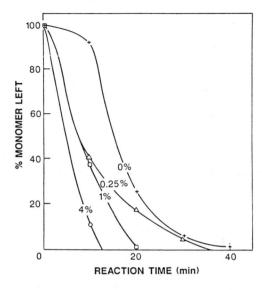

Figure 5.4 Graphs of polymerization of (n-HexSiMe)$_n$ by sodium condensation from n-HexSiMeCl$_2$, with various amounts of 15-crown-5 ether added, showing the catalytic effect of crown ether on the initiation of polymerization. Reprinted with permission from S. Gauthier and D. J. Worsfold, *Macromolecules* **1989,** *22,* 2213. Copyright 1989 American Chemical Society.

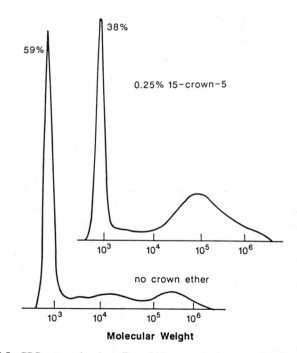

Figure 5.5 GPC curves showing effect of 15-crown-5 ether on molecular weight distribution and yield of (n-HexSiMe)$_n$. Reprinted with permission from S. Gauthier and D. J. Worsfold, *Macromolecules* **1989,** *22,* 2213. Copyright 1989 American Chemical Society.

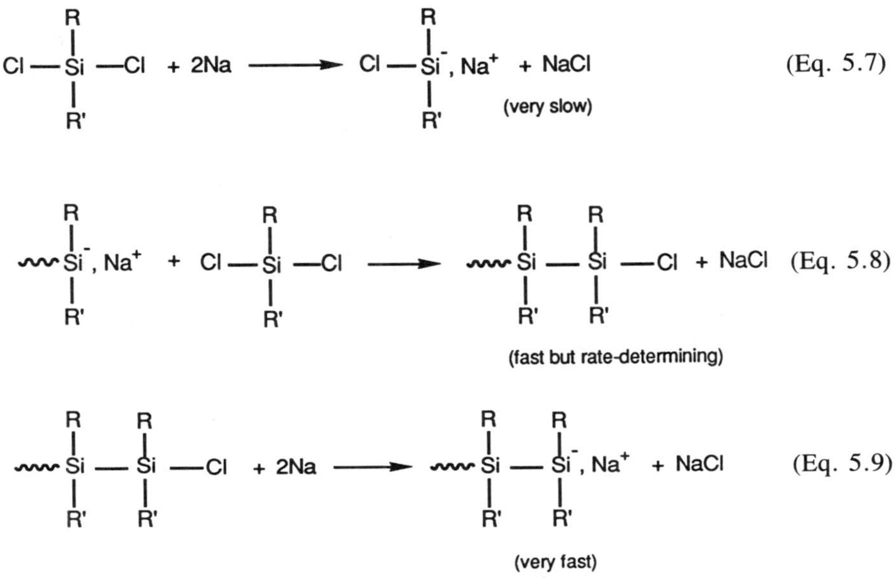

$$\underset{\underset{R'}{|}}{\overset{\overset{R}{|}}{Cl-Si-Cl}} + 2Na \longrightarrow \underset{\underset{R'}{|}}{\overset{\overset{R}{|}}{Cl-Si^-,Na^+}} + NaCl \qquad \text{(Eq. 5.7)}$$

(very slow)

$$\underset{\underset{R'}{|}}{\overset{\overset{R}{|}}{\text{∿∿}Si^-,Na^+}} + \underset{\underset{R'}{|}}{\overset{\overset{R}{|}}{Cl-Si-Cl}} \longrightarrow \underset{\underset{R'}{|}}{\overset{\overset{R}{|}}{\text{∿∿}Si}}-\underset{\underset{R'}{|}}{\overset{\overset{R}{|}}{Si-Cl}} + NaCl \quad \text{(Eq. 5.8)}$$

(fast but rate-determining)

$$\underset{\underset{R'}{|}}{\overset{\overset{R}{|}}{\text{∿∿}Si}}-\underset{\underset{R'}{|}}{\overset{\overset{R}{|}}{Si-Cl}} + 2Na \longrightarrow \underset{\underset{R'}{|}}{\overset{\overset{R}{|}}{\text{∿∿}Si}}-\underset{\underset{R'}{|}}{\overset{\overset{R}{|}}{Si^-,Na^+}} + NaCl \qquad \text{(Eq. 5.9)}$$

(very fast)

Scheme 5.1

thereby reduce the induction period. In addition, the catalyst most probably complexes the sodium cations of the Si^- Na^+ ion pairs, changing the product distribution.

The mechanism in Scheme 5.1 can be elaborated further, as shown in Scheme 5.2. The reaction of chlorosilanes with sodium probably proceeds by an initial single electron transfer to form an anion radical, which loses chloride rapidly to form a silyl radical. The fact that $Si-H$ termination is found for some chains[16] and that 5-hexenylphenyldichlorosilane gives mainly cyclized products resulting from addition to the $C=C$ double bond[17] both suggest that radicals are present at some stage. The latter may be reduced in another electron-transfer step to give a silyl anion, which might either be bound to the sodium surface or be present in solution as an ion pair. The anion could then react with an $Si-Cl$ bond as in reaction **8** of Scheme 5.1; this is very probably the major reaction leading to chain extension. However, the silyl radical might be directly involved in chain extension by combination with another radical, or could add to a carbon-carbon double bond, or accept a hydrogen atom from the solvent to give an $Si-H$-terminated chain.

The evidence presented so far does not exclude other intermediates which might be present during the polymerization, such as silylenes (R_2Si), disilenes ($R_2Si=SiR_2$), or small, strained polysilane rings, which might react with silyl anions or radicals to give chain extension. We are, then, faced with a situation where there are probably several mechanisms for polymerization, some of which may take place at or near the sodium surface and others in solution. The balance among competing mechanisms may depend on solvent, temperature, sodium sur-

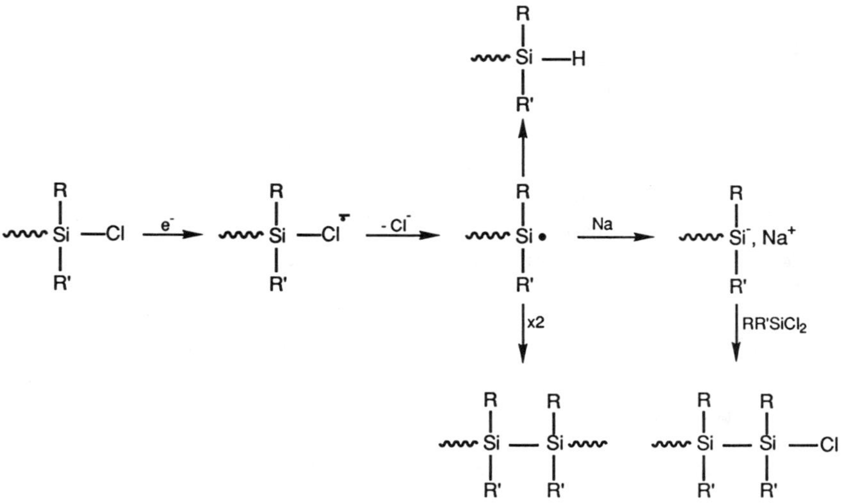

Scheme 5.2

face area, additives such as crown ethers, and the nature of the organic groups attached to silicon. For instance, arylalkyl dichlorosilanes undergo condensations much more rapidly than do dialkyldichlorosilanes, by mechanism(s) of polymerization which appear to be somewhat different.[15] At this time it is possible to optimize reaction conditions for synthesis of a particular polysilane polymer, but development of an inclusive theory of polysilane polymerization by alkali metals will require more experimentation.*

5.3.2 Disilene or Silylene Polymerization

Polyolefins are, of course, usually synthesized by the catalyzed polymerization of alkenes. Why is an analogous route, polymerization of disilenes, $R_2Si=SiR_2$, not employed to prepare polysilanes? The reason is paradoxical. The energy barrier to polymerization of doubly bonded silicon compounds is simply too low, so that in most cases they polymerize (or oligomerize) immediately when they are generated!

Photolysis of cyclic silanes or linear trisilanes in hydrocarbon glasses at 77K leads to divalent silicon species, silylenes (eq. 5.10).[19] When the hydrocarbon is warmed slightly to "anneal" the glass, the silylenes combine to give disilenes. Upon further warming to melt the hydrocarbon, polymerization of the disilene usually takes place. If the groups R and R' are moderately large the product is exclusively a cyclic oligomer. One well-studied example is tetra-*iso*-propyldisilene, which cyclodimerizes in high yield to the stable four-membered ring.[20]

---

*A recent communication[18] reports an efficient synthesis of polysilanes from dichlorosilanes and sodium in ethylether, with 15-crown-5 ether as a catalyst.

$$(RR'Si)_n \text{ or } RR'Si(SiMe_3)_2 \xrightarrow[77K]{h\upsilon} [RR'Si:] \xrightarrow{\Delta} RR'Si{=}SiRR' \xrightarrow{\Delta} (RR'Si)_n$$

(Eq. 5.10)

$$iPr_2Si{=}Si\,iPr_2 \xrightarrow{\Delta} \begin{array}{c} iPr_2Si{-}Si\,iPr_2 \\ | \qquad | \\ iPr_2Si{-}Si\,iPr_2 \end{array}$$

(Eq. 5.11)

These reactions illustrate a limitation on the possible structures for polysilane high polymers, no matter how they are prepared. Even when the substituent groups on silicon are small, the cyclic oligomers are thermodynamically more stable than the high polymers.[21] A high polymer may be synthesized in such cases, but only by reactions which do not proceed to equilibrium. As the size of the substituent groups becomes larger, the stability of the polymer becomes so much less than that of its cyclic oligomers that only the latter are formed. Thus polysilanes can be made containing one methyl group and one branched-chain alkyl group, that is, $(i\text{-}PrSiMe)_n$, but with two branched alkyl groups, the cyclic oligomer is so strongly favored that no polymer is observed, as in the case of $(i\text{-}Pr_2Si)_4$.*[20]

With even larger substituent groups, the disilene becomes the thermodynamically preferred form, and can then be isolated.[22] The best known example is tetramesityldisilene (Structure 5.5), ordinarily synthesized as shown in eq. (5.12). The stable, isolable disilenes have an elaborate and beautiful chemistry, but they cannot be polymerized.[23]

$$\left(\text{-}\bigcirc\text{-}\right)_2 Si(SiMe_3)_2 \xrightarrow[-(Me_3Si)_2]{h\upsilon} \left(\text{-}\bigcirc\text{-}\right)_2 Si{=}Si\left(\text{-}\bigcirc\text{-}\right)_2 \xleftarrow{h\upsilon} \left[\left(\text{-}\bigcirc\text{-}\right)_2 Si\right]_3$$

5.5

(Eq. 5.12)

Even though polymerization of room-temperature-stable disilenes is not possible, masked or transient disilenes may be useful sources of polysilane polymers in some cases. Sakurai has shown that high polymers are produced by reaction of anions with disilabicyclooctadienes. Although these molecules are thermal or photochemical sources of disilenes, the polymerization reaction most likely proceeds anionically, with elimination of the disilene fragment as a new silyl anion which can continue the reaction chain (eq. 5.13).[24] The polymers produced in this way may have properties different from those of ordinary polysilanes made by sodium coupling. In particular, disilacyclooctadienes (Structure 5.6) have been used to

---

*Similar, and even more stringent, limitations apply to olefin polymers. In general it is not possible to synthesize polyolefins with more than two alkyl substituents per two carbon atoms. With heavier substitution the polymer becomes unstable relative to the monomer.

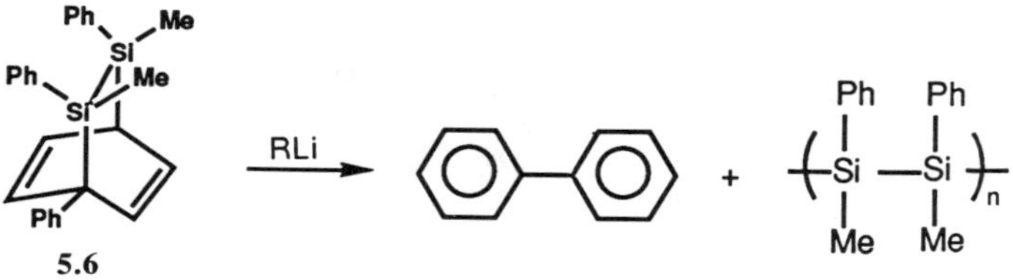

(Eq. 5.13)

make regularly ordered polysilane copolymers, as explained in Section 5.9.4. Stereoisomeric disilacyclooctadienes might also serve as precursors to stereoregular polysilanes.

Synthesis of polysilane polymers from silylenes, RR'Si, may also be useful in special cases. As shown in eq. (5.14), dimethylsilylene can be conveniently generated by thermolysis of 1,2-dimethoxytetramethyldisilane (Structure 5.7). In the absence of reagents to trap the silylene, the ultimate product is poly(dimethyl)-silylene, which probably arises from polymerization of tetramethyldisilene. Other polysilane polymers might be produced by the same method, varying the substitution at silicon in the starting disilane.

$$MeOSiMe_2SiMe_2OMe \xrightarrow{275°C} (MeO)_2SiMe_2 + [Me_2Si:]$$

$$[Me_2Si=SiMe_2] \longrightarrow (Me_2Si)_n$$

(Eq. 5.14)

5.3.3 Dehydrogenative Coupling

As an alternative route for the synthesis of polysilanes, dehydrogenation of diorganosilanes is attractive (eq. 5.15):

$$n\ RR'SiH_2 \xrightarrow{catalyst} H(SiRR')_nH + (n-1)H_2 \qquad (Eq.\ 5.15)$$

This method might be more easily controlled than the usual alkali metal condensation of dichlorosilanes, especially if the dehydrogenation could be brought about catalytically by transition metals. A general method for dehydrogenative polymerization of silanes is not yet available, but quite promising discoveries have been made in this area, especially by Harrod and his coworkers.[25,26]

A key finding was that phenylsilane is converted rapidly and quantitatively to an oligomer with a degree of polymerization of about 10, when it is treated at room temperature with dimethyltitanocene, $Cp_2Ti(CH_3)_2$ (Cp = cyclopenta-

dienyl). The best catalysts appear to be dialkyltitanocenes, Cp_2TiR_2, or zirconocene alkyls or hydrides, Cp_2ZrR_2 or Cp_2ZrH_2. Weaker activity as dehydrogenative
coupling agents is shown by cyclopentadienyl complexes of V, Th, and U. With
zirconium-containing catalysts degrees of polymerization up to ~ 20 have been
observed. Polymerization takes place only for primary silanes, $RSiH_3$; with diorganosilanes $RR'SiH_2$, the products are dimers, $RR'SiHSiHRR'$. Arylsilanes are
coupled more rapidly than are alkylsilanes; reactivities decrease in the following
sequence: $PhSiH_3$ > p-$TolylSiH_3$ > $PhSiMeH_2$ > $PhCH_2SiH_3$ > n-$HexSiH_3$
> cyclo-$HexSiH_3$.

With phenylmethylsilane and cyclohexylsilane, the dimer is produced rather
than polymer. The steric constraints on the reaction appear to be severe, as is also
indicated by the fact that pentamethylcyclopentadienyl Ti and Zr complexes are
poorer catalysts than are the analogous Cp complexes.

The polymerization is accelerated by the addition of an olefin such as cy-
clohexene, which undergoes hydrogenation as the reaction takes place. With titanocene catalysts the oligomerization is unchanged except for the increase in rate.
However, zirconocene compounds are also catalysts for hydrosilylation, which
involves the addition of Si—H to olefins. As a result, the polymer may contain
cyclohexyl as well as phenyl groups:

$$PhSiH_3 + \bigcirc \!\!\!\!= \xrightarrow{Cp_2ZrMe_2} H\!\!-\!\!\Big(\underset{H}{\overset{Ph}{Si}}\Big)_{\!n}\!\!\Big(\underset{\textcircled{S}}{\overset{Ph}{Si}}\Big)_{\!m}\!\!-\!\!H \qquad (Eq. 5.16)$$

The mechanism of dehydrogenative polymerization is not understood. Interesting hydrogen-bridged molecules such as Structures 5.7 and 5.8 have been isolated from the reaction mixture, but these do not appear to be the catalytically
active species.[25] Harrod proposes that the reaction proceeds through the formation

5.7   5.8

of intermediate Structure 5.10 from 5.9; 5.10 could then lose H_2 to give the active
titanium-silylene species Structure 5.11 (Scheme 5.3). The latter could add $RSiH_3$
across the Ti=Si bond as shown; repetitive addition and H_2 elimination would
build up the polymer chain. Another possible pathway to the polymeric products
has been suggested recently,[27] proceeding through a series of sigma bond
metatheses (Scheme 5.4).

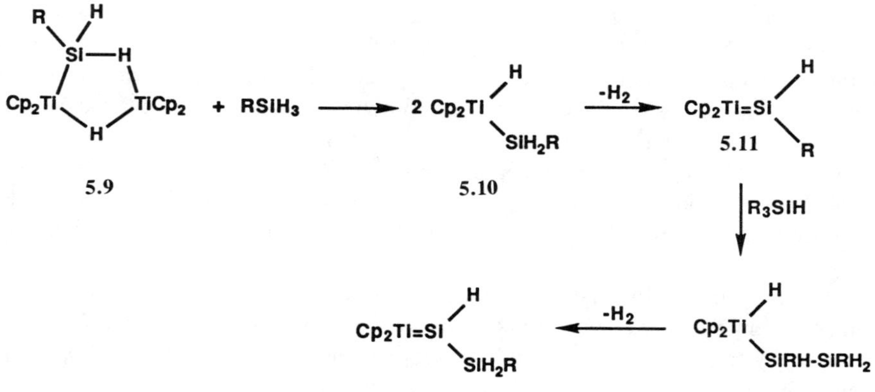

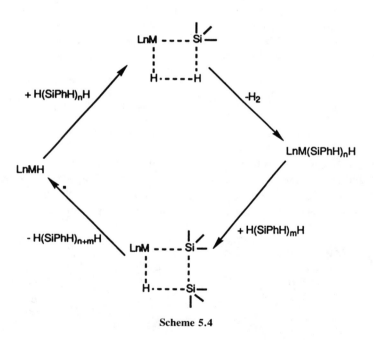

Scheme 5.3

Scheme 5.4

Dehydrogenative coupling by titanium and zirconium complexes is at least as effective for germanium as for silicon hydrides; in fact dialkylgermanes, RR′GeH₂, are oligomerized rather than dimerized by these catalysts. Much remains to be learned about dehydrogenative coupling of silanes, but with further development this route may provide a useful alternative to the usual polymerization by sodium condensation.

5.3.4 Other Polymerization Reactions

As explained in Chapter 4, silicone polymers are usually made by ring-opening polymerization of cyclosiloxane rings. However, because polysilane polymers are unstable relative to cyclosilane rings, ring-opening polymerization cannot generally be carried out. There remains the possibility that *strained* cyclosilane rings may be used in ring-opening polymerization under kinetically controlled conditions. In fact, the cyclotetrasilane $(Me_2Si)_4$ is observed to polymerize to $(Me_2Si)_n$ on long standing. Poly(phenylmethyl)silylene has been synthesized by ring-opening polymerization of the strained cyclosilane $(PhMeSi)_4$, prepared as shown in Scheme 5.5.[28] Probably other polysilanes could be made by similar polymerization of strained cyclosilanes.

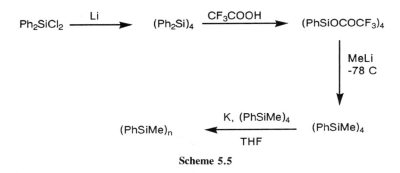

Scheme 5.5

As an alternative to condensation of dichlorosilanes by alkali metals, it is possible to synthesize polysilanes by the reaction of preformed dilithio compounds with dichlorosilanes. An early synthesis of a copolymer by this route[10] is illustrated in eq. (5.17) (see also Section 5.9.2). Finally, poly(phenylmethylsilylene) has been made by thermal decomposition of a silyl-mercury polymer[29] (eq. 5.18).

$$Li(Ph_2Si)_5Li \ + \ Cl(Me_2Si)_5Cl \ \longrightarrow \ (Ph_2Si)_n (Me_2Si)_m \qquad\qquad \text{(Eq. 5.17)}$$

$$^-PhMeSiHg^- \ \xrightarrow{\ \Delta\ } \ (PhMeSi)_n$$

$$\text{(Eq. 5.18)}$$

5.4 CHEMICAL MODIFICATION OF POLYSILANES

The presence of functional groups in polysilane polymers is limited by the vigorous conditions of the sodium condensation reaction usually used in synthesis. Any functions which react with molten sodium metal must therefore be introduced after the polymer is synthesized. Although relatively little work has yet been done, a few methods have been developed for modifying the structures of polysilanes, and many others should be possible.

Halogen atoms can be introduced into alkenylpolysilanes by HBr or HCl addition reactions, as shown for the 3-cyclohexenyl-ethyl polymers in eq. (5.19). Mild Lewis acid catalysis is required.[30] Chlorine atoms have also been introduced into arylpolysilanes by chloromethylation, in which $-CH_2Cl$ groups are substituted onto phenyl rings in the polymer.[31]

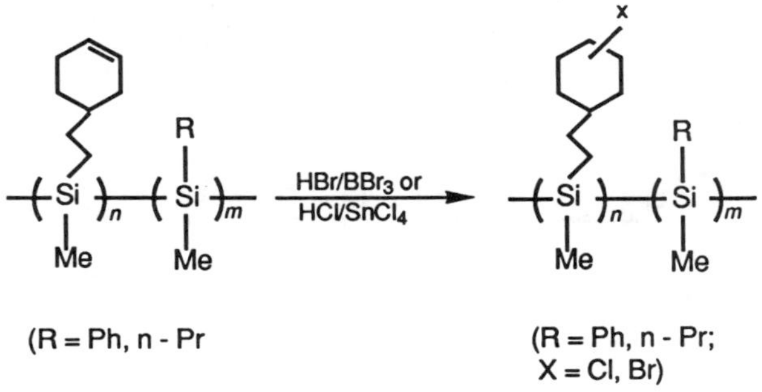

$$(Eq. 5.19)$$

(R = Ph, n - Pr   (R = Ph, n - Pr;
                                                                    X = Cl, Br)

Another approach to functionalized polysilanes is to use protected functional groups. This method was employed to synthesize polysilanes bearing phenolic $-OH$ groups, providing an alkali-soluble polysilane (Scheme 5.6).[32] The protected OH group was introduced as a trimethylsiloxy substituent, which was not affected by the sodium metal during the condensation. Methanol solvolysis later converted the siloxy substituent to the free phenol. Because the polymer is alkali soluble, it can be used as a photoresist which is developed with aqueous solutions (see Section 5.10.2). The phenol function can be further converted to an ester group by reacting it with an acid chloride or to a carboxylic acid by treating it with a diacid chloride followed by hydrolysis.[33]

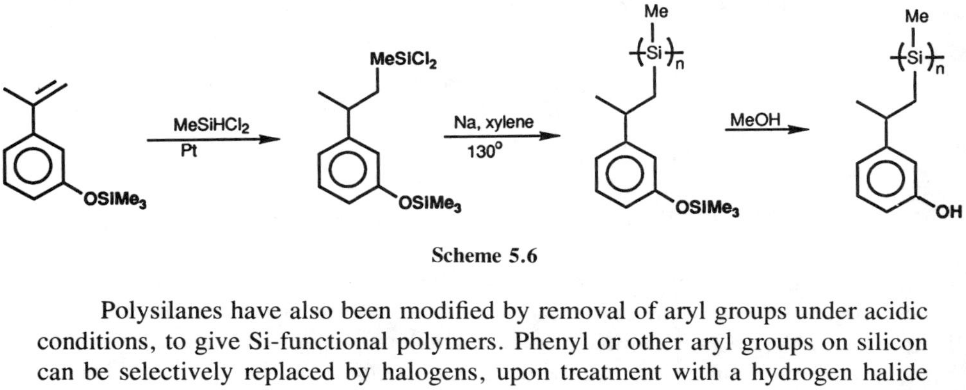

Scheme 5.6

Polysilanes have also been modified by removal of aryl groups under acidic conditions, to give Si-functional polymers. Phenyl or other aryl groups on silicon can be selectively replaced by halogens, upon treatment with a hydrogen halide and a Lewis acid. This reaction has been employed with phenyl-containing poly-silanes to replace phenyl groups by chlorine (eq. 5.20).[34]

The fraction of phenyl groups replaced, p/n in eq. (5.20), can be varied depending upon the amount of HCl used and the reaction conditions. Silicon-chlo-

$$(\text{Eq. 5.20})$$

rine linkages in the resulting polymer are quite reactive toward nucleophilic dis-placement, as expected. Thus treatment of the polymer with excess n-butyl-lithium replaces most of the chlorine atoms by n-butyl groups. However it is quite difficult to replace all the chlorine, and if Si—Cl bonds remain, the polymer undergoes hydrolysis and subsequent cross-linking upon exposure to air.

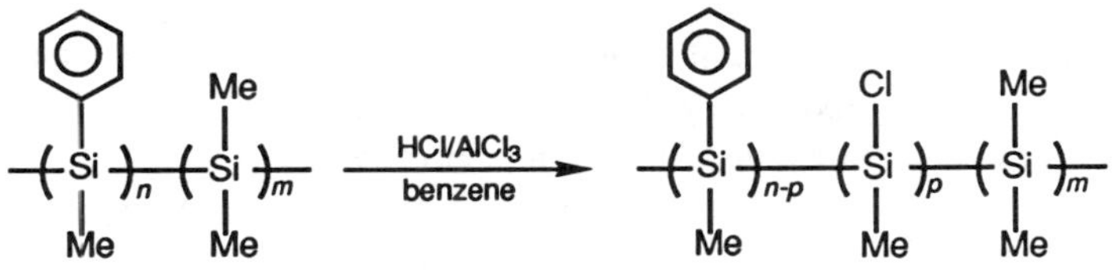

A more versatile substituted polysilane is obtained when trifluoromethane sulfonic (triflic) acid, CF_3SO_3H, is used to replace phenyl substituents.[35] Treat-ment of $(PhMeSi)_n$ with triflic acid in dichloromethane leads to rapid replacement of up to 80% of the phenyl groups by triflate. Some cleavage of Si—Si bonds and consequent degradation of the polymer takes place at high levels of substitution, but the molecular weight is apparently little changed by replacement of up to 30% of the phenyl groups. p-Methoxyphenyl substituents are replaced even more rap-idly than phenyl. The triflate groups on silicon are highly reactive and can be replaced, for example, by reaction with alcohols, to give methoxy, ethoxy or *tert*-butoxy–substituted polysilanes. The triflated polymer catalyzes cationic polymer-ization of tetrahydrofuran, leading to comblike graft copolymers with a polysilane backbone and polytetrahydrofuran side chains. Likewise, reaction of the triflate-containing polysilane with methyl methacrylate forms a comblike copolymer by group transfer polymerization. Other copolymerization reactions should be possi-ble.

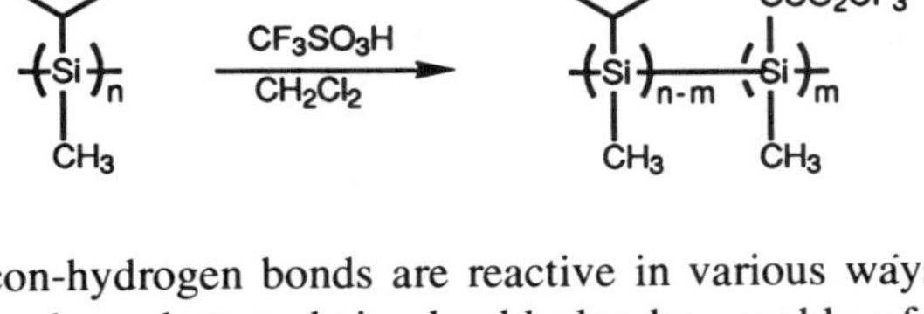

$$(\text{Eq. 5.21})$$

Since silicon-hydrogen bonds are reactive in various ways, polymers con-taining Si—H in the polymer chain should also be capable of further chemical modification. To date these polymers have been used mainly for cross-linking, as explained in Section 5.8.

5.5 PHYSICAL PROPERTIES OF POLYSILANES

The properties of polysilanes depend very much on the nature of the organic groups bound to silicon. Some polymers are so highly crystalline as to be insoluble and infusible. These include the $(Me_2Si)_n$ originally prepared by Burkhard, $(MeEtSi)_n$ and $(Et_2Si)_n$, as well as $(Ph_2Si)_n$ and most poly(diarylsilylene)s.* Decreasing the symmetry reduces the crystallinity and leads to the formation of tractable polymers. Thus $(PhMeSi)_n$ and most poly(arylalkylsilylene)s are resinous solids which melt when heated and are soluble in organic solvents.

Crystallinity is also reduced as the alkyl chains become longer, leading to soluble polymers. The higher poly(methyl-alkylsilylene)s, $(MeSiR)_n$ where R is n-propyl or larger, are all soluble, meltable solids. Within this family the glass transition temperatures decrease to a minimum at R = n-hexyl, which has $T_g \sim -75°C$. Polymers with R = n-butyl to n-dodecyl are all rubbery elastomers at room temperature. The symmetrical dialkylsilylene polymers, $(R_2Si)_n$ where R is n-propyl or larger, are also soluble solids. These polymers have been carefully studied in the solid state and show interesting phase transitions which will be described further in the paragraphs that follow.

Crystallinity is also lower for copolymers than for homopolymers that contain similar side groups. For example, the copolymer $(Me_2Si)_n(Ph_2Si)_m$ is soluble and meltable, although the corresponding homopolymers are not.[37] Among the polysilane copolymers, only the $(PhMeSi)_n(Me_2Si)_m$ "polysilastyrene" copolymer family has been studied in any detail. X-ray powder patterns for polymers with different copolymer compositions show that the crystallinity is at a minimum near a $PhMeSi:Me_2Si$ ratio of 1.

In the synthesis of polysilane copolymers, the different kinds of silicon atoms may be unevenly distributed between the oligomeric and high polymeric fractions. The composition of the polymer may therefore differ from the initial feed ratio of different dichlorosilanes. To determine the polymer composition, it is convenient to employ proton NMR spectroscopy. Figure 5.6 shows the ^1H NMR of a sample of "polysilastyrene."[38] There are only two resonances, a broad peak at δ 7.0 for all the aryl protons and an extremely broad peak near δ 1.0 for the methyl protons. By integrating the area under these resonances, the ratio of aryl protons to methyl protons can be determined, and hence the ratio of PhMeSi to Me_2Si groups in the polymer may be calculated.

A crucially important property of all polymers is of course their molecular weight. For polysilanes as well as other polymers, the molecular-weight distribution is most easily investigated by size exclusion chromatography, often called gel permeation chromatography (GPC). As mentioned earlier, polysilanes prepared by the usual route, sodium condensation of dichlorosilanes, are commonly found to

---

*To obtain soluble poly(diarylsilylene)s, it is necessary to attach long-chain alkyl groups to the aromatic rings, as in $[(p\text{-}n\text{-hexylphenyl})_2Si]_n$.[36]

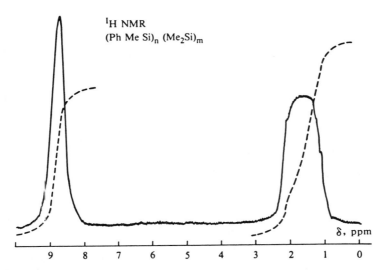

Figure 5.6 Proton NMR spectrum of "polysilastyrene." Dashed lines show integration of phenyl and methyl proton areas, allowing determination of the composition of the copolymer.

have bimodal, or even trimodal, molecular weight distributions.* For example, the gel permeation chromatogram for polysilastyrene shown in Figure 5.7 contains at least two rather well-defined regions of molecular weight. If necessary, the high- and low molecular weight polymers may be separated fairly well by fractional precipitation. In a typical procedure the polymer is dissolved in a good solvent like THF or toluene, and is precipitated by the slow addition of a poor solvent such as 2-propanol. The material with higher molecular weight is less soluble and precipitates first.

Although a molecular weight scale is shown in Figure 5.7, it is surely inaccurate for the polysilane polymer. The scale is based on calibration of the GPC instrument with samples of polystyrene of known molecular weight. A polysilane, and indeed any other polymer than polystyrene, will behave differently in the GPC experiment, so that the "molecular weights" obtained by GPC are only relative.

Accurate molecular weight determinations have been obtained for some polysilanes by light scattering, but because this technique is much more time consuming than GPC, only a few samples have yet been studied.[39] The results are shown in Table 5.2, compared with "molecular weights" from the GPC experiment. The molecular weights determined by light scattering are higher than the GPC values, by factors which depend on the particular structure of the polysilane.

*As explained in the section on synthesis, the molecular weight distribution becomes more nearly monomodal when the polysilane synthesis is carried out in the presence of cation complexing agents or is aided by ultrasound.

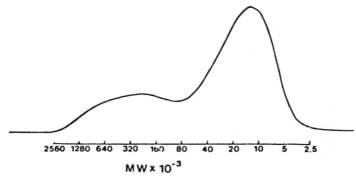

Figure 5.7   GPC curve for polysilastyrene, showing bi- or polymodal molecular weight distribution.

TABLE 5.2   Properties of Polysilanes from Light-Scattering Studies

Polymer	Solvent	M_w^{LS}	M_w^{GPC}	M_w/M_n	R_G, nm	C_∞
$(n\text{-PrSiMe})_n$	THF	210,000	180,000	3.1	31	19
$(cyclo\text{-HexSiMe})_n$	$c\text{-}C_6H_{12}$	2,540,000	500,000	2.7	76	14
$(n\text{-Hex}_2Si)_n$	hexane	6,100,000	1,900,000	2.3	108	21
$(n\text{-Hex}_2Si)_n$	THF	6,300,000	1,900,000	2.3	92	20
$(n\text{-Oct}_2Si)_n$	THF	320,000	2,600,000	2.4	100	30
$(PhSiMe)_n$	THF	46,000	19,000	4.2	21	64
$[(p\text{-}n\text{-BuC}_6H_4)_2Si]_n$	THF	450,000	500,000	—	60	70

M_w^{LS}: weight average M from light-scattering measurement.

M_w^{GPC}: weight average "M" from gel permeation chromatography, relative to polystyrene standards.

R_G: radius of gyration. C_∞: characteristic ratio.

The light-scattering behavior of those polysilanes studied indicates that they form random coils in solution, but are slightly extended and stiffened compared with typical polyolefins. One measure of chain flexibility is the characteristic ratio C_∞ which is also shown in the table. The values of C_∞ for most polysilanes of about 20 are larger than those for typical hydrocarbon polymers (~ 10), indicating that the polysilanes are somewhat less flexible than polyolefins. However, poly(diarylsilylene)s are much more rodlike and inflexible, with persistence lengths of more than 100 Å.[39]

5.6 ELECTRONIC PROPERTIES AND CONFORMATION

5.6.1 Electron Delocalization

The polysilanes (and related polygermanes) are different from all other high polymers, in that they exhibit *sigma-electron delocalization*. This phenomenon leads to special physical properties: strong electronic absorption, conductivity, photoconductivity, photosensitivity, and so on, which are crucial for many of the technological applications of polysilanes. Other polymers, such as polyacetylene and polythiophene, show electron delocalization, but in these materials the delocalization involves pi-electrons.

In most polymers, organic or inorganic, delocalization of sigma-electrons is not evident. Yet we know from modern chemical bonding theory that all electrons can be regarded as delocalized. Why is delocalization in the sigma-electron system not observed, say, for polyethylene?

polysilane, σ-delocalized   polyacetylene, π-delocalized

Part of the reason has to do with the ionization energies for the sigma-electrons. For many covalent bonds (C—C, C—H, C—O, Si—O, etc.), the ionization potentials of the sigma-electrons are high. Because the sigma-electrons are tightly held, it ordinarily makes little difference whether they are considered to be delocalized or not, and a model assuming localized sigma-electrons may be used. But for electrons in Si—Si sigma bonds, the ionization energies are much lower, indeed often less than those of pi electrons in olefins. For example, compare the first ionization potentials for $CH_3CH_2CH_2CH_3$, 10.6 eV; $H_2C=CH—CH=CH_2$, 9.1 eV; and $Me(SiMe_2)_4Me$, 8.0 eV.[41] Because the sigma-electrons in polysilanes are so loosely bound, a localized model is no longer appropriate.*

It is also important that the interaction between adjacent silicon orbitals is relatively large. It is convenient to think about this interaction as taking place between neighboring Si—Si bonds, leading to an energy splitting.[41] Corresponding to the filled Si—Si σ bonding orbitals there are Si—Si σ* antibonding orbitals, which are also split by σ-resonance. As the number of silicon atoms increases, the energy gap between the σ HOMO and σ* LUMO becomes smaller. Eventually, the set of interacting filled orbitals combine to form a valence band, while the unfilled orbitals combine to generate an empty conduction band (Figure 5.8).

When polysilanes absorb energy in the ultraviolet region, electrons are promoted from the σ valence band to the σ* conduction band.[42] Absorption maxima for polysilanes fall between 300 and 400 nm; typical ultraviolet spectra for some polysilanes in solution are shown in Figure 5.9. Because this σ → σ* transition is "permitted," the electronic absorptions are intense, with extinction coefficients ϵ between 5000 and 10,000 per Si—Si bond. Consistent with the splitting portrayed

*For an excellent qualitative discussion of the differences between Si—Si and C—C bonding, see J. Michl, *Acct. Chem. Res.* **1990**, 23 , 127.

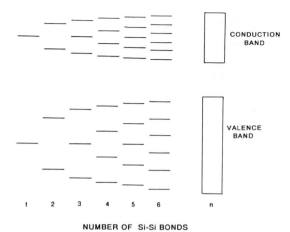

NUMBER OF Si-Si BONDS

Figure 5.8 Schematic diagram showing splitting of filled and unfilled energy levels for a polysilane as the length of the chain increases, leading to formation of a valence band and a conduction band.

in Figure 5.8, as the number of silicon atoms increases, the energy of the electronic transition decreases. The absorption wavelength therefore increases and eventually reaches a limit when the number of silicon atoms reaches ~ 30[43,44], as shown in Figure 5.10. However, the σ-σ^* separation depends on the conformation of the

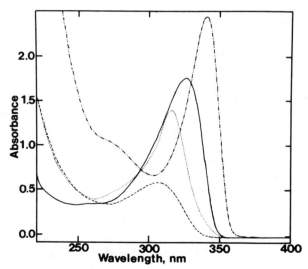

Figure 5.9 Ultraviolet absorption spectra of some polysilanes. $-\cdot-\cdot-$ (PhSiMe)$_n$, ——— (cyclo-HexSiMe)$_n$, $\cdot\cdot\cdot\cdot$ (n HexSiMe)$_n$, ----- (n DodecylSiMe)$_n$. Reprinted with permission from R. West, J. *Organometal. Chem.* **1986,** *300,* 327.

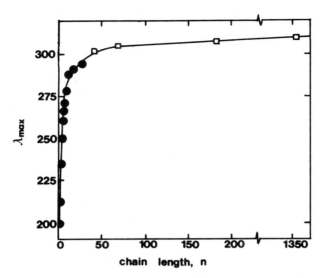

Figure 5.10 Ultraviolet absorption maxima as a function of chain length for alkyl-polysilanes. ●, permethyl compounds; ○, (*n* DodecylSiMe)$_n$. Reprinted by permission from P. Trefonas et al., *J. Polym. Sci., Polym. Lett. Ed.* **1983**, *21*, 823. Copyright © 1983 John Wiley & Sons, Inc.

polysilane chain, and therefore the absorption wavelength λ_{max} also depends strongly on the spatial arrangement of the chain. Theoretical[42,44,45] as well as experimental evidence indicates that λ_{max} increases as the number of *trans*-Si—Si—Si—Si conformations increases.[45,46]

For poly(di-*n*-alkylsilylene)s in solution at ordinary temperatures, the limiting value of λ_{max} is 305–315 nm. For polysilanes that contain branched alkyl groups, λ_{max} shifts to longer wavelength, for example, λ_{max} = 326 nm for (cyclohexyl-SiMe)$_n$, probably because more *trans* conformations are present. Many polysilanes show strong reversible *thermochromism*, with λ_{max} usually moving to longer wavelength as the temperature decreases. This effect probably results from an increase in the proportion of *trans* conformations at lower temperature. In some cases, depending on the nature of the alkyl groups and the solvent, a discontinuous wavelength shift occurs over a relatively narrow temperature range, as shown in Figure 5.11.[48] This surprising result shows that two conformational isomers must coexist in solution; the transformation has been explained as due to a coil-to-rod transformation or to crystallization of the side chains, either of which would lead to a larger proportion of *trans*-junctions in the low-temperature form.

Arylmethylpolysilanes (ArSiMe)$_n$ absorb between 325 and 350 nm, depending on the nature of the aryl substituent. This long-wavelength shift probably results mainly from interactions between the aryl π and polysilane σ electrons, as has been observed for aryldisilanes, but conformational effects may also play a part. Most diarylpolysilanes show λ_{max} near 400 nm, a shift which seems too large

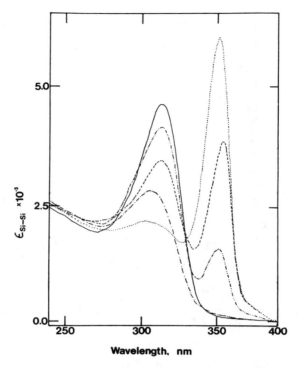

Figure 5.11 Ultraviolet absorption versus temperature for $(n\text{-Bu}_2 \text{Si})_n$, showing abrupt shift to longer wavelength as temperature is decreased. Reprinted by permission from P. Trefonas et al., *Organometallics*, **1985**, *4*, 1318. Copyright © 1985 John Wiley & Sons, Inc.

to be explained by σ-π mixing alone. These polymers are probably rodlike, containing long all-*trans* segments enforced by the diaryl substituents, as suggested from other evidence.[40]

A particularly interesting example is poly[bis(*p-n*-butoxy-phenyl)silane], which is atypical among diarylsilane polymers in having λ_{max} near 325 nm at room temperature. However, when a solution of this polymer is heated, a new band grows in at 409 nm, similar to the other poly(diarylsilylenes). It has been suggested that this polymer has a highly kinked backbone at low temperatures because of unfavorable dipole interactions in the *trans*-conformation. When the temperature is raised, the polymer unkinks to give a mainly *trans* arrangement, and the absorption band shifts accordingly.[49]

The thermochromic changes observed for so many polysilanes result from equilibration between different conformations. From molecular mechanics calculations on polymer segments, it appears that the preference for *gauche*, *trans*, or mixed conformations is likely to be small.[50] The conformations of solid polysilanes will be considered in the next section.

5.6.2 Solid Polysilanes—Electronic Properties and Conformation

The best studied polysilane is undoubtedly poly(di-*n*-hexylsilylene) (PDHS).[48,50] Interest in this polymer arose from the discovery that it undergoes a striking change in ultraviolet absorption, at 42°C. Above this temperature PDHS absorbs at 317 nm, similar to the same polymer in solution at room temperature. Below 42°C a new phase grows in which shows λ_{max} = 372 nm (Figure 5.12).

Extensive studies of PDHS by x-ray diffraction, electron diffraction, and vibrational and NMR spectroscopy have shown that the low-temperature form is highly crystalline with an all-*trans*, zigzag arrangement of the polymer chain. Below the transition temperature, crystallization of the side chains takes place, which enforces the *trans*-conformation for the main polymer chain. Other polysilanes which behave similarly are poly(di-*n*-heptylsilylene) and poly(di-*n*-octylsilylene); both show thermochromism and an all-*trans* structure below their transition temperatures.

X-ray studies indicate that solid $(Et_2Si)_n$ and $(n\text{-}Pr_2Si)_n$ also have all-*trans* conformations.[51] However, as shown in Table 5.3, the intermediate compounds, $(n\text{-}Bu_2Si)_n$ and $(n\text{-}pentyl_2Si)_n$, have quite a different structure.[46,52] These polymers, which absorb near 315 nm and are not thermochromic, adopt a coiled arrangement with a twist of ~30° at each silicon atom. A representation of the resulting 7:3 helix is shown in Figure 5.13. A remarkable property of $(n\text{-}Bu_2Si)_n$ is that it is

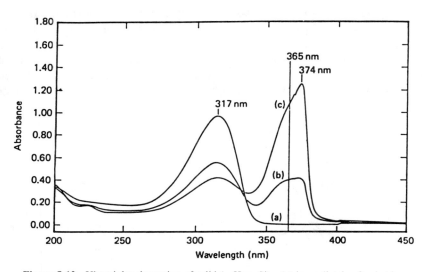

Figure 5.12 Ultraviolet absorption of solid $(n\text{-}Hex_2\ Si)_n$. (a) immediately after baking at 100°C; (b) and (c) after standing at 25°C, showing growth of the 374 nm absorption. Reprinted with permission from R. D. Miller et al., *J. Am. Chem. Soc.* **1985**, *107*, 2172. Copyright 1985 American Chemical Society.

TABLE 5.3   Structures of Some Polysilane Polymers at 25°C.

Polymer	Structure	Torsion $<$, °	Polymer	Structure	Torsion $<$, °
$(Me_2Si)_n$	TT	180	$(nBuSinHex)_n$	mesophase	—
$(nPr_2Si)_n$	TT	180	$(nHex_2Si)_n$	TT	180
$(nBu_2Si)_n$	7/3 helix	154	$(nOct_2Si)_n$	TT	180
$(nPen_2Si)_n$	7/3 helix	154	$[nC_{14}H_{29})_2Si]_n$	TGTG'	180, ~90

piezochromic; application of moderate pressure converts this polymer irreversibly to the all-*trans* form.[53]

Polysilanes with longer-chain alkyl substituents also show complex conformational behavior. The C_{14} polymer, $[(n-C_{14}H_{29})_2Si]_n$, apparently adopts yet another conformation in which *trans* and *gauche* linkages alternate.[50]

When more than one type of alkyl group is present, the polysilanes show much reduced crystallinity. Polymers such as $(n-BuSiMe)_n$ and $(n-HexSiMe)_n$ are rubbery and amorphous at room temperature and have glass transition temperatures far below 0°C. If the two alkyl groups are only slightly different, as in $(n-BuSi-n-Hex)_n$, intermediate behavior is found; crystallinity is lost at room temperature, but partial ordering is preserved.[54] This leads directly to our next topic, liquid crystallinity in polysilanes.

5.6.3 Mesophases and Liquid Crystalline Polysilanes

For $(n-Hex_2Si)_n$, the stable phase below 42°C is crystalline with the all-*trans* configuration. What happens when this classic polymer is warmed above the transition temperature? Recent studies show that the high-temperature form of $(n-Hex_2Si)_n$ is a liquid crystal-like *mesophase*, in which crystallinity is lost but some ordering remains.[55] The polymer chains are arranged like parallel columns in a hexagonal lattice (Figure 5.14).[55] The arrangement is like that in nematic

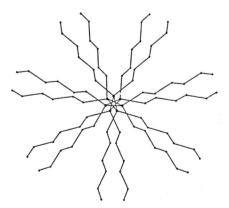

Figure 5.13 View of 7/3 helical conformation for $(n-Hex_2 Si)_n$. Larger dots represent Si atoms, smaller dots represent C atoms. Reprinted with permission from R. D. Miller et al., Chapter 4 in *Inorganic and Organometallic Polymers*, ACS Symposium Series 360, copyright 1988 American Chemical Society.

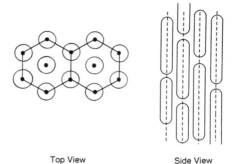

Figure 5.14 Proposed arrangement of polysilane molecules in the hexagonal liquid-crystal mesophase.

Top View   Side View

liquid crystals. Similar nematic-type mesophases have been found for other organic polymers, including certain siloxanes and polyphosphazenes.

From x-ray diffraction studies, the interchain distance in the high-temperature phase of $(n\text{-Hex}_2\text{Si})_n$ is 13.5 Å. From the lattice constant and reasonable assumptions about bond lengths and angles, it is possible to calculate the projection of the Si—Si distance along the axis of the column. The arrangement which best fits the data is that the molecules are tightly coiled into an all-*gauche* helix! If the model is correct, the conformational change that occurs at the transition temperature for $(n\text{-Hex}_2\text{Si})_n$ is profound. The *n*-hexyl side chains become disordered, and the fully extended polymer molecules wind into a tight spiral.

Other polysilanes also show liquid crystal-like mesophases, and in fact this behavior is common for polysilanes with different alkyl substituents. An example is $(n\text{-BuSin-Hex})_n$, which undergoes a transition to a partially crystalline form at $-30°C$. From this temperature up to $+200°C$, it is birefringent and shows sharp x-ray reflections assignable to a hexagonal lattice (Figure 5.15), consistent with a columnar liquid crystalline phase.[54] Similar behavior is observed for copolymers such as $(n\text{-Pr}_2\text{Si})_n(n\text{-Hex}_2\text{Si})_m$, which exists in a similar liquid-crystalline mesophase from below 0° to above 200°C.

5.6.4 Electrical Conductivity and Photoconductivity

The conductivity of polysilanes provides striking evidence for delocalization of σ-electrons. Pure polysilane polymers are insulators with conductivities $< 10^{-12}$ ohm^{-1} cm^{-1}, but treatment with oxidizing agents such as AsF_5, SbF_5, or H_2SO_4 renders them electrically conducting.[9] Only preliminary experiments have been reported to date, but with cross-linked polysilastyrene and AsF_5 conductivity as high as 0.5-V ohm^1 cm^{-1} has been seen. This value is toward the upper end of the range for semiconductors.

The conductivity of polysilanes is believed to arise from transfer of an electron from the silicon chain to the oxidizing agent, leaving a "hole" (actually a polysilane cation radical) which may be delocalized along the polysilane chain.

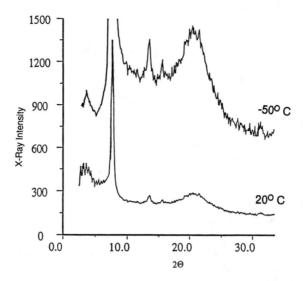

Figure 5.15 X-ray diffractogram for films of (n BuSin-Hex)$_n$ at $-50°$ and 20°C. The strong peak near $2\theta = 7.5°$ corresponds to the interchain spacing of ~ 12 Å; the weaker peaks at higher 2θ values indicate that the polymer has a hexagonal lattice.

Electrons must hop from one polymer chain to another to account for the bulk conductivity.

Polysilane polymers also show photoconductivity when irradiated with ultraviolet light,[56] and are excellent charge transport materials in electrophotography.[57] In a typical charge transport experiment, electrical charges are generated at the surface of a polysilane film from a photoactive layer such as selenium (Figure 5.16) and moved through the polymer film by an electrical field. Polymers studied included (PhSiMe)$_n$, (PhSiEt)$_n$, polysilastyrene, (n-OctSiMe)$_n$, and (cyclo-Hex-

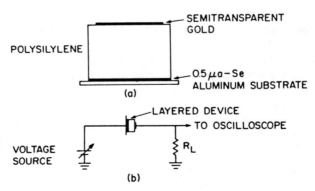

Figure 5.16 Schematic drawing of experiment used to determine hole conductivity of polysilanes. Reprinted by permission from M. Stolka et al., *J. Polym. Sci., Polym. Chem. Ed.*, **1987**, *25*, 823. Copyright © 1988 John Wiley & Sons, Inc.

TABLE 5.4   Charge Mobility and Activation Energy for Charge Transport for Electrophotographic Polymers

Polymer	Hole Mobility cm^2 V^{-1} s^{-1}	Activation Energy, eV
Polyvinyl carbazole	10^{-8}	0.6
Amine-doped polycarbonate	10^{-4}	0.3–0.7
(PhSiMe)$_n$	$>10^{-4}$	0.28
Amine-doped (PhSiMe)$_n$	10^{-3}	—

SiMe)$_n$. All these polysilanes transport charge, and their efficiency is almost independent of the nature of the substituent groups. In each case the conductivity is entirely by hole transport; that is, positive but not negative charges appear to migrate through the polysilane layer under the influence of an electric field. The polysilanes show charge mobilities as high as 5×10^{-3} units, and activation energies for charge transport even lower than those of materials now used in electrophotography (Table 5.4).

As charge transport materials, the polysilanes are unique in that the active sites are on the polymer backbone itself. In other electrophotographic materials, charged sites are on pendent groups, as in poly(vinyl carbazole), or on molecules mixed with the polymer, as in polycarbonate film doped with aromatic amines. "Doping" of the polysilane with aromatic amines increases the hole mobility by almost another order of magnitude, as shown from the table.[58]

In addition to photoconductivity, polysilanes have been found to exhibit marked nonlinear optical properties,[51,59,60] suggesting that they may eventually be useful in laser and other optical technology. The third-order nonlinear susceptibility, $\chi_{(3)}$, is a measure of the strength of this effect. The nonlinear properties of polysilanes, like the absorption spectra, seem to be dependent on chain conformation and are enhanced for polymers having the all-*trans* arrangement (Table 5.5). The value of 11×10^{-12} esu observed for (n-Hex$_2$Si)$_n$ below its transition temperature is the largest ever observed for a polymer which is transparent in the visible region.

TABLE 5.5   Third-Order Nonlinear Susceptibilities for Polysilanes and Some Related Polymers[50]

Polymer	T°C	χ^3, esu $\times 10^{12}$
(PhSiMe)$_n$	23	1.5
	25	7.2, 4.2
(n-Hex$_2$Si)$_n$	23	11.0
	50	2.0
(n-Hex$_2$Ge)$_n$	−11	6.5
	23	3.3
(p-C$_6$H$_4$CH=CH$_2$)$_n$	25	7.8
Polydiacetylenes, various		1–400

5.7 PHOTODEGRADATION OF POLYSILANES

Upon irradiation with ultraviolet light, most polysilanes undergo chain scission into smaller fragments. When polysilanes $(RR'Si)_n$ are photolyzed with UV light ($\lambda = 254$ nm) in the presence of triethylsilane, large amounts of $Et_3SiSiRR'H$ are produced. This product arises from the insertion of silylene into the $Si-H$ bond of triethylsilane, as shown in eq. 5.22.

$$Et_3Si\text{-}H + RR'Si: \longrightarrow Et_3Si\text{-}SiRR'\text{-}H \qquad \text{(Eq. 5.22)}$$

Upon exhaustive photolysis the other major product is the disilane, $HSiRR'SiRR'H$. From these and other experiments it appears that the major reactions in photolysis of polysilanes are homolysis to give silyl radicals (eq. 5.23) and silylene elimination (eq. 5.24):[61,62] Some chain scission with transfer of a substituent also takes place (eq. 5.25).

$$\overset{\displaystyle R}{\underset{\displaystyle R'}{\wedge\!\wedge\!\wedge Si}}-\overset{\displaystyle R}{\underset{\displaystyle R'}{Si \wedge\!\wedge\!\wedge}} \xrightarrow{h\upsilon} 2 \wedge\!\wedge\!\wedge \overset{\displaystyle R}{\underset{\displaystyle R'}{Si}}\bullet \qquad \text{(Eq. 5.23)}$$

$$\overset{\displaystyle R}{\underset{\displaystyle R'}{\wedge\!\wedge\!\wedge Si}}-\overset{\displaystyle R}{\underset{\displaystyle R'}{Si}}-\overset{\displaystyle R}{\underset{\displaystyle R'}{Si \wedge\!\wedge\!\wedge}} \xrightarrow{h\upsilon} \overset{\displaystyle R}{\underset{\displaystyle R'}{\wedge\!\wedge\!\wedge Si}}-\overset{\displaystyle R}{\underset{\displaystyle R'}{Si \wedge\!\wedge\!\wedge}} + RR'Si: \qquad \text{(Eq. 5.24)}$$

$$\overset{\displaystyle R}{\underset{\displaystyle R'}{\wedge\!\wedge\!\wedge Si}}-\overset{\displaystyle R}{\underset{\displaystyle R'}{Si}}-\overset{\displaystyle R}{\underset{\displaystyle R'}{Si \wedge\!\wedge\!\wedge}} \xrightarrow{h\upsilon} \overset{\displaystyle R}{\underset{\displaystyle R'}{\wedge\!\wedge\!\wedge Si}}-R + :\overset{\displaystyle R}{\underset{\displaystyle R'}{Si}}-\overset{\displaystyle }{\underset{\displaystyle R'}{Si \wedge\!\wedge\!\wedge}} \qquad \text{(Eq. 5.25)}$$

Chain shortening results from all of these reactions, but especially from eqs. 5.23 and 5.25. When 254 nm uv light is used, photolysis ceases at the disilane stage because uv radiation is no longer absorbed. With uv photons of lower energy (>300 nm), elimination of silylene (reaction 5.24) does not take place, and the major initial step is homolytic scission.[62] Secondary reactions also occur, leading in the case of $(n\text{-Hex}_2Si)_n$ to the formation of $Si-H$ bonds and $Si=C$ double bonds. There is also evidence for the formation of persistent silyl radicals of the type $-SiRR'-\dot{S}iR-SiRR'-$.

Photolysis ceases at the disilene stage where the UV light is no longer observed. With UV photons of lower energy ($\lambda > 300$ nm), elimination of silylene (reaction 5.24) does not take place, and the major initial step is homolytic scission.[62]

When aryl substituent groups are present, some cross-linking takes place as well as chain scission.[46] Groups containing $C=C$ double bonds are even more effective at bringing about cross-linking. This leads directly to the topic of cross-linking in polysilanes.

5.8 CROSS-LINKING

For some uses it is important to form bonds that link different polysilane chains, to transform soluble, meltable polysilanes into insoluble resins. This process is vital if the polysilanes are to be used as precursors to silicon carbide ceramics, since, if cross-linking is not carried out, most of the polymer is volatilized before thermolysis to silicon carbide can take place. Several methods have therefore been developed to bring about cross-linking of polysilanes.[34,63]

5.8.1 Photolysis of Alkenyl-Substituted Polysilanes

As mentioned in the preceding section, when vinyl or other alkenyl groups are present as substituents, photolysis leads to cross-linking, which dominates over chain scission. Probably the cross-linking results when silyl radicals formed in the initial photolysis add to vinyl carbon atoms on neighboring chains, but other mechanisms are also possible.

5.8.2 Free-Radical Cross-linking with Polyunsaturated Additives

The same reactions can occur when the $C=C$ double bonds are present on a molecule which is simply mixed with the polysilane and can then serve as a cross-linking agent. This provides a very general method for cross-linking of any polysilane. The radicals necessary to cause cross-linking can be formed by photolysis of the polysilane or by heating the mixture with a free radical initiator such as azobis(isobutyronitrile), AIBN. To be useful as cross-linking agents, additive molecules must include two or more unsaturated groups. Compounds which have been used include tetravinylsilane; trivinylphenylsilane; the hexavinyldisilyl compounds $(CH_2=CH)_3Si(CH_2)_nSi(CH=CH)_3$, $n = 2$, 4, or 6; 1,5-hexadiene; and triallylbenzene-1,3,5-tricarboxylate.

In photochemical cross-linking, the silyl radicals formed in the initial cleavage reaction probably add to the $C=C$ double bonds. This addition produces cross-links and also generates new carbon radicals which may cause further cross-linking reactions (eq. 5.26).

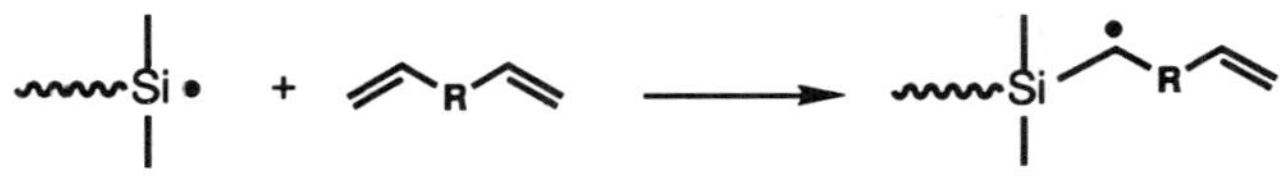

$$\text{(Eq. 5.26)}$$

Using thermal radical initiators, the first steps may be either addition of initiator radicals to the double bonds of the polyunsaturated compound, or abstraction of hydrogen from organic groups on the polysilane. For instance, loss of hydrogens from methyl groups on the polysilane would lead to formation of cross-links (eq. 5.27).

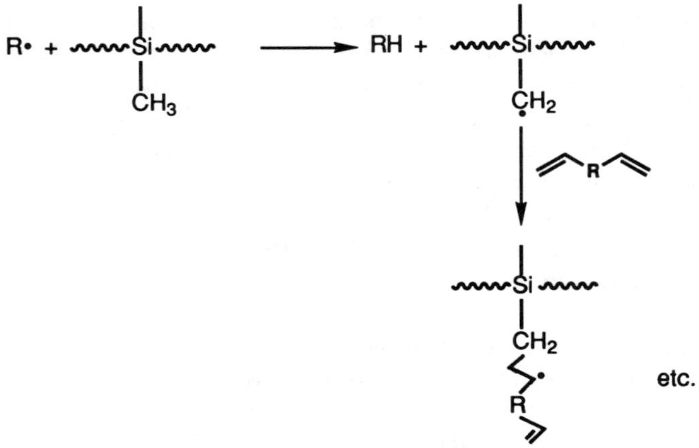
R• +
Si
CH3
RH +
Si
CH2
R
Si
CH2
R
etc.

(Eq. 5.27)

5.8.3 Chemical Cross-linking of Hydrosilylene Polymers, "Room Temperature Vulcanizing"

It is possible to synthesize polysilane polymers that contain hydrogen, as in phenylhydrosilylene-based polymers. The polymerization and workup must however be carried out quite carefully so that the Si—H groups are not lost through hydrolysis. The polymers were synthesized using an excess of sodium, so that the chains are mainly anionically terminated (eq. 5.28). $Ph_2SiMeCl$ was added as a chain blocking agent. The polymers, now terminated with Ph_2MeSi groups, were isolated under neutral conditions to avoid solvolysis of any Si—H linkages. The resulting polymers have rather low molecular weights, with about 10–20 silicon atoms in the chain.[63,64] In addition to the homopolymer shown in eq. (5.28), copolymers of PhSiH with other silylene groups (Structure 5.12) have been synthe-

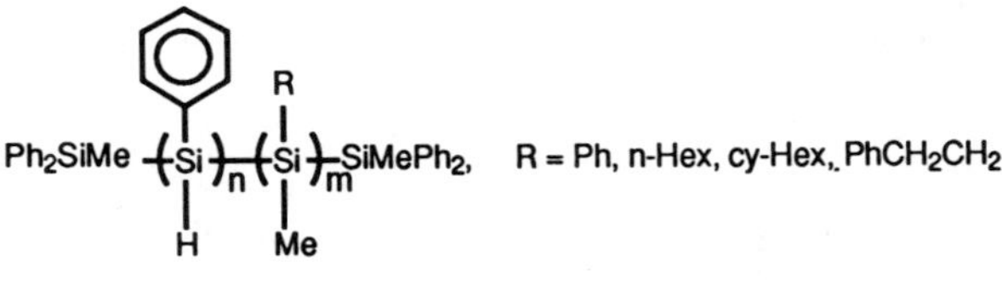
R
Ph2SiMe
Si
n
Si
m
SiMePh2,
R = Ph, n-Hex, cy-Hex,. PhCH2CH2
H
Me

5.12

sized. These polymers are low-melting solids or in some cases (R = n-hexyl) thick fluids at room temperature.

$$PhSiHCl_2 \xrightarrow{\text{Na}} {}^{-}SiPhH \text{—}(SiPhH)_{\overline{n}} PhHSi^{-}$$

$$\xrightarrow{Ph_2SiMeCl} Ph_2SiMe \text{—}(SiPhH)_{\overline{n}} SiMePh_2$$

(Eq. 5.28)

The well-known hydrosilylation reaction is effective for bringing about cross-linking of these polymers. Trivinylphenylsilane was used as the cross-linking agent, in the presence of a trace of chloroplatinic acid as a catalyst (eq. 5.29), but in practice any compound containing several vinyl groups could be used. The mixture of reactants forms a viscous fluid or paste which sets to a firm rubbery solid as cross-linking takes place. By this process "room temperature vulcanization" of polysilanes can be brought about, analogous to the RTV polymerization of silicone elastomers.[64]

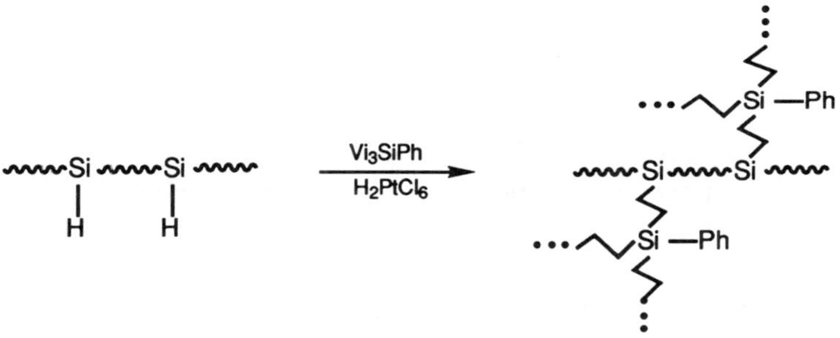

$$\text{~~~Si~~~Si~~~} \xrightarrow[H_2PtCl_6]{Vi_3SiPh} \text{~~~Si~~~Si~~~}$$

(Eq. 5.29)

5.9 STRUCTURES OF POLYSILANES

5.9.1 Configurations and Stereochemistry

Polysilane homopolymers in which all the side groups are identical, such as $(n\text{-}Bu_2Si)_n$, exhibit no stereoisomeric effects since a plane of symmetry can be drawn through each silicon atom. In polymers bearing two different substituents on each silicon, for example, $(n\text{-}BuSiMe)_n$, each silicon is a stereogenic center, and the relative configurations of other silicon atoms becomes significant. Consider a polysilane with silicon atoms bearing substituents A and B. The relationship between neighboring atoms is meso (or M) if they have the same configuration, racemic (or R) if they have opposite configurations. In general, polymers which contain stereogenic centers are called isotactic if all the relative configurations are

M, syndiotactic if all relative configurations are R, and atactic if the configurations are random.[65]

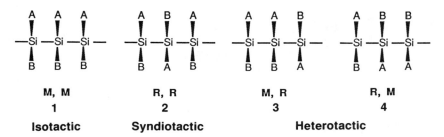

For a triad of silicon atoms, there are four possibilities: M,M (isotactic), R,R (syndiotactic), and M,R and R,M (heterotactic); the last two are equivalent since the polymer chain may be "read" in either direction. If there is no stereochemical preference, as in an atactic polymer, the amounts of isotactic, syndiotactic, and heterotactic triads would be 1:1:2, the heterotactic arrangement being twice as probable. If the relative configuration of two nearest neighbors on each side is considered, there are 10 independent arrangements:

RRRR	RMMR
RRRM = MRRR	MRRM
RRMR = RMRR	MMMR = RMMM
RRMM = MMRR	MMRM = MRMM
RMRM = MRMR	MMMM

Various methods may be used to examine configurations of polysilanes, but ^{29}Si NMR spectroscopy has been the most useful. Silicon-29, like carbon-13, has spin 1/2 and a relatively low abundance, 4.7%. NMR spectroscopy using ^{29}Si has been very important for the characterization of siloxane polymers, and is proving to be equally useful for polysilanes.

The ^{29}Si NMR spectra of solutions of several dialkylpolysilanes is shown in Figure 5.17. For $(n\text{-}Hex_2Si)_n$, where there is no possibility for stereoisomerism, the resonance is a single sharp line showing that all the silicon atoms have identical environments. But for the $(n\text{-}alkyl\text{-}SiMe)_n$ polymers, which contain stereogenic silicon atoms, multiple resonances appear. The symmetrical, five- to seven-line patterns can be interpreted as arising from completely atactic polymers, having a statistical distribution of relative configurations at the two nearest-neighbor silicon atoms on each side.[66,67]

For arylmethyl polymers the results are quite different. For the best studied example, $(PhSiMe)_n$, the ^{29}Si NMR pattern consists of three broad lines, each showing some evidence of further splitting (Figure 5.18). With aryl substituents, the relative configuration of the nearest-neighbor silicon atoms may become more important in determining the chemical shift, so the three lines probably correspond

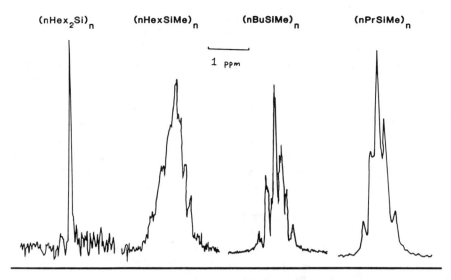

Figure 5.17 ²⁹Si NMR spectra for some alkylpolysilanes. Reprinted by permission from A. Wolff et al., *J. Polym. Sci., Polym. Chem. Ed.*, **1988**, *26*, 701. Copyright © 1988 John Wiley & Sons, Inc.

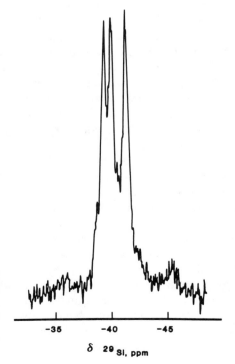

Figure 5.18 ²⁹Si NMR spectrum of (PhSiMe)_n. The three principal lines have been associated with the triad arrangements, rr, mm, and rm. Reprinted by permission from A. Wolff et al., *J. Polym. Sci., Polym. Chem. Ed.*, **1988**, *26*, 713. Copyright © 1988 John Wiley & Sons, Inc.

to RR, MM, and RM relative configurations. However, the lines are not in the expected $1:1:2$ ratio for an atactic polymer; instead the intensity ratio is about $3:3:4$, suggesting that there is partial tacticity in $(PhSiMe)_n$.[68] The patterns for other arylalkyl polysilanes differ, but in general two or three broad resonances are found. None of the aryl polymers studied so far has given a symmetrical pattern like those for the alkylpolysilanes.

Fully stereoregular polysilanes have not been reported. Evidently the sodium condensation reaction used in synthesis leads either to atactic polymers or to ones which do not show complete tacticity. However, other methods of polymerization, such as polymerization of disilenes, dehydrogenative coupling, or ring-opening polymerization, may produce stereoregular polysilanes, and this is one important reason to search for new methods of synthesis.

5.9.2 Ordering in Copolymers

Polysilane copolymers may have several kinds of arrangements. A copolymer that contains two kinds of silicon atoms might be (1) randomly arranged, or (2) blocklike, with long runs of silicon atoms of just one kind, or (3) ordered, with the two kinds of silicon atoms alternating along the chain.

Ordering in polysilanes is conveniently studied by ^{29}Si NMR spectroscopy. The usual synthesis by sodium condensation produces either random or blocklike copolymers, depending on the relative reactivities of the chlorosilanes. Random copolymers are obtained for similar chlorosilanes, for example $n\text{-}BuSiMeCl_2$ and $n\text{-}HexSiMeCl_2$, but if the chlorosilanes have rather different rates of reaction, one will be consumed preferentially to generate blocklike copolymers. As an example, the ^{29}Si NMR spectrum for $(PhSiMe)_n(n\text{-}HexSiMe)_m$ is shown in Figure 5.19.

The spectrum shows a peak near that for pure $(n\text{-}HexSiMe)_n$, as well as the three-line pattern characteristic of $(PhSiMe)_n$, with only weak resonances falling between these two values. The NMR indicates a highly blocklike structure, with long runs of similar silicon atoms; this arrangement results because $PhMeSiCl_2$ is very much more reactive than $n\text{-}HexSiMeCl_2$.

The ^{29}Si NMR spectrum of polysilastyrene, $(PhSiMe)_n(Me_2Si)_m$, indicates that this polymer also has a rather blocklike structure. A much less blocklike arrangement in a polymer with the same composition, $(PhSiMe-SiMe_2)_n$, has been obtained by sodium condensation of the dichlorodisilane $ClSiMe_2-PhMeSiCl$.[68] Note that $(PhSiMe-SiMe_2)_n$ is neither a random nor a fully ordered copolymer (it is partially ordered, since no runs of more than two PhSiMe or Me_2Si units occur). The ^{29}Si NMR spectrum, and other physical properties, of the two polymers are quite different.

The first fully ordered polysilane copolymers have recently been synthesized, by two different routes. Careful condensation of a 1,3-dibromotrisilane produced the AABAAB-type copolymer shown in eq. (5.30),[69] and an ABAB-type copolymer was obtained by anionic polymerization (eq. 5.31).[70] Both polymers show only two narrow lines in their ^{29}Si NMR spectra, which indicates that they

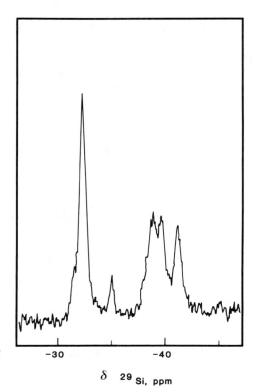

Figure 5.19 ^{29}Si NMR spectrum of (PhSiMe)$_n$ (n Hex$_2$ Si)$_n$, a blocklike co-polymer.

-30 -40

δ ^{29}Si, ppm

have an ordered structure. These and similar polysilanes may show unusual electronic behavior related to their ordered structure, and are hence of theoretical and perhaps practical interest.

$$\text{BrSiMe}_2\text{Si(n-Hex)}_2\text{SiMe}_2\text{Br} \xrightarrow{\text{Na}^0} \begin{array}{c} \text{Me nHex Me} \\ | \quad | \quad | \\ \text{+Si—Si—Si+}_n \\ | \quad | \quad | \\ \text{Me nHex Me} \end{array}$$

(Eq. 5.30)

(Eq. 5.31)

Another approach to ordering in polysilanes involves cocondensation of a dilithiopolysilane oligomer with an α,ω-dichloropolysilane, to give a block copolymer, as shown in Scheme 5.7.[71] In principle this method might yield fully ordered polymers, but there is evidence that $Li(Ph_2Si)_5Li$ undergoes redistribution to give $(Ph_2Si)_4$ and shorter $Si-Si$ chains, and the actual degree of ordering of copolymers made by this route has not been investigated.

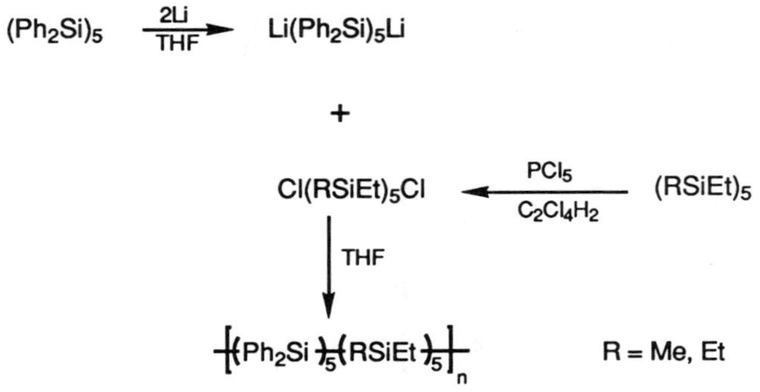

$$(Ph_2Si)_5 \xrightarrow[\text{THF}]{2Li} Li(Ph_2Si)_5Li$$

$$+$$

$$Cl(RSiEt)_5Cl \xleftarrow[C_2Cl_4H_2]{PCl_5} (RSiEt)_5$$

$$\downarrow \text{THF}$$

$$+\!\!\left[(Ph_2Si)_5(RSiEt)_5\right]_n\!\!+ \qquad R = Me, Et$$

Scheme 5.7

5.9.3 Network Polysilanes

Most of the research on polysilanes has been concerned with linear polymers with two organic groups on each silicon, $(RR'Si)_n$. Recently, however, polymers with only a single organic substituent per silicon, $(RSi)_n$, have been investigated in several laboratories.[72,74] These "polysilyne" polymers were first prepared at AT&T Bell Laboratories by reduction of alkylchlorosilanes in pentane with an emulsion of sodium-potassium alloy, produced by ultrasound irradiation. For alkyl groups larger than ethyl, the resulting polymers are soluble in organic solvents such as THF and can be cast into transparent yellow films.

Polymers which have been described include the alkylpolysilynes with R = n-propyl, n-butyl, isobutyl, n-hexyl, cyclohexyl, β-phenethyl, as well as the arylsilyne polymers with R = phenyl and p-tolyl. Molecular weights are typically 10,000 to 20,000. All the polysilynes are visibly colored, yellow or orange, and their electronic spectra differ greatly from those of linear polysilanes (Figure 5.20).

Polysilynes might adopt an unsaturated structure $(RSi=SiR-SiR=Si-SiR)$, with alternating single and double bonds, analogous to the structure of polyacetylenes. Such polymers would be very interesting, but are disfavored because of the relative weakness of the $Si=Si$ double bonds. All the evidence indicates that the polysilynes have instead a network structure with cross-links as shown in Figure 5.21. In the three-dimensional network each silicon is linked to three other silicon atoms by sigma bonds. The polysilynes are therefore intermediate

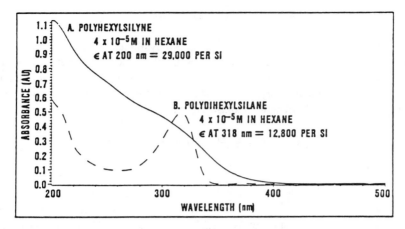

Figure 5.20 Ultraviolet spectra of network polymers $(n\text{-HexSi})_n$, compared with $(n\text{-Hex}_2\text{ Si})_n$. Reprinted with permission from Bianconi et al., *Macromolecules* **1989**, 22, 1697. Copyright 1989 American Chemical Society.

between the linear polysilanes, where each silicon is bonded to two others, and elemental silicon in which each silicon forms four sigma bonds to other Si atoms. It follows that the polysilynes should exhibit enhanced electron delocalization, as shown for instance in their absorption spectra. Relatively little is yet known about the polysilynes, but it is likely that they will be increasingly important in the future.

Figure 5.21 The network structure proposed for $(n\text{-HexSi})_n$. Large spheres represent Si atoms. Reprinted with permission from Bianconi et al., *Macromolecules* **1989**, 22, 1697. Copyright 1989 American Chemical Society.

5.9.4 Copolymers with Other Monomers

A few examples of copolymers between polysilanes and other polymer units have been synthesized, although their properties have scarcely been investigated. Silylene-olefin block copolymers can be made by adding monomers such as styrene to anionically terminated polysilanes. Anionic polymerization of the olefin takes place leading to copolymers such as Structure 5.13.

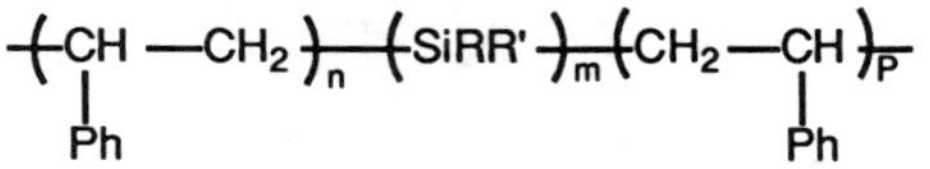

$$\left(\begin{matrix} CH - CH_2 \\ | \\ Ph \end{matrix}\right)_n \left(SiRR'\right)_m \left(\begin{matrix} CH_2 - CH \\ | \\ Ph \end{matrix}\right)_p$$

5.13

Initiation of alkene polymerization by polysilanes as photocatalysts also produces polyolefins containing polysilane blocks at one end. Copolymers of polysilanes with polysiloxanes have been obtained by oxidation of polysilanes with the peroxide, m-chloroperbenzoic acid, which inserts oxygen atoms into Si—Si bonds. The oxidation sites tend to cluster, leading to blocklike, rather than random, silanesiloxane copolymers (eq. 5.32).[75]

$$\left(SiRR'\right)_n \xrightarrow[MCPBA]{[O]} \left(SiRR'\right)_m O \left(SiRR'O\right)_p \left(SiRR'\right)_q \left(SiRR'O\right)_r \cdots$$

(Eq. 5.32)

Alternating copolymers of disilanyl groups with aromatic rings, and with C≡C triple bonds, have been prepared. The aryl copolymers are synthesized by sodium condensation of bis(chlorodialkylsilyl)benzene (eq. 5.33). Because these polymers are light sensitive, they are of interest as photoresist materials for microlithography.[76]

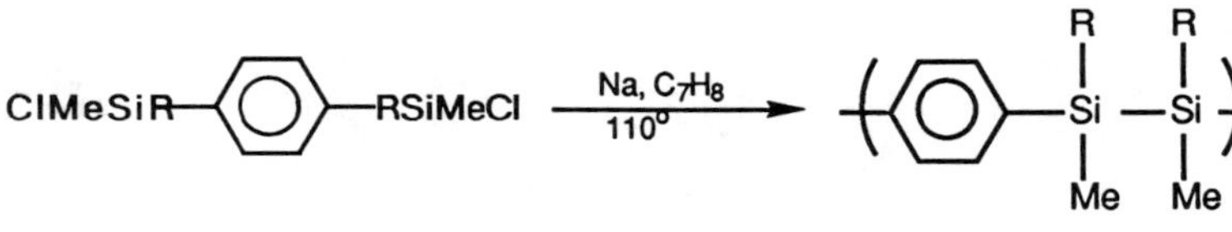

$$ClMeSiR-\left\langle\bigcirc\right\rangle-RSiMeCl \xrightarrow[110^\circ]{Na, C_7H_8} \left(\left\langle\bigcirc\right\rangle - \begin{matrix} R \\ | \\ Si \\ | \\ Me \end{matrix} - \begin{matrix} R \\ | \\ Si \\ | \\ Me \end{matrix}\right)_n$$

R = Et, Ph

(Eq. 5.33)

Alternating copolymers of disilanyl with ethynyl units have been obtained both by condensation and by ring-opening polymerization.[77] As shown in Scheme 5.8, the lithium and magnesium derivatives of the intermediate diethynyldisilane (Structure 5.14) react quite differently with dichlorodisilanes. The lithium compounds yield polymer (Structure 5.16) while the magnesium compounds give mainly the strained eight-membered rings (Structure 5.15). Bases, such as acetylide anions, open the rings and provide another route to the polymers.

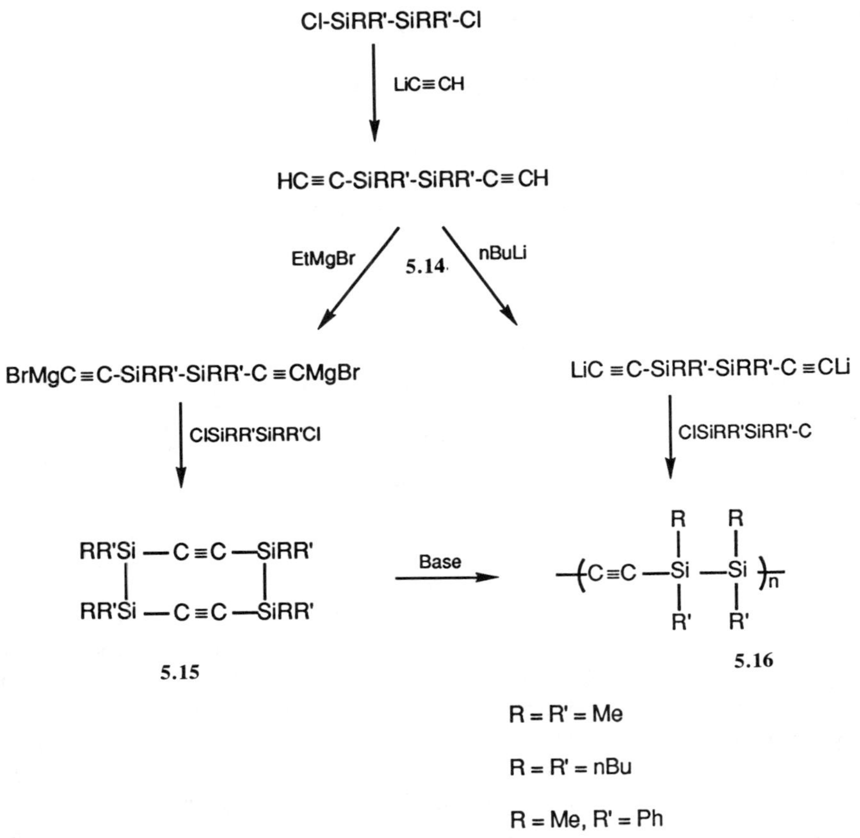

Cl-SiRR'-SiRR'-Cl

$\downarrow$ LiC≡CH

HC≡C-SiRR'-SiRR'-C≡CH

EtMgBr ⟋   **5.14**   ⟍ nBuLi

BrMgC≡C-SiRR'-SiRR'-C≡CMgBr   LiC≡C-SiRR'-SiRR'-C≡CLi

$\downarrow$ ClSiRR'SiRR'Cl $\downarrow$ ClSiRR'SiRR'-C

RR'Si —C≡C —SiRR'   ⟶   R   R
    | | **Base** | |
RR'Si —C≡C —SiRR'   —(C≡C —Si —Si)ₙ
                                             | |
                                            R' R'

5.15   **5.16**

R = R' = Me

R = R' = nBu

R = Me, R' = Ph

Scheme 5.8

Although these polymers have not yet been carefully studied, they seem to show electron delocalization by σ-π interaction, involving both the π electron of the ethynyl groups and the σ electrons of the Si—Si bonds. They may then provide a third class of electron-delocalized polymers, besides the π-delocalized conduction polymers such as polyacetylene and the σ-delocalized polysilanes.

Finally, it should be noted that random copolymers between polysilanes and polygermanes have been synthesized by cocondensation of dichlorosilanes and dichlorogermanes (eq. 5.34).[78] Pure polygermane polymers, $(RR'Ge)_n$, have similarly been prepared.[78,79] They exhibit electronic properties similar to those of the polysilanes.

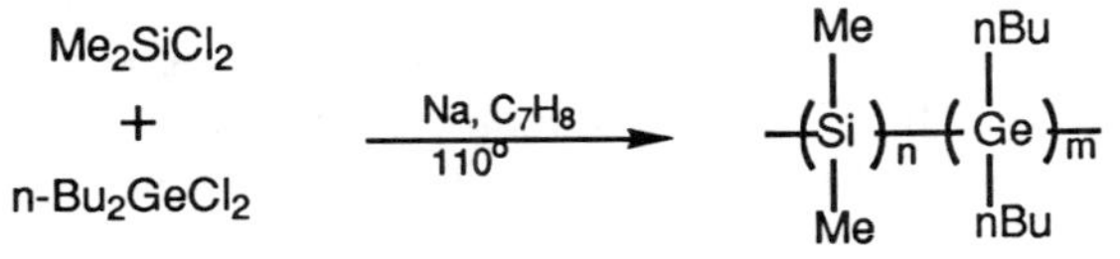

$$Me_2SiCl_2 + n\text{-}Bu_2GeCl_2 \xrightarrow[110°]{Na, C_7H_8} \text{—(Si)}_n\text{(Ge)}_m\text{—} \qquad (\text{Eq. } 5.34)$$

with Me, Me on Si and nBu, nBu on Ge

5.10 TECHNOLOGY OF POLYSILANES

5.10.1 Precursors to Silicon Carbide

The discovery by Yajima that polysilanes could be pyrolyzed to silicon carbide was mentioned in the introduction.[4] In this process, either $(Me_2Si)_n$ or the cyclic oligomer $(Me_2Si)_6$ are synthesized from Me_2SiCl_2 and are then heated to near 450°C (Scheme 5.9).

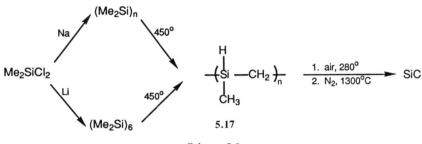

Scheme 5.9

This discovery has now been commercialized by the Nippon Carbon Co. for the production of NICALON™ silicon carbide fibers. In this process, methylene groups become inserted into many of the Si—Si bonds to give a polycarbosilane polymer with the idealized Structure 5.17. The actual structure of the polycarbosilane is more complex, including cross-links through CH groups attached to three silicon atoms. This intermediate polymer is fractionated by precipitation from a solvent to give material with molecular weight ~ 1800, which can be spun into fibers. These are heated in air to form a skin of SiO_2 on the surface, making the fibers rigid so that they will not melt when pyrolyzed. Further heating to > 800°C converts the polycarbosilane to amorphous silicon carbide; at still higher temperatures, crystallites of β-SiC begin to grow in the ceramic, reinforcing and strengthening the fibers. Maximum tensile strength is attained near 1300°C. This technology can also be used for the manufacture of monolithic silicon carbide objects. Modifications of the process include the addition of polyborodiphenylsiloxane as a catalyst and the inclusion of $Ti(O-i-Pr)_4$ to give (after pyrolysis) ceramic fibers containing titanium and oxygen as well as silicon and carbon. A more complete description of this technology can be found in an excellent review.[80]

Other linear polysilanes can also be used as ceramic precursors. Solid objects of silicon carbide are manufactured by Nippon Soda Co. from polysilastyrene and finely divided silicon carbide powder.[81] These materials are blended, injection-molded to the desired shape using conventional equipment and then pyrolyzed to give silicon carbide with 75–80% of the theoretical maximum density. The 20–25% of void space limits the ultimate strength of the ceramic somewhat, but there is a compensating advantage. The vacant space is present as tiny bubblelike "microvoids," which limit the growth of cracks. As a result, this material is the first silicon carbide which can be further shaped by machining. Polysilastyrene, when photocross-linked, can also be pyrolyzed to form silicon carbide fibers or coatings.

More complex, cross-linked polymers containing silicon-silicon bonds have also been used as silicon carbide precursors. The mixture of methylchlorodisilanes formed in the direct process for synthesis of Me_2SiCl_2 contains the three compounds shown in eq. 5.35. When these are heated with a phosphonium salt catalyst, they undergo a redistribution reaction to give a polycyclic, cross-linked oligomer (Structure 5.18) having the approximate composition shown, which undergoes pyrolysis to silicon carbide in high yield.[82] In addition a variety of cross-linked polymers have been obtained from alkali metal condensation of mixtures $CH_2{=}CHSiMeCl_2$ with Me_2SiCl_2, $MeSiHCl_2$, and Me_3SiCl, which produce silicon carbide in good conversion upon pyrolysis.[83]

$$\left.\begin{array}{c} MeCl_2SiSiCl_2Me \\ + \\ MeCl_2SiSiClMe_2 \\ + \\ Me_2ClSiSiMe_2Cl \end{array}\right\} \xrightarrow[250^{\circ}C]{nBu_4P^+Cl^-} \left[(Me_2Si)_3(MeSi)_{17}Cl_5 \right] \xrightarrow{\Delta} SiC$$

$$\mathbf{5.18}$$

(Eq. 5.35)

5.10.2 Polysilanes as Photoresists

Modern electronic and communications industry depends upon microlithography for the manufacture of microchips and other components. In this process, an image is transferred to a photosensitive layer, called a photoresist, by exposing it through a mask. There are two types of photoresists, as shown in Figure 5.22. Upon irradiation, the photoresist either becomes cross-linked (a negative-working photoresist) or becomes degraded (a positive-working photoresist). Most microlithography at present is done with positive-working photoresists.

As microelectronic components have grown more complex, it has become necessary to repeat this process as many as 25 times to transfer all the necessary information to the microchip. In order to cover existing topography on the chip, it is often better to use a bilayer resist system, as shown in Figure 5.23. The silicon wafer, with its existing topography, is first coated with a relatively thick layer of a polymer which is not photosensitive, as a planarizing layer. This is covered with a very thin (0.05 to 0.2 μm) layer of photoresist material. After exposure and development, the image is transferred through the planarizing layer by etching it

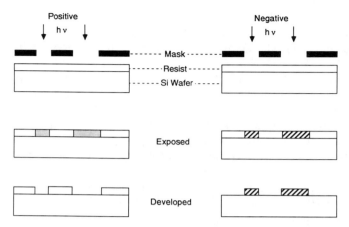

Figure 5.22   Schematic drawing showing the mode of operation of positive and negative photoresists.

away with an oxygen plasma. The photoresist layer must not be removed by the oxygen plasma, and this leads to one important advantage of polysilanes as photoresists. During exposure to the plasma, polysilanes (and other silicon-containing polymers) form a coating of silicon dioxide, which protects the underlying polymers from attack by the oxygen ions. The bilayer process makes it possible to generate steep-sided, narrow features desired in modern microchip fabrication.

The polysilanes are useful as photoresists[13] because they undergo scission when exposed to ultraviolet light.[46] Both the extinction coefficient and the wavelength of maximum absorption decrease with increasing exposure, so that polysilanes can be "bleached" photochemically, as illustrated for $(n\text{-HexSiMe})_n$ in Figure 5.24. Dialkylsilane polymers are generally better as positive photoresists than

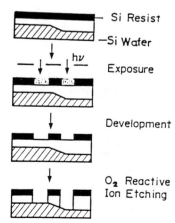

Figure 5.23   Shows the use of a polysilane photoresist in a bilayer construction, above an inert layer which is removed by oxygen ion etching.

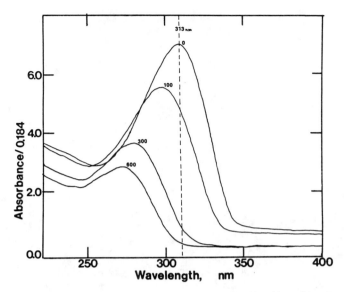

Figure 5.24 Ultraviolet spectra of (n-HexSiMe)$_n$ showing bleaching of the absorption band with increasing doses of UV radiation. Reprinted by permission from P. Trefonas et al., *J. Polym. Sci., Polym. Chem. Ed.*, **1983**, *22*, 822. Copyright © 1983 John Wiley & Sons, Inc.

arylalkyl or diaryl polymers, because the photoscission is not accompanied by cross-linking for the dialkyl materials.[85]

At present most microlithography is carried out using visible light. However, there is an increasing need for ever-closer spacing of features on microchips, to increase speed of operation as well as to permit further miniaturization. At the present maximum resolution of 0.5 to 1.0 μm, diffraction effects from the light are limiting. For the next increase in resolution, it will be necessary to use light of shorter wavelength. The polysilanes are well suited for this next generation of photoresists, since they are active in the ultraviolet region. A bilayer composition made using a (cyclo-HexSiMe)$_n$ photoresist, with resolution of 0.45 μm, generated at IBM-Almaden Research Laboratories is shown in Figure 5.25.[86]

Polysilanes also behave as promising photoresists at higher energy, in the x-ray region, and can be employed for electron beam imaging without the use of a mask. In addition, some polysilanes can be developed without the use of a solvent, using UV laser photolysis. Such photoresists are said to be "self-developing."[87]

5.10.3 Polysilanes as Photoinitiators

Photolysis of polysilanes produces silyl radicals, which can be used to induce free radical reactions. Because the silyl radicals can add to carbon-carbon double

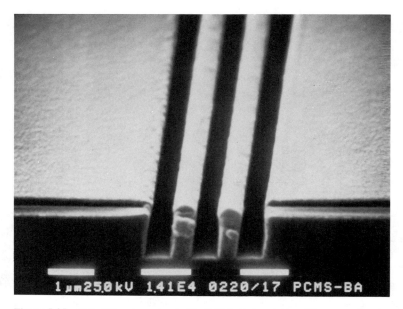

Figure 5.25   Test chip, using a (cyclo-HexSiMe) photoresist in a bilayer construction, after exposure, development and etching. Resolution is about 0.45 μm. Photo courtesy of R. D. Miller.

bonds and begin the formation of polymer chains, polysilanes can be used as radical photoinitiators.[88,89] Polysilane photoinitiation has been demonstrated for numerous acrylate monomers and has been studied in some detail for styrene. Various polysilanes can be used to initiate the polymerization of the many vinyl compounds that are susceptible to free radical chain growth. Of the polymers investigated, using irradiation at 254 nm, $(PhSiMe)_n$ and its copolymers were the most effective, followed by $(PhC_2H_4SiMe)_n$ and its copolymers, while pure alkyl polymers such as $(cyclo-HexSiMe)_n$ are the least suitable.

As photoinitiators the polysilanes are rather inefficient. Although the quantum yield for formation of radicals from polysilanes is high, only about one photon in 10^3 leads to radical chain formation. The low efficiency is somewhat mitigated by the fact that the extinction coefficients of the polysilanes are higher. The net result is that, compared to conventional photoinitiators, such as benzoin methyl ether, the photoinitiation efficiency of polysilanes is lower by about a factor of 10.

The unique advantage of the polysilanes, which makes them of interest in spite of their low efficiency, is that they are less susceptible to inhibition by oxygen than other photoinitiators. As described in Section 5.8.1, photolysis of polysilanes leads initially to elimination of silylenes (which do not initiate radical chains) and scission to silyl radicals. Many of the radicals are probably lost through recombination; many others under secondary processes that lead ultimately to stable silyl radicals and to Si=C compounds (silenes). Both of these may act to scavenge oxygen during the photoinitiation.

Industrial photoprocesses, such as photopolymerization of films and coatings, must ordinarily be carefully protected from oxygen, and this is both difficult and expensive. Thus because of their relative resistance to the effects of oxygen, the polysilanes may be useful either as photoinitiators or as additives in photoinitiated reactions.

5.11 ADDITIONAL READING

Further information about the entire field of polysilanes, cyclic, linear, and high polymeric, can be found in R. West, Chapter 19, "Polysilanes," in *The Chemistry of Organosilicon Compounds*, S. Patai and Z. Rappoport, eds., John Wiley, New York, 1989. An in-depth review of polysilane high polymers emphasizing spectroscopic and other physical characterization and containing extensive tables of data is "Polysilane High Polymers," R. D. Miller and J. Michl, *Chem. Reviews* **1989**, *89*, p 1357. Many useful chapters on various aspects of polysilane chemistry appear in the book, *Silicon-Based Polymer Science*, J. M. Zeigler and F. W. G. Fearon, eds., Advances in Chemistry Series 224, American Chemical Society, Washington, DC, 1990.

5.12 REFERENCES

1. Rochow, E. G. *An Introduction to the Chemistry of the Silicones*, 2nd ed.; John Wiley: New York, 1951, p 61.

2. Kipping, F. S. *J. Chem. Soc.* **1924**, *125*, 2291.

3. Burkhard, C. A. *J. Am. Chem. Soc.* **1949**, *71*, 963.

4. Kumada, M.; Tamao, K. *Adv. Organometal. Chem.* **1968**, *6*, 19.

5. West, R. *Pure Appl. Chem.* **1982**, *54*, 1041.

6. Brough, L. F.; Matsumura, K.; West, R. *Angew. Chem. Int. Ed. Engl.* **1979**, *18*, 955; Brough, L. F.; West, R. *J. Am. Chem. Soc.* **1981**, *103*, 3049.

7. Yajima, S.; Hayashi, J.; Omori, M. *Chem. Lett.* **1975**, 931; Yajima, S.; Okamura, K.; Hayashi, J. Ibid. **1975**, 1209.

8. Mazdyasni, K. S.; West, R.; David, L. D. *J. Am. Ceram. Soc.* **1978**, *61*, 504.

9. West, R.; David, L. D.; Djurovich, P. I.; Stearley, K. L.; Srinivasan, K. S. V.; Yu, H. *J. Am. Chem. Soc.* **1981**, *103*, 7352.

10. Wesson, J. D.; Williams, T. C. *J. Polym. Sci., Polym. Chem. Ed.* **1980**, *18*, 959.

11. Trujillo, R. E. *J. Organomet. Chem.* **1980**, *198*, C27.

12. Trefonas III, P.; Djurovich, P. I.; Zhang, X-H.; West, R.; Miller, R. D.; Hofer, D. *J. Polym. Sci., Polym. Lett. Ed.* **1983**, *21*, 819.

13. Gauthier, S.; Worsfold, D. J. *Macromolecules* **1989**, *22*, 2213.

14. Kim, H. K.; Matyjaszewski, K. *J. Am. Chem. Soc.* **1988**, *110*, 3321; Fujino, M.; Isaka, H. *J. Chem. Soc. Chem. Commun.* **1989**, 466.

15. Worsfold, D. J. In *Inorganic and Organometallic Polymers*; Zeldin, M.; Wynne, K. J.; Allcock, H. R., Eds.; ACS Symposium Series; American Chemical Society: Washington, DC, 1988; Vol. 360, pp 101–111.

16. Zeigler, J. M. *Polym. Prepr.* (Am. Chem. Soc., Div. Polym. Chem.) **1986**, *27*, 109; Zeigler, J. M.; Harrah, L. A.; Johnson, A. W. ibid. **1987**, *28*, 424.

17. Kim, H. K.; Matyjaszewski, K. *Polym. Prepr.* **1989**, *30*, 119.

18. Cragg, R. H.; Jones, R. G.; Swain, A. C.; Webb, S. J. *J. Chem. Soc., Chem. Commun.* **1990**, 1147.

19. Michalczyk, M. J.; Fink, M. J.; De Young, D. J.; Carlson, C. W.; Welsh, K. M.; West, R.; Michl, J. *Silicon, Germanium, Tin & Lead Cmpds.* **1986**, *9*, 75.

20. Matsumoto, H.; Arai, T.; Watanabe, H.; Nagai, Y. *J. Chem. Soc., Chem. Commun.* **1984**, 724.

21. Brough, L. F.; West, R. *J. Organomet. Chem.* **1980**, *194*, 139.

22. Michalczyk, M. J.; West, R.; Michl, J. *J. Am. Chem. Soc.* **1984**, *106*, 821.

23. West, R. *Angew. Chem. Int. Ed. Engl.* **1987**, *26*, 1201.

24. Sakamoto, K.; Obata, K.; Hirata, H.; Nakajima, M.; Sakurai, H. *J. Am. Chem. Soc.* **1989**, *111*, 7641.

25. Aitken, C. T.; Harrod, J. F.; Samuel, E. *J. Am. Chem. Soc.* **1986**, *108*, 4059. *Can. J. Chem.* **1986**, *64*, 1677.

26. Harrod, J. F. In *Inorganic and Organometallic Polymers*; Zeldin, M.; Wynne, K. J.; Allcock, H. R., Eds.; ACS Symposium Series; American Chemical Society: Washington, DC, 1988; Vol. 360, pp 89–100; Campbell, W. H.; Hilty, T. K.; Yurga, L. *Organometallics* **1989**, *8*, 2615.

27. Woo, H.-G.; Tilley, T. D. *J. Am. Chem. Soc.* **1989**, *111*, 3757; Tilley, T. D. *Comments Inorg. Chem.* **1990**, *10*, 37.

28. Cypryk, M.; Gupta, Y.; Matyjaszewski, K. *J. Am. Chem. Soc.* **1991**, *113*, 1046.

29. Maxka, J.; Mitter, F. K.; Powell, D. R.; West, R. *Organometallics* **1991**, *11*, 660.

30. Stuger, H.; West, R. *Macromolecules* **1985**, *18*, 2349.

31. Ban, H.; Sukegawa, K.; Togawa, S. *Macromolecules* **1987**, *20*, 1775.

32. Horiguchi, R.; Onishi, Y.; Hayase, S. *Macromolecules* **1988**, *21*, 304.

33. Hayase, S.; Horiguchi, R.; Onishi, Y.; Ushiroguchi, T. *Macromolecules* **1989**, *22*, 2933.

34. West, R. *J. Organomet. Chem.* **1986**, *300*, 327.

35. Matyjaszewski, K.; Chen, Y. L.; Kim H. H. In *Inorganic and Organometallic Polymers*; Zeldin, M.; Wynne, K. J.; Allcock, H. R., Eds.; ACS Symposium Series 360; American Chemical Society: Washington, DC, 1988, pp 78–88.

36. Miller, R.; Sooriyakumaran, R. *J. Polym. Sci., Polym. Lett. Ed.* **1987**, *25*, 321.

37. Zhang, Z.-H.; West, R. *J. Polym. Sci., Polym. Lett. Ed.* **1985**, *23*, 497.

38. West, R.; David, L. D.; Djurovich, P. I.; Yu, H.; Sinclair, R. *Am. Cer. Soc. Bull.* **1983**, *62*, 825.

39. Cotts, P. M.; Miller, R. D.; Trefonas III, P. T.; West, R.; Fickes, G. *Macromolecules* **1987**, *20*, 1046.

40. Cotts, P. M.; Miller, R. D.; Sooriyakumaran, R. In *Silicon-Based Polymer Science*; Zeigler, J. M.; Fearon, F. W. G., Eds.; Adv. Chem. Ser. 224, 1990; pp 397–412.

41. Bock, H.; Ensslin, W. *Angew. Chem. Int. Ed. Engl.* **1971**, *10*, 404.

42. Takeda, K.; Teramae, H.; Matsumoto, N. *J. Am. Chem. Soc.* **1986**, *108*, 8186.

43. Trefonas III, P.; West, R.; Miller, R. D.; Hofer, D. *J. Polym. Sci., Polym. Lett. Ed.* **1983**, *21*, 832.

44. Klingensmith, K.; Downing, J. W.; Miller, R. D.; Michl, J. *J. Am. Chem. Soc.* **1986**, *108*, 7438.

45. Michl, J.; Downing, J. W.; Karatsu, T.; Klingensmith, K. A.; Wallraff, G. M.; Miller, R. D. In *Inorganic and Organometallic Polymers*; Zeldin, M.; Wynne, K. J.; Allcock, H. R., Eds.; ACS Symposium Series; American Chemical Society: Washington, DC, 1988; Vol. 360, pp 61–77.

46. Miller, R. D.; Rabolt, J.; Sooriyakumaran, R.; Fleming, W.; Fickes, G. N.; Farmer, B. L.; Kuzmany H. In *Inorganic and Organometallic Polymers*; Zeldin, M.; Wynne, K. J.; Allcock, H. R., Eds.; ACS Symposium Series; American Chemical Society: Washington, DC, 1988; Vol. 360, pp 43–60.

47. Harrah, L. A.; Zeigler, J. M. *Macromolecules* **1987**, *20*, 2037.

48. Schilling, F. C.; Bovey, F. A.; Lovinger, A. J.; Zeigler, J. M. In *Silicon-Based Polymer Science*; Zeigler, J. M.; Fearon, F. W. G., Eds.; Adv. Chem. Ser. 224, 1990; pp 341–378.

49. Miller, R. D.; Sooriyakumaran, R. *Macromolecules* **1988**, *21*, 3179.

50. Miller, R. D.; Michl, J. *Chem. Rev.* **1989**, *89*, 1359.

51. Lovinger, A. J.; Davis, D. D.; Schilling, F. C.; Bovey, F. A.; Zeigler, J. M. *Polym. Commun.* **1989**, *30*, 356.

52. Schilling, F. C.; Lovinger, A. J.; Zeigler, J. M.; Davis, D. D.; Bovey, F. A. *Macromolecules* **1989**, *22*, 3055.

53. Schilling, F. C.; Bovey, F. A.; Davis, D. D.; Lovinger, A. J.; MacGregor, R. B.; Walsh, C. A.; Zeigler, J. M. *Macromolecules* **1989**, *22*, 4645.

54. Asuke, T.; West, R. *Macromolecules* **1991**, *24*, 343.

55. Weber, P.; Guillon, D.; Skoulios, A.; Miller, R. D. *J. Phys. France* **1989**, *50*, 795; Lovinger, A. J.; Schilling, F. C.; Bovey, F. A.; Zeigler, J. M. *Macromolecules* **1986**, *19*, 2660.

56. Kepler, R. G.; Zeigler, J. M.; Harrah, L. A.; Kurtz, S. R. *Phys. Rev. B* **1982**, *35*, 2818.

57. Stolka, M.; Yuh, H.-J.; McCrane, K.; Dai, D. M. *J. Polym. Sci., Polym. Chem. Ed.* **1987**, *25*, 823; Stolka, M.; Abkowitz, M. *J. Non-Cryst. Solids* **1987**, *97*, 1111.

58. Yokoyama, K.; Notsu, S.; Yokoyama, M. *J. Chem. Soc., Chem. Commun.* **1990**, 805.

59. Kajzar, F.; Messier, J.; Rosilio, C. *J. Appl. Phys.* **1986**, *60*, 3040.

60. Baumert, J. C.; Bjorklund, G. C.; Jundt, D. H.; Jurich, M. C.; Looser, H.; Miller, R. D.; Rabolt, J.; Sooriyakumaran, R.; Swalen, J. D.; Twieg, R. J. *Appl. Phys. Lett.* **1988**, *53*, 1147.

61. Trefonas, P.; Miller, R.; West, R. *J. Am. Chem. Soc.* **1985**, *107*, 2737.

62. Karatsu, T.; Miller, R. D.; Sooriyakumaran, R.; Michl, J. *J. Am. Chem. Soc.* **1989**, *111*, 1140.

63. West, R.; Zhang, X.-H.; Djurovich, P. I.; Stuger, H. In *Science of Ceramic Chemical Processing*; Hench, L. L.; Ulrich, D. R., Eds.; John Wiley: New York, **1986**; pp 337–344.

64. Zhang, Z. H.; West, R., unpublished studies.
65. Randall, J. C. *Polymer Sequence Determination*; Academic Press: New York, 1977; pp 1–40; Bovey, F. A. *Acct. Chem. Res.* **1968**, *1*, 175.
66. Schilling, F. C.; Bovey, F. A.; Zeigler, J. M. *Macromolecules*, **1986**, *19*, 239.
67. Wolff, A. R.; Maxka, J.; West, R. *J. Polym. Sci., Polym. Chem. Ed.* **1988**, *26*, 713.
68. Wolff, A. R.; Nozue, I.; Maxka, J.; West, R. *J. Polym. Sci., Polym. Chem. Ed.* **1988**, *26*, 701.
69. Menescal, R.; West, R., *Macromolecules* **1990**, *23*, 4492.
70. Sakamoto, K.; Yoshida, M.; Sakurai, H., *Macromolecules* **1990**, *23*, 4494.
71. Wesson, J. P.; Williams, T. C. *J. Polym. Sci., Polym. Chem. Ed.* **1981**, *19*, 65.
72. Bianconi, P. A.; Schilling, F. C.; Weidman, T. W. *Macromolecules* **1989**, *22*, 1697.
73. Matyjaszewski, K.; Kim, H. K. *Polym. Bull.* **1989**, *22*, 253.
74. Yuan, C.-H.; Menescal, R.; West, R., unpublished studies.
75. Trefonas III, P.; West, R. *J. Polym. Sci., Polym. Lett. Ed.* **1985**, *23*, 469.
76. Ishikawa, M.; Nate, K. In *Inorganic and Organometallic Polymers*; Zeldin, M.; Wynne, K. J.; Allcock, H. R., Eds.; ACS Symposium Series; American Chemical Society: Washington, DC, 1988; Vol. 360, pp 209–223.
77. Iwahara, T.; Hayase, S.; West, R. *Macromolecules* **1990**, *23*, 1298.
78. Trefonas, P.; West, R. *J. Polym. Chem., Polym. Chem. Ed.* **1985**, *23*, 2099.
79. Miller, R. D.; Sooriyakumaran, R. *J. Polym. Chem., Polym. Chem. Ed.* **1987**, *25*, 111.
80. Baney, R.; Chandra, G. In *Encyclopedia of Polymer Science and Engineering*, 2nd ed.; John Wiley: New York, 1988; Vol. 13, p 312.
81. Nichara, K.; Yamamoto, T.; Arima, J.; Takemoto, R.; Suganana, K.; Watanabe, R.; Nishikawa, T.; Okamura, M. *J. Polym. Sci., Polym. Chem. Ed.*, in press.
82. Baney, R. H.; Gaul, Jr., G. H.; Hilty, T. K. *Organometallics* **1983**, *2*, 859.
83. Schilling, Jr., C. L. *Brit. Polym. J.* **1986**, *18*, 355.
84. Miller, R. D. In *Silicon-Based Polymer Science*; Zeigler, J. M.; Fearon, F. W. G., Eds.; Adv. Chem. Ser. 224, 1990; pp 413–458.
85. Miller, R. D.; Hofer, D.; Fickes, G. N.; Willson, C. G.; Marinero, E.; Trefonas II, P.; West, R. *Polym. Engr. Sci.* **1986**, *26*, 1129.
86. Miller, R. D.; Wallraff, G.; Clecak, N.; Sooriyakumaran, R.; Michl, J.; Karatsu, T.; McKinley, A. J.; Klingensmith, K. A.; Downing, J. *Polym. Mater. Sci. Eng.* **1989**, *60*, 49.
87. Zeigler, J. M.; Harrah, L. A.; Johnson, A. W. *SPIE, Advances in Resist Technology and Processing* **1985**, *539*, 166.
88. West, R.; Wolff, A. R.; Peterson, D. J. *J. Radiat. Curing* **1986**, *13*, 35.
89. Wolff, A. R.; West, R. *Applied Organomet. Chem.* **1987**, *1*, 7.

6

Miscellaneous Inorganic Polymers

6.1 INTRODUCTION

The polymers discussed in this chapter are simply those that do not fit into one of the earlier, more general categories.[1-20] They are generally polymers that are not yet of great commercial importance, often because they are so new that much more research needs to be done before they can be utilized effectively.

6.2 OTHER PHOSPHORUS-CONTAINING POLYMERS

Elemental phosphorus itself exists in several polymeric forms. If the white allotropic form, which consists of P_4 tetrahedral molecules, is put under high pressure, preferably at elevated temperatures, it can be catalytically converted to other modifications.[8] It first becomes red, then violet, then black as the degree of polymerization increases. These materials are very difficult to characterize because of branching and the formation of cyclics. In the extreme limit, the structure approaches that of graphite, and shows good electrical conductivity.[7] No evidence exists at all for the formation of high molecular weight, linear chains of elemental phosphorus.

Like other elements important in inorganic chemistry, phosphorus is used mainly in combination with other elements in a polymeric chain.[21] If present in too large a proportion, there are two major disadvantages, namely, rather high cost and susceptibility to oxidative hydrolytic reaction. However, smaller amounts of phosphorus can be greatly advantageous, imparting flame retardancy, improved adhesion to metals, and ion-binding characteristics. Phosphorus-containing polymers are therefore extensively used as flame retardants for fabrics, for adhesives, and for ion-exchange resins.

One of the most direct ways for the introduction of phosphorus into a polymer is by direct phosphorylation, for example, by reaction of the polymer with phosphorus trichloride and oxygen. This has been done with many classic thermoplastics such as polyolefins and acrylics, common thermosetting materials such as the polyester resins and epoxies, polysaccharides such as cellulose derivatives, and biopolymers such as proteins.[21] However, these materials are better considered to be modified organic polymers, rather than as inorganic or semi-inorganic polymers.[11]

Phosphorus is thought to form chains with side group organic units (with repeat unit $-PR_2-$), analogous to the polysilanes, but the molecular weights are very low.[16] Phosphorus-nitrogen backbone polymers having the saturated repeat unit $-RPONR'-$ (as distinct from phosphazenes) have also been prepared, but these species have been studied very little.

Some phosphorus-sulfur compounds have also been prepared, but thus far these are oligomeric cagelike structures rather than linear or branched chains. Examples are P_4S_3, P_4S_5, P_4S_7, and P_4S_{10}.[7]

The so-called phoryl resins, or aromatic polyphosphates, can be prepared from phosphodichloridates ($ROPOCl_2$) and aromatic diols. Their repeat units are shown in Figure 6.1.[7] Although they have some attractive properties, such as good transparency and hardness, they lack long-term hydrolytic stability. This problem has been partially overcome by elimination of the phosphorus-oxygen bonds, as for example, in the poly(phosphinoisocyanates). Their structure is shown in Figure 6.2.[7] It is also possible to form poly(metal phosphinates) with repeat unit $-M(OPR_2O)_2-$ by allowing a metal alkoxide to react with a phosphinic acid.[21] Typical metal atoms are aluminum, cobalt, chromium, nickel, titanium, and zinc.[21] Polymeric phosphine oxides can be prepared by the reactions

$$x\ RPCl_2 + nx\ CH_2{=}CR'R'' \longrightarrow [-PRCl_2(CH_2CR'R'')_n-]_x \quad (Eq.\ 6.1)$$

$$[(-CH_2CR'R'')_n-]_x + x\ H_2O \longrightarrow [-PR(O)(CH_2CR'R'')_n-]_x$$
$$+\ 2x\ HCl \quad (Eq.\ 6.2)$$

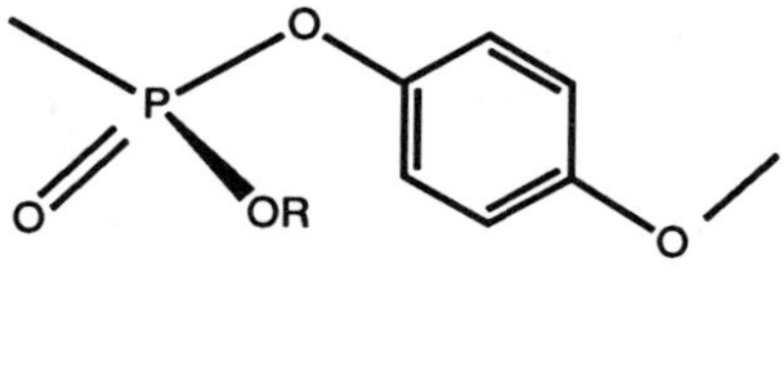

Figure 6.1 An aromatic polyphosphate, or "phoryl resin."

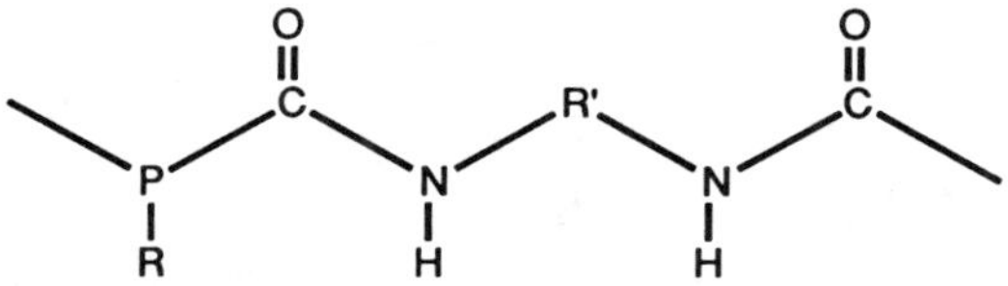

Figure 6.2   A poly(phosphinoisocyanate) chain.

The materials obtained are extremely stable,[21] because the electron pair of the phosphine structure has been donated to an oxygen atom. A final series of chains of this type are the poly(phosphoryldimethylamides), the structures of which are shown in Figure 6.3. They have interesting elastomeric properties, but presumably are hydrolytically sensitive.[7]

Some of the very newest inorganic polymers of any type also contain phosphorus atoms.[22] They are derived from the basic phosphazene structure described in Chapter 3 and are obtained by the ring-opening polymerization of heterocyclic compounds in which one of the disubstituted phosphorus atoms is replaced by another moiety. Specifically, introduction of a carbon atom can yield poly(carbophosphazenes), with the repeat unit shown in Figure 6.4. Alternatively, replacement with a sulfur atom can yield a poly(thiophosphazene), with the repeat unit shown in Figure 6.5. Relatively little is known about these polymers at the present time.[22]

Another series of polymers that contain phosphorus are the polyphosphates, which have structures very similar to silicates. Some of these are well known as

Figure 6.3   A poly(phosphoryldimethyl-amide) chain.

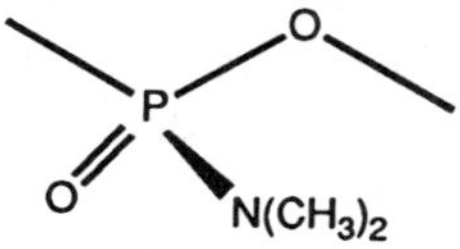

Figure 6.4   A poly(carbophosphazene) chain.

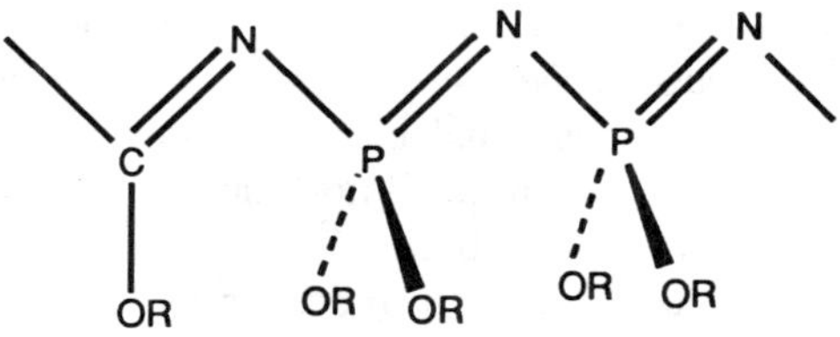

Figure 6.5   A poly(thiophosphazene) chain.

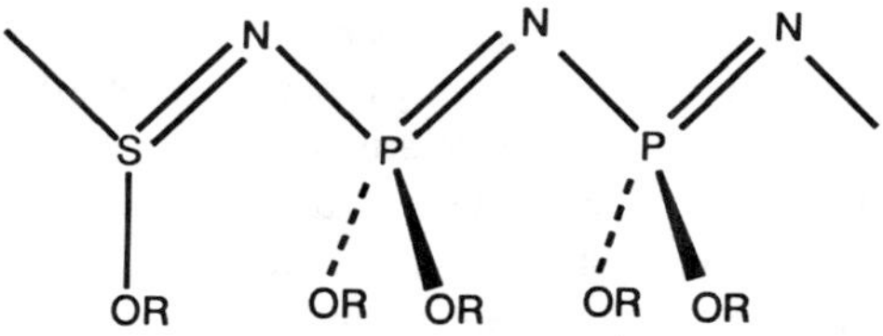

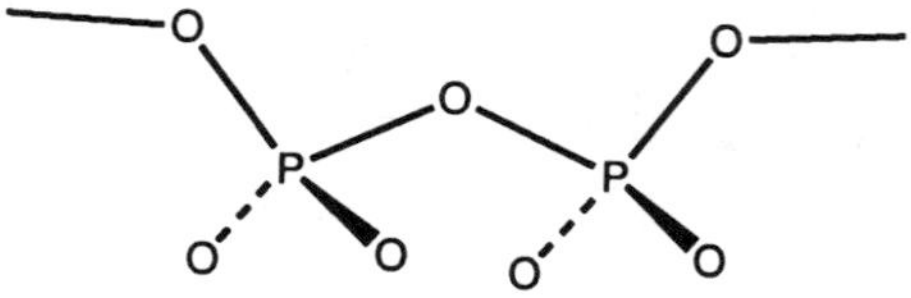

Figure 6.6 A polyphosphate chain.

"Grahams's salt," "Kurrol sodium salt," and "Maddrell salt."[8] Their linear chains are illustrated in Figure 6.6. They have not been studied extensively, possibly because they readily undergo molecular reorganization and hydrolytic cleavage to low molecular weight materials. In fact, most of the high molecular weight materials are mixtures of linear oligomeric chains, cyclics, and cross-linked network segments. Also present are strongly bonded countercations. Some of these materials exhibit chelating properties, and are therefore used in metal fabrication and protection.[21] However, the polyphosphates are unusually interesting chain molecules, in that they are totally inorganic, are polyelectrolytes, and contain repeat units of great importance in biopolymeric materials such as the polynucleotides. The skeletal P—O bonds have a length of 1.62 Å, with bond angles of 130° at the O atoms and 102° at the P atoms.[9,23] This inequality in skeletal bond angles is a frequent occurrence in inorganic polymers, as already mentioned, and can have a profound effect on the configurational characteristics of the chains. Specifically, this structural feature causes the all-*trans* conformation of the molecule (often the form of lowest conformational energy) to approximate a closed polygon. Since the difference in bond angles is 28°, it takes approximately 13 repeat units to complete the 360° required for the chain to close on itself. These chains are quite similar in this regard to the polysiloxane chains discussed in Chapter 4.

The potential barrier that hinders rotations about the skeletal bonds in polyphosphates is relatively small, probably less than 1 kcal mol^{-1}. Nonetheless, the rotational isomeric state model has been used for this chain, since it represents a reasonable approximation to the continuous range of rotational angles. Conformational energy calculations, including both steric and coulombic interactions, indicate that *gauche* states are of essentially the same energy as *trans* states when neighboring states remained *trans*. However, consecutive *gauche* states of either like or opposite sign are almost entirely suppressed, by steric interferences, coulombic repulsions, or both.[9,23]

Although this chain model is rather simple, it gave a good account of the relatively large values of the characteristic ratio (7.1 and 6.6) reported for lithium polyphosphate and sodium polyphosphate ("Graham's salt"), respectively. In brief, there are numerous *gauche* states present along the chain backbone, because of their relatively low energy. They generally do not occur in pairs (which are compact) because of the relatively high energies of such conformations. They represent departures from the all-*trans* form of the chain, which has a very small end-to-end distance. It is their presence that explains the relatively high spatial extension of the polyphosphate chain.

Theoretical and experimental studies have also been carried out on sodium phosphate.[23] Theoretical values for molar cyclization constants for the oligomeric molecules were calculated using this rotational isomeric state model. Specifically, calculations were made of the probability that a polyphosphate chain of the specified length would cyclize. More specifically, the calculations gave the total number of conformations with terminal atoms in sufficiently close proximity to form the final bond completing the cyclic molecule. The calculations were carried out using direct enumeration techniques and, alternatively, using a Gaussian distribution for the end-to-end distances. The calculated results were in satisfactory agreement with experiment.[23]

Arsenic is thought to occur in some related polymeric chains, possibly ladderlike in structure. The backbones are described as being made up entirely of arsenic atoms, and the molecules have the empirical formula $[AsCH_3]_x$.[16,24] This polymer can be prepared by the reaction of iodine with $(AsCH_3)_5$, but the range of colors and physical forms obtained suggest that the material may not be uniform in composition. Antimony is also thought to form ladder structures similar to those described for arsenic. The polymers are formed by the reaction of alkylstilbenes, $RSbH_2$, with a halogen source.[16,24] Bismuth forms very few compounds of any type with Bi—Bi bonds, although there has been a report of a polymer with the structure $(C_6H_5Bi)_x$.[16,24]

6.3 OTHER SILICON-CONTAINING POLYMERS

It should be mentioned first that a number of minerals, and glass itself, contain silicon and consist of polymeric structures.[11] Glass is a highly irregular material that consists of rings and linear chains of silicate units in complicated three-dimensional arrangements. On the other hand, some minerals consist of single chains or double chains in which negatively charged oxygen atoms are neutralized by positively charged metal cations. It is sometimes possible to make such materials more tractable by breaking open the cross-links and inserting nonionic, nonpolar groups in their place.[14] Nonetheless, these materials generally are not synthesizable, characterizable, and processable in the way that most organic or inorganic-organic polymers are.

The most important class of silicon-containing polymers that have not yet been covered are the polysilazanes, shown in Figure 6.7. These polymers, or precursors to them, are generally prepared by the reaction of organic-substituted chlo-

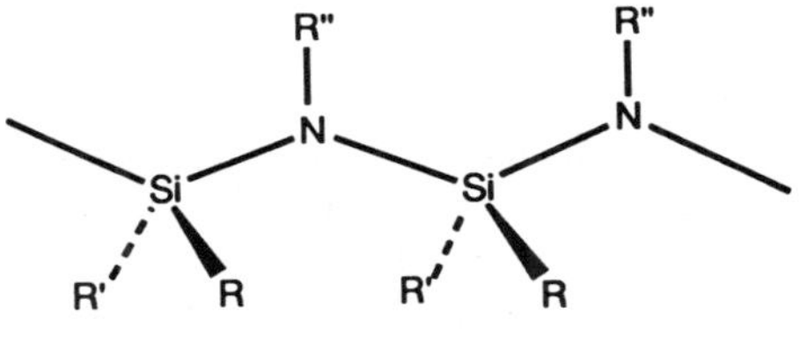

Figure 6.7 A polysilazane chain.

rosilanes with ammonia or amines as is shown in eq. (6.3).[25]

$$R_3SiCl + 2R_2'NH \longrightarrow R_3SiNR_2' + R_2'NH_2^+Cl^- \qquad (Eq. 6.3)$$

Such molecules can also be formed by the dehydrocoupling reaction catalyzed by bases or transition metals:

$$R_3SiH + R_2'NH \longrightarrow R_3SiNR_2' + H_2 \qquad (Eq. 6.4)$$

Another possibility is to use a deamination/condensation reaction, as shown in eq. (6.5). As a final alternative, a redistribution reaction, as

$$2R_2Si(NHR')_2 \longrightarrow (R'NH)R_2SiN(R')SiR_2(NHR') + R'NH_2 \qquad (Eq. 6.5)$$

illustrated in eq. (6.6), can be employed.

$$2RSiCl_2 + (R_2SiNH)_3 \longrightarrow ClR_2SiNHSiR_2Cl + Cl(R_2SiNH)_2SiR_2Cl$$

$$(Eq. 6.6)$$

The application of these approaches to the formation of the corresponding polymers can be illustrated by the reaction[25]

$$CH_3NH_2 + H_2SiCl_2 \longrightarrow [-H_2SiNCH_3-]_x + 2HCl \qquad (Eq. 6.7)$$

with molecular weights in the vicinity of 10^3 g mol^{-1}. Such oligosilazanes can be used in a dehydrocoupling reaction catalyzed by a transition metal to obtain higher degrees of polymerization.

It is also possible to form novel polysilazanes by dehydrocyclodimerization of the ammonolysis product of CH_3SiHCl_2, or coammonolysis products of this silane with other chlorosilanes.[26] Ring compounds that contain the desired Si and N atoms may also be used as starting materials, followed by hydrolysis, thermolysis, polymerization, or copolymerization to convert them into silazane-type materials. Examples are cyclic bis (silyl) imidates, N-acylcyclosilazoxanes, N-arylcyclosilazoxanes, and N,N'-diarylcyclosilazanes.[27]

More complicated materials can be derived from the polysilazane structure. Some, like poly(methyldisilylazane), are highly cross-linked, polycyclic structures that are presumably also cagelike. These polymers are soluble in some solvents. Hence, GPC measurements have given some information on their molecular weights and molecular-weight distributions. ^{29}Si and ^{13}C NMR have also been used to clarify some of the features mentioned.[28] The primary use of these polymers is as preceramic materials,[29] which, when pyrolyzed, yield silicon nitride, Si_3N_4. The major general problems in this area are (1) use of a polymer that is tractable (soluble, meltable, or malleable), (2) generation of molecular weights high enough or samples sufficiently cross-linked for the polymer to hold its shape during the pyrolysis step, (3) obtaining a high weight percent conversion to the ceramic, and (4) avoidance of reactions that could lead to excessive porosity. A specific problem in the pyrolysis to silicon nitride is the necessity to remove all of the extraneous carbon atoms left over from the organic groups. These would affect the properties of the resultant silicon nitride product.[19,25,30]

A variety of other ceramics are prepared by pyrolysis of preceramic polymers.[25,30] Some examples are silicon carbide, SiC, tungsten carbide, WC, aluminum nitride, AlN, and titanium nitride, TiN. In some cases, these materials are
obtained by simple pyrolysis in an inert atmosphere or under vacuum. In other
cases, a reactive atmosphere such as ammonia is needed to introduce some of the
atoms required in the final product.

6.4 POLYGERMANES

Germanium falls directly below silicon in the periodic table and, as can be seen
from Figures 6.8 and 6.9, the chain structure of the polygermanes is very similar
to that of the polysilanes, discussed in Chapter 5. Not surprisingly, therefore, their
properties are also very similar to those of the polysilanes.[31-36]

The earliest preparations of poly(organogermanes) yielded oligomers when
phenyl R side groups were present. Some homopolymers with n-butyl side groups
have been prepared, and a few copolymers having silane repeat units, also are
known. In these, a greater variety of side groups were employed. Monomers are
typically prepared by Grignard reactions with, for example, $GeCl_4$. This is illustrated in eqs. (6.8) and (6.9).

$$GeCl_4 + RMgX \longrightarrow R_4Ge \qquad\qquad (Eq. 6.8)$$

$$R_4Ge + GeCl_4 \longrightarrow R_2GeCl_2 \qquad\qquad (Eq. 6.9)$$

The resulting product, R_2GeCl_2, can then undergo a Wurtz coupling reaction in
the presence of sodium, as in the preparation of the polysilanes. For this specific
monomer, the molecular weights of the polymers can be quite high, with number-
average values above half a million. Polydispersity indices are also rather high
(around two or greater). When R_2GeBr_2 is used, the molecular weights have tended
to be relatively low, under 10,000 g mol^{-1}, and the distributions seem to be nar-

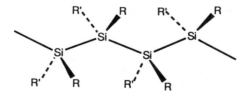

Figure 6.8   A polysilane chain.

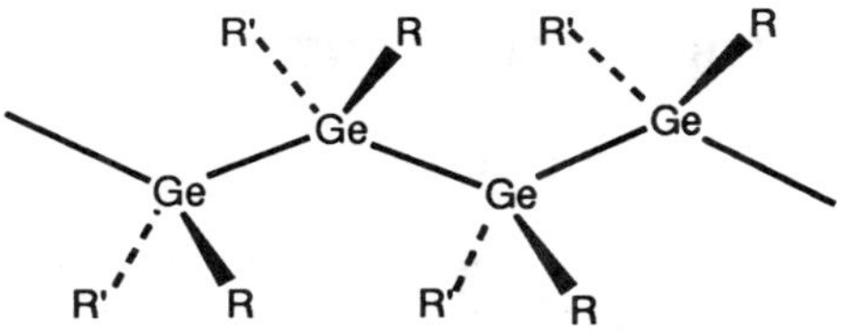

Figure 6.9   A polygermane chain.

row, with polydispersity indices of 1.1 to 1.3. The thermal stability is reported to be quite good.[32,33]

An alternative, recently reported route to polygermanes is the catalyzed dehydrogenative coupling of germanium hydrides.[36] Titanocene and zirconene derivatives are efficient catalysts, yielding high rates and high conversions. Dimethyltitanocene is particularly effective, but yields different products with different starting germanes. Specifically, primary germanes undergo coupling to give three-dimensional networks, presumably because of the reactivity of the backbone hydrogen atoms. Secondary germanes couple, apparently to form linear chains, but of rather low molecular weight.[36]

There has been a great deal of interest in the UV-visible spectroscopy of the polygermanes, particularly in comparison with the analogous chains that have silicon or tin backbones.[32-34] Both conventional and Raman spectroscopy have been employed. One interesting observation is that the symmetrically disubstituted poly(di-*n*-alkylgermanes) exhibit thermochromic transitions at temperatures below those of their polysilane analogs. Another is the conclusion that in poly(di-*n*-hexylgermane) the side chains adopt *trans*-planar conformations as in the polysilane counterpart. The two chains are also similar in that both backbones can, under certain circumstances, also adopt planar zigzag conformations.

Intramolecular energy calculations have been made on polygermane chains, so additional information on their conformational preferences is available.[35] More specifically, MNDO/2 molecular orbital calculations have been carried out as a function of rotations about the backbone bonds. For poly(dimethylgermane), the results predict a broad energy minimum located at the *trans* conformation, with two symmetrical, steeper and somewhat shallower minima near the usual *gauche* locations. The results are very similar to those for poly(dimethylsilane), except that the barriers are considerably lower.[35] This is apparently due to the fact that Ge—Ge bonds are approximately 0.30 Å longer than Si—Si bonds. Unfortunately, no relevant experimental data that might be used to test the validity of the calculations are available at the present time.

Like polysilanes, polygermanes can be decomposed and volatilized by exposure to radiation of the appropriate frequency. For this and other reasons, they have been considered for microlithographic applications.[32,35] Because they are more difficult and more expensive to prepare, and have no obvious advantages over polysilanes, they have found little use in this type of application.

There has also been some interest in the preparation of poly(germanoxanes) with repeat units of the type $[-GeR_2O-]$.[8]

6.5 POLYMERIC SULFUR AND SELENIUM

6.5.1 Polymeric Sulfur

This polymer is obviously totally inorganic, and can be made directly by the ring-opening polymerization of rhombic sulfur, which consists of eight-membered sulfur rings. Rhombic sulfur is a brittle, crystalline solid at room temperature.

Heating to 113°C causes it to melt to a reddish-yellow liquid of relatively low viscosity. Above approximately 160°C, the viscosity increases dramatically because of the free radical polymerization of the cyclic molecules into long, linear chains.[6, 8, 14, 19] At this point, a degree of polymerization of approximately 10^5 is obtained. If the temperature is increased to above approximately 175°C, depolymerization occurs, as evidenced by a decreasing viscosity. A similar type of depolymerization occurs with polysiloxanes. In thermodynamic terms, the negative $-T\Delta S$ term overcomes the positive ΔH term for chain depolymerization. (The temperature at which the two terms are just equal to one another is called the "ceiling temperature" for the polymerization.)

In any case, if this polymerized form of elemental sulfur is quenched (cooled rapidly), it becomes a solid. This solid is glassy at very low temperatures, but becomes highly elastomeric above its glass transition temperature of approximately $-30°C$.[6, 8, 14, 19] The situation is complicated by the presence of unpolymerized S_8 molecules which would certainly act as plasticizers. So far, attempts to cross link the elastomeric form into a network structure suitable for stress-strain measurements have not been very successful. The polymer is unstable at room temperature, gradually crystallizing, and eventually reverting entirely to the S_8 cyclics.

The ability of sulfur to form chains is also important in the sulfur vulcanization of diene elastomers such as natural rubber. In this case, strings of sulfur atoms of varying length form the cross-links that tie one chain to another. In this sense, they can be thought of as the short chains of a short-chain–long-chain bimodal network, as was described in Chapter 4.

Cyclic Se_8 molecules have also been polymerized by heating.[19] These chains are described in the following section.

Sulfur chains are fascinating since their structure is the simplest possible: all the chain atoms are identical and bear no substituents or side chains.[23] A sketch of the chain is shown in Figure 6.10. The bond lengths are 2.06 Å, the bond angles are 106°, and the rotational states are located at $\pm 90°$. The rotational barrier is approximately 10 kcal mol^{-1} and the van der Waals radius of the sulfur atom is 1.80 Å. The sulfur chain in the crystalline state is a helix in which all of the rotational angles are of the same sign and are 96°, very close to the value expected from the minimum conformational energy.[23]

The distances between neighboring atoms in all conformations give rise to favorable (attractive) interactions. These attractions are particularly strong in the two most compact conformational pairs, $\pm 90°$, $\pm 90°$, and in none of the conformations are severe steric overlaps present. These circumstances are very different from those in typical organic polymers. These conformational preferences should make sulfur chains extremely compact, with very small unperturbed dimensions. Conformational analyses, in fact, indicate that the characteristic ratio of polymeric sulfur should be less than unity (which is the "freely jointed" value).[23]

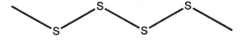

Figure 6.10 The polymeric sulfur chain.

Because of the instability of sulfur chains, it has not yet been possible to obtain an experimental value of the characteristic ratio with which to test the theoretical predictions. However, experimental values of the dipole moments are available for alkyl-terminated chains of 2, 3, and 4 sulfur atoms. Comparison of these experimental results with values calculated using the same conformational analyses does give good agreement, thus lending support to the proposed model.[23] An accurate experimental value is known for the entropy of cyclization of polymeric sulfur back to the cyclic octamer. Use of the conformational information mentioned, in an appropriate cyclization theory does give a good account of this entropy change. Additional calculated results have been carried out to elucidate the conformational preferences of a variety of small molecules that contain $S-S$ bonds, the relative stabilities of cyclic sulfur molecules of varying ring size, and the self-depression of the melting point of elemental atomic sulfur arising from the presence of sulfur cyclics.

It is also possible to prepare sulfanes, RS_xR, which are relatively long chains of sulfur capped with R end groups.[7] In one synthesis, a thiosulfate solution is acidified with a strong acid, to yield a hydrogen-terminated sulfur chain. Other end units such as alkyl groups, aryl groups, halogens, and mono- and disulfonic acids are also known. Some polymers of this type have been characterized and used in applications such as coating additives.

A variety of copolymers have been synthesized by the direct copolymerization of S_8 with other comonomers.[37] For example, S_8 undergoes an equilibrium copolymerization with Se_8, as shown in the reaction

$$S_8 + Se_8 \longrightarrow [-S_x Se_y -] + S_{8-x} Se_x \qquad \text{(Eq. 6.10)}$$

Cyclic species $S_{8-x} Se_x$ are formed as well as the linear copolymer

$$[-S_x Se_y -].$$

Reaction of sulfur with arsenic is more complicated than can be explained by this simple difunctional equilibrium. In particular, the trifunctional As atoms give rise to a three-dimensional network structure.[37]

Small amounts of phosphorus can be used to vulcanize polymeric sulfur. Larger amounts yield the P_4S_4 "bird-cage" molecules mentioned in the earlier section on "Other Phosphorus-Containing Polymers," together with phosphorus-sulfur copolymers. The copolymers are hydrolytically unstable.[37]

6.5.2 Polymeric Selenium

Selenium exists in the form of a cyclic octamer, which can rearrange to long, polymeric chains. As can be seen from Figure 6.11, the selenium chain is very similar to that of sulfur. The bond angles are 104°, essentially the same as those in polymeric sulfur, and rotational states are also thought to occur at the same angles, $\pm 90°$.[23] However, the skeletal bonds are 2.34 Å long, which is longer than the $S-S$ bonds in the polymeric sulfur chain. This long bond length more

Figure 6.11 The polymeric selenium chain.

than offsets the repulsions expected from the increase in van der Waals radius (to 1.90 Å), and this chain should be even more flexible. If so, it might have a lower glass transition temperature than polymeric sulfur. Actually, this is hard to predict purely on the basis of bond flexibility, since interchain interactions are generally also important (or "plasticizers" simply would not work!).

The crystalline state conformation of polymeric selenium is a helix in which all the rotational angles are of the same sign, with a value of $\pm 78°$. This is fairly close to the expected value of $\pm 90°$.[23] The barrier to rotation about the Se—Se bond is thought to be roughly the same as that about S—S bonds. No reliable experimental results are available on the statistical properties of these chains. Some values of the glass transition temperature of polymeric selenium have been reported,[15] and these could at least provide a measure of the dynamic flexibility of the chains. However, these results are probably compromised by the presence of cyclic selenium molecules that act as plasticizer.

Not surprisingly, tellurium forms some polymeric materials analogous to those formed by sulfur and selenium.[15]

6.6 OTHER SULFUR-CONTAINING POLYMERS

The most important polymer in this category is probably poly(sulfur nitride), also known as polythiazyl,[12,37,38] which is shown schematically in Figure 6.12. The synthesis of this sulfur-nitrogen polymer starts from the eight-membered cyclic tetramer, $[SN]_4$. The tetramer itself can be prepared by the reaction of ammonia with elemental sulfur, sulfur fluorides, or sulfur chloride. It is a crystalline solid with a sublimation temperature of 178°C. It is also prone to detonation. In the vapor state it can be catalytically converted with metallic silver at elevated temperatures to the cyclic dimer which, in turn, polymerizes in the solid state to the $[SN]_x$ polymer.[12,37,38] Several days are normally required for the polymerization. The mechanism is thought to involve ring cleavage, with the two resulting radicals combining with the nearest neighbors bracketing them in the crystal structure.[19,39] Because a sublimation step is involved in the polymerization process, there has been considerable interest in epitaxial chain growth, where the polymerization occurs on surfaces of known and controlled structure. In practice, the polymer is depolymerized at approximately 150°C to a gaseous S_4N_4 isomer. When this iso-

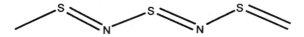

Figure 6.12 The polymeric sulfur nitride, or polythiazyl, chain.

mer in the vapor state comes into contact with the desired surface at a lower temperature, it spontaneously repolymerizes to give the $[SN]_x$ film.[12]

The polymer is highly crystalline, which facilitates structure analysis, but has thwarted attempts to measure its properties in solution. The preparation of single crystals has been useful with regard to the characterization of its structure. Its crystal structure is known, but not much is known about its molecular weight. The S—N and N—S bonds are essentially identical in length, with a bond order that is intermediate between those expected for normal single and double bonds.[12] Conformationally, the chains are in nearly planar, crank-shaft arrangements. Electron microscopy shows the crystals to be composed of layers of fibers parallel to the long axis of the crystal. They are soft and malleable, and have a golden metallic luster. They conduct electricity like a metal down to very low temperatures, with highly anisotropic conductivity in the case of the oriented chains.[19] The conductivity increases by several orders of magnitude as the temperature is decreased from room temperature to a few degrees Kelvin. At 0.26 K, $[SN]_x$ becomes a superconductor. The electrical conductivity is thought to require both the charge delocalization that is possible along the chains, and the highly ordered chain arrangements present in the crystalline state. Brominated polythiazyl chains are also being studied in this regard.[39] They are prepared by direct bromination of the polymer at ambient temperatures, and have the composition $[SNBr_y]_x$, where y is approximately 0.4.

The chains were found to be stable in an inert atmosphere, but decomposed above approximately 140°C, into sulfur, nitrogen, and low-molecular-weight sulfur-nitrogen molecules. Unfortunately the polymer oxidizes easily, which may cause loss of some of its most interesting properties.[19] Its insolubility and failure to melt without decomposition greatly complicate its processing into fibers or films. However, it has been used extensively as an electrode material. A final difficulty is the explosive nature of many sulfur-nitrogen compounds. The situation could possibly be simplified if sulfur-nitrogen chains could be prepared with substituents linked to the S atoms.[14] These and other problems are being studied at the present time.

For example, considerable interest exists in the preparation of sulfur-nitrogen polymers that have organic substituents on the sulfur atoms.[37] This could help alleviate the intractability problem mentioned, and could also give rise to a series of polymers that parallel the phosphazenes in their structural variability. A series of polymers with S—N backbones of this type has been prepared, but with oxygen atoms as some of the substituents. The basic structure has the repeat unit —RS(=O)(N)— or —FS(=O)(N)—, but they have not yet been studied in detail. The fluorine-containing polymer, however, is known to be a tough elastomer which is unaffected by water, acids, or bases up to 100°C.[7]

Some of the most useful series of polymers in this category are the polysulfides.[40] Their repeat units consist of a variable number of skeletal sulfur atoms separated by organic groups. The organic groups can be hydrocarbon sequences, as in $-(CH_2)_mS_n-$, ether moieties, as in $-(CH_2)_mOS_n-$, or units such as in $-(CH_2)_mOCH_2OS_n-$. In these notations, m and n are typically 2 to 4. The early materials of this type were high molecular weight solid elastomers called "Thiokols," but the most important products today are lower molecular weight liquids with mercaptan end groups that permit them to be converted into an elastomer in situ. They have excellent radiation resistance as well as very good adhesion to some metals and to glass. This latter characteristic makes them particularly useful as adhesives, sealants, and coatings.[40]

Because of the inorganic nature inherent in the strings of sulfur atoms in their backbones, these polymers have excellent resistance to hydrocarbon solvents that can swell common hydrocarbon elastomers. They also have good resistance to moisture, oxygen, and ozone; good weatherability; and better than average low temperature properties.[40] In general, the higher the sulfur content, the better the properties of this type. However, the sequences of skeletal sulfur atoms also present problems, as can be anticipated from the polymerization-reorganization behavior mentioned earlier with regard to polymeric sulfur itself. Because of this problem, the polysulfide elastomers have poor high-temperature stability and high-compression set (permanent deformation or "creep" under stress). The presence of the skeletal sulfur atoms and the sulfur atoms in the mercaptan end groups can cause odor problems as well, although most of this has been blamed on trace impurities.

Materials prepared with liquidlike properties, as mentioned earlier, have molecular weights of only a few thousand. A variety of chemical curing agents are used in cross-linking them, with PbO_2 and chromates being good examples.[40]

These polymers are typically prepared by reaction of sodium polysulfide with a formal. Bis(2-chloroethylformal) is commonly used, but some *tri*functional chlorinated hydrocarbon with its associated higher functionality is typically also added to promote branching. The sodium polysulfide is usually made by the direct reaction of sodium hydroxide with sulfur at temperatures near 120°C, as shown in the equation[40]

$$6 \text{ NaOH} + 2\,(x + 1)\,\text{S} \longrightarrow 2\,\text{Na}_2\text{S}_x + \text{Na}-(\text{S})_x-\text{Na} + 3\,\text{H}_2\text{O}$$

(Eq. 6.11)

The polysulfide then reacts by a condensation polymerization with the difunctional and trifunctional chloro compounds. It is also possible to react cyclic S_8 with a mercaptan-capped molecule or dithiol in a condensation reaction in the presence of a basic catalyst. An example is the reaction with ethanedithiol[37]

$$x \text{ HS}(\text{CH}_2)_2 \text{SH} + (xy/8)\,\text{S}_8 \longrightarrow [-(\text{CH}_2)_2\text{S}_y-] + \text{H}_2\text{S} \quad \text{(Eq. 6.12)}$$

Another preparative technique involves the ring-opening polymerization of cyclic polysulfides.[37] For example, 1-oxa-4,5-dithiacycloheptane polymerizes very rapidly at room temperature when treated with an aqueous sulfide:

$$x \;\; \longrightarrow \;\; [-(\text{CH}_2)_2\text{O}(\text{CH}_2)_2\text{S}_2-]_x \quad \text{(Eq. 6.13)}$$

The related monomer, 1,3-dioxa-6,7-dithiacyclononane, can be polymerized in the same manner. A variety of catalysts, both anionic and cationic, are effective. Similarly, 1,2-dithiacycloalkanes can be polymerized with traces of $AlCl_3$:

$$x \quad \begin{pmatrix} (CH_2)_m \\ | \\ S\text{-}S \end{pmatrix} \quad \longrightarrow \quad [-(CH_2)_m S_2-]_x \qquad \text{(Eq. 6.14)}$$

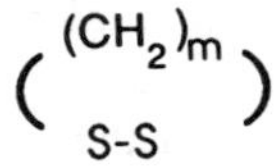

Polymers that have several values of m in the range 3–13 have been prepared. The estimated ring strain energies of these monomers show a good correlation with measured heats of polymerization.

Some of the corresponding cyclic trisulfides can also be polymerized, often to high-molecular-weight materials. For example, 1-ethyl-2,3,4,-trithiacyclopentane can be polymerized[37]

$$x \quad \begin{array}{c} C_2H_5 \quad S \\ \diagdown \diagup \\ S \\ | \\ S \end{array} \quad \longrightarrow \quad [-CH_2CH(C_2H_5)S_3-]_x \qquad \text{(Eq. 6.15)}$$

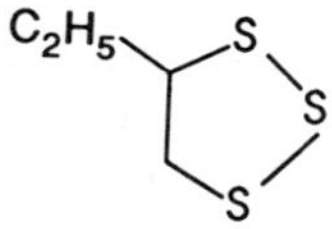

to produce an elastomer that has a molecular weight of several hundred thousand. Another related polymerization involves cyclic trithiaderivatives of tetrafluoroethylene, prepared by sulfuration of the olefin in the gas phase:[37]

$$x \; F_2C{=}CF_2 + (3n/8)\; S_8 \quad \longrightarrow \quad x \quad \begin{array}{c} F_2C \diagup S \\ | \quad\quad S \\ F_2C \diagdown S \diagup \end{array} \qquad \text{(Eq. 6.16)}$$

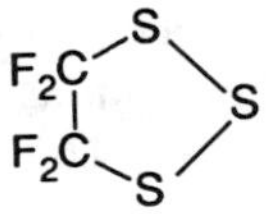

This monomer can then be polymerized in the presence of base

$$x \quad \begin{array}{c} F_2C \diagup S \\ | \quad\quad S \\ F_2C \diagdown S \diagup \end{array} \quad \longrightarrow \quad [-CF_2CF_2S_3-]_x \qquad \text{(Eq. 6.17)}$$

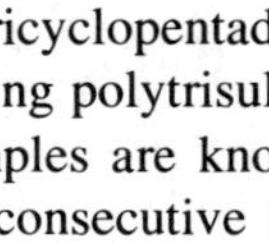

to yield a crystalline polymer of relatively high molecular weight. The sulfuration method can also be used to prepare ortho-fused 1,2,3-trithia derivatives of norbornene, and di- and tricyclopentadiene, which can then be polymerized anionically to the corresponding polytrisulfides.[37]

Only a few examples are known of polymerizations to cyclic species that have more than three consecutive S atoms in the repeat unit. One example is poly(tetrafluoroethylene tetrasulfide), which can be prepared in a manner similar to that shown in eq. (6.16).[37]

The cyclic S_8 monomer can be copolymerized directly with unsaturated monomers.[37] This reaction is illustrated by the equation

$$x \; CH_2{=}CR'R'' + (xy/8)\; S_8 \quad \longrightarrow \quad [-CH_2CR'R''S_y-] \qquad \text{(Eq. 6.18)}$$

Examples of organic comonomers studied to date are methyl acrylate, methyl methacrylate, and vinyl acetate. The number of S atoms thus introduced into the repeat unit is highly variable and changes with the nature of the diene, the temperature, feed composition, and method of initiation. However, typical values are generally in the range 3–8.[37] The reaction of sulfur with the double bonds in diene elastomers, such as natural rubber and cis-1,4-butadiene, also falls into this category. Not surprisingly, similar reactions occur between S_8 and small-molecule dienes. Also, copolymerizations between S_8 and aryl alkynes have been reported, with an example of the organic comonomer being phenylacetylene.[37]

Cyclic S_8 can also be copolymerized with cyclic mono- and polysulfides such as thiiranes and trithiolanes.[37] The result is a polysulfide with an augmented number of S atoms in the repeat unit.

S—S bonds are important in biological polymers. For example, one of the ways in which proteins are held in very precisely controlled arrangements is through disulfide linkages, and this is particularly important in enzymes and antibodies.[37] Not surprisingly, a great deal of interest exists in carrying out reactions in the laboratory that simulate the function of these biopolymers. Disulfide spacers have also been used to attach biologically active groups to a variety of polymer backbones.

Some general comments can also be made with regard to the physical properties of some of these polymers. For example, the properties of polysulfides generally show a very strong dependence on composition.[40] The glass transition temperature T_g is of particular interest. Specifically, polysulfides in which the organic groups are a simple $-(CH_2)_m-$ hydrocarbon sequence have relatively high values of T_g, typically around $-25°C$. If the organic chain units are ether groups, $-(CH_2)_mO-$, the values of T_g are lower, typically around $-50°C$.[40,41] This is consistent with the known increase in flexibility that results from the presence of dicoordinate oxygen atoms.[23] The formal group, $-(CH_2)_mOCH_2O-$, would thus be expected to introduce even more flexibility, and this is found to be the case. These polymers show the lowest values of T_g in this series, around $-75°C$. Lengthening of the hydrocarbon chain tends to decrease the T_g, which is consistent with the very low glass transition temperature (probably around $-125°C$) found for polyethylene.[41] On the other hand, lengthening of the sulfur chain tends to increase T_g. This is also to be expected, in view of the relatively high value of T_g for polymeric sulfur ($-30°C$).

The tendency for crystallization also varies with structure, in particular with the lengths and nature of the organic sequences. Crystallization during storage at low temperatures can cause such an elastomer to harden, and to become less processable. Crystallization can be avoided entirely by making the chain structure highly irregular, by using a mixture of the organic reactants instead of a single species. One such mixture consists of butyl formal and butyl ether.[40]

A number of sulfur-containing polymers do not contain S—S bonds, but are simply analogs of the corresponding oxygen-containing polymers long known in classical organic polymer chemistry. Some examples are poly(thioesters), poly(thiocarbonates), and poly(thiocarbamates). Their structures are shown in Fig-

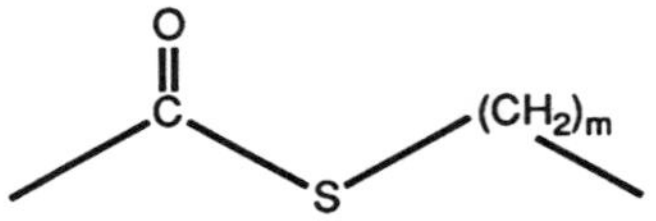

Figure 6.13 A poly(thioester) chain.

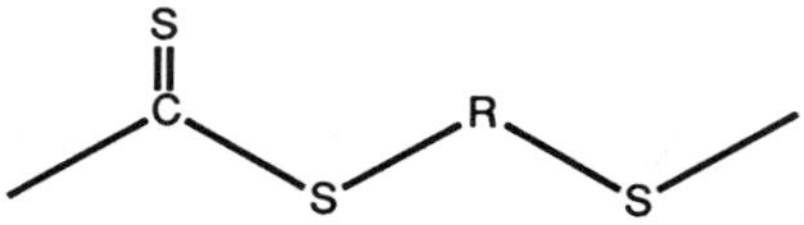

Figure 6.14 A poly(thiocarbonate) chain.

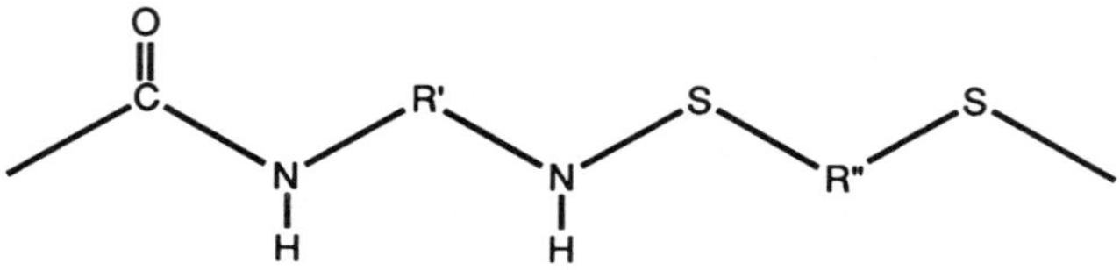

Figure 6.15 A poly(thiocarbamate) chain.

ures 6.13 through 6.15. Poly(thioethers) that have structures $[-(CH_2)_mS-]_x$ also exist. These are analogs of the much-studied polyoxides $[-(CH)_mO-]_x$.[23] None of the polymers in these series possesses much inorganic character and are best considered as parts of the organic series, as mentioned earlier.

6.7 BORON-CONTAINING POLYMERS

One of the simplest polymers containing boron consists of chains of boron fluoride, of repeat unit $-BF-$. It can be prepared by the reaction of elemental boron with boron trifluoride at high temperatures. The polymer is a rubbery elastomer, but it has been little studied because of its hydrolytic instability and tendency to ignite spontaneously in air.[7] However, a variety of other structures are formed with a number of metals, for example chains with Fe, ladders with Ta, sheets with Ti, tetragonal structures with U, and cubic structures with Ca and Ar.[7] Boron also forms chains that are analogs of poly(dimethylsiloxane), with repeat unit $-BCH_3-O-$.[7]

One important class of boron-containing polymers are those which contain the carborane cage, of structure $-CB_{10}H_{10}C-$. This group is prepared by the addition of acetylene to pentaborane or decaborane, $-B_{10}H_{14}-$. The resultant o-carborane rearranges to m-carborane, as shown in Figure 6.16.[19] Reactive sites can be introduced by conversion of this product to m-LiCB$_{10}$H$_{10}$CLi, by reaction with butyl lithium. For example, it has been incorporated between siloxane chain units by reaction with a dichlorosiloxane. This gives increased stability to the resulting copolymeric elastomer. Although the transition temperature of this copolymer, like any other, depends on composition, the copolymers produced in this way

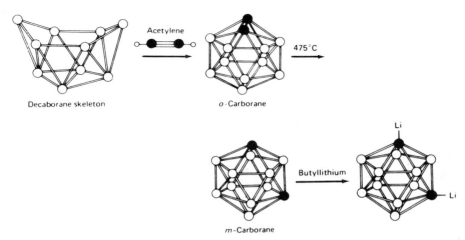

Figure 6.16   Carborane-type structures.

can have very high softening temperatures. Such materials have useful properties, at least for limited periods of time, at temperatures as high as 500°C. They also have relatively low glass transition temperatures, around −30°C, as required in many high-performance applications.[12]

The decaborane cage structure has also been used with some comonomers, specifically diamines, to give relatively high molecular weight polymers. Fibers formed from these chains can be pyrolyzed to give products that consist largely of boron nitride, BN.

The pentaborane cage structure $-B_5H_9-$ has been used as a side group in the preparation of vinyl-type polymers, but only of relatively low molecular weight. Pyrolysis of this material gives primarily boron carbide, B_4C.

Another example is the poly(borazine) chain, with repeat unit $-BRNR'-$.[42,43] The synthesis of tractable, unbranched polymers of this type is complicated by the formation of borazines, which are very stable cyclic molecules $[(BRNR')_3]$. This problem can possibly be overcome through the use of cyclic mono- and diborylamines.[42] Some of these polymers are thermally stable, up to approximately 500°C, but are easily hydrolyzed. Their most important use is probably as precursors to boron nitride ceramics.[43] A related polymer is polyborazylene, which is a cyclolinear boron-nitrogen analog of polyphenylene. The simplest structure proposed, which corresponds to $[-B_3N_3H_4-]_x$, is shown in Figure 6.17. Two final examples are boron nitride itself, $-BN-$,[44,45] and boron phosphide. Boron nitride can be prepared by heating borax ($Na_2B_4O_7$) in the presence of ammonium chloride at 1000°C.[7] It is thought to have a planar structure similar to that of graphite, as is illustrated in Figure 6.18. It can be converted into a cubic, diamondlike structure called borazon. It is resistant to hydrolysis and oxygen, and can be harder than diamond.[8] Also, it can be used at temperatures as high as 2000°C. It is used mainly as a refractory material, abrasive, or insulator.

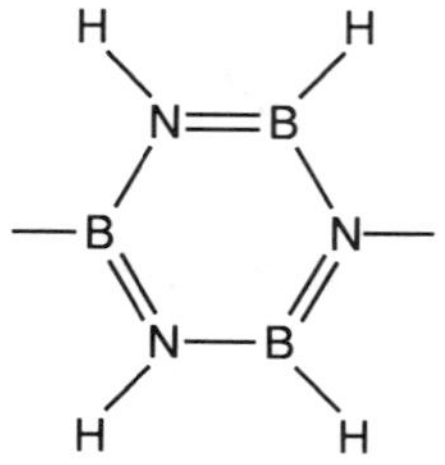

Figure 6.17 The boron nitride chain.

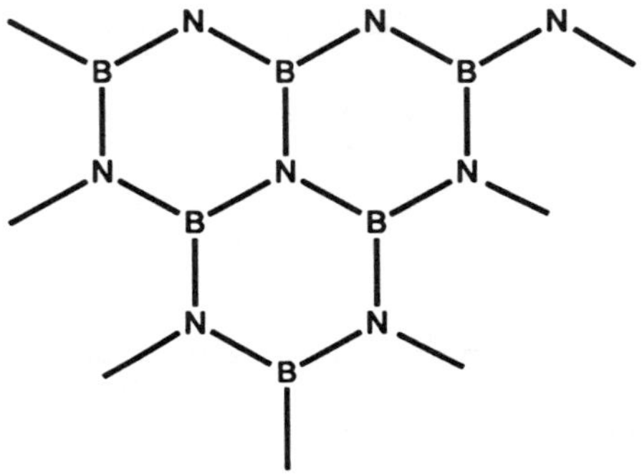

Figure 6.18 Boron nitride.

It is also possible to prepare all-carbon polymers of closely related structure. For example, pyrolysis of polyacrylonitrile, $(-CH_2CHCN-)_x$, first results in cyclization of some of the $-CN$ side chains.[19] Prolonged pyrolysis yields very pure graphitic material. It is very strong and has a high thermal stability. In the form of fibers, it can be used for reinforcement in high-performance composites.

6.8 ALUMINUM-CONTAINING POLYMERS

The reactions of metallic aluminum with an alcohol can generate oligomeric structures that contain Al—O bonds. Of greater interest, however, is the reaction of $AlCl_3$ with the salt of an organic acid to give aluminum dialkanoates. These have chainlike structures, and a range of molecular weights. These polymers are used as "drag-reduction" agents either in water or organic fluids. Drag-reduction agents reduce the resistance of fluid flow through pipes. This is very useful in fire fighting, oil pumping, and so on.

The reactions of trialkylaluminum reagents with an organic acid can yield a poly(acyloxyaloxane) or poly(aluminoxane) with the structure shown in Figure 6.19. Although the molecular weights are quite low, and the chains may have considerable branching and cross-linking,[7] the polymers can be melt spun into fibers, which themselves can be pyrolyzed to alumina, Al_2O_3. Other applications include their use as drying or gelling agents and as lubricants.

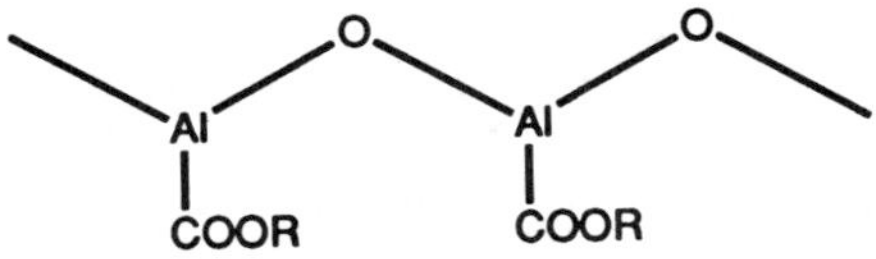

Figure 6.19 A polyacyloxyaloxane
chain.

Another type of polymer in this category is represented by Figure 6.19, in which the O atoms are replaced by NH groups. These aluminum-nitrogen chains are thermally and hydrolytically unstable and are sensitive to acids and bases.[7] However, this polymer can also be drawn into fibers, which can be converted to aluminum nitride, AlN, in the presence of ammonia.

6.9 TIN-CONTAINING POLYMERS

Most polymers with tin atoms in the chain backbone can be represented by the stannoxane structure shown in Figure 6.20, where R represents a variety of organic groups. They are typically prepared by an interfacial polymerization between a diorganotin dihalide and a difunctional organic molecule. The nitrogen analogs, in which NH groups replace the O atoms in this structure, have also been prepared. The most notable characteristic of this class of polymers is their good thermal stability.

A variety of oligomers with novel "drum" and "ladder" structures have also been prepared. They have the formulas $[R'Sn(O)O_2CR]_6$ and $[(R'Sn(O)O_2CR)_2R'Sn(O_2CR)_3]_2$ and contain four- and six-membered rings. They can be synthesized by the reaction of stannoic acid with a carboxylic acid, but the reaction of $RSnCl_3$ with the silver salt of a carboxylic acid can also be used. If diphenylphosphinic acid or dicyclohexylphosphinic acid is used instead of the carboxylic acid, oxygen-capped clusters or cubic and butterfly structures are obtained.[46]

Also of interest are polymers that have tin atoms in their side chains. Typically these polymers consist of vinyl chains with $-COOSnR_3$ side groups, as illustrated in Figure 6.21. Such side groups have also been much used in copolymers, for example, in poly(tri-*n*-butyltin methacrylate/methyl methacrylate).[47] These materials have been used as antifouling paints on ships and off-shore oil platforms. They are believed to work by hydrolysis to liberate Sn ions, which deter the growth of marine organisms. The R_3SnOH formed in the hydrolysis and some

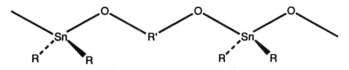

Figure 6.20 A polymer with tin atoms in the chain backbone.

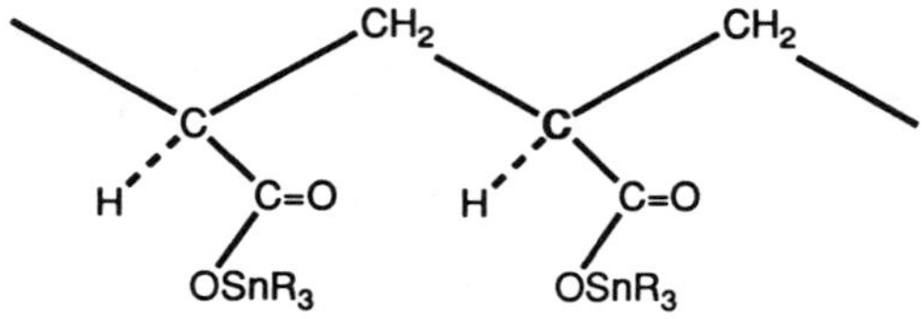

Figure 6.21 A polymer with tin atoms in the side groups, bonded through easily hydrolyzable Sn—O bonds.

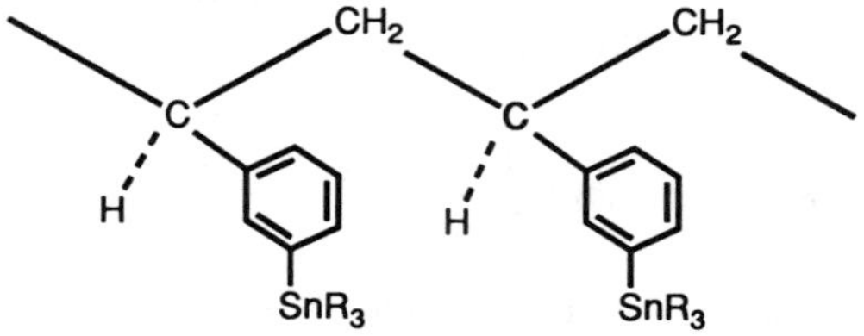

Figure 6.22 A polymer with tin atoms in the side groups, bonded through relatively stable Sn—C bonds.

of its degradation products are tenaciously adsorbed by clays. However, the tin ions can have considerable toxicity and can accumulate in bodies of water when ships are in harbor for extended periods of time. The use of tin polymers for this application is now being discouraged.

A new type of polymer with tin atoms in the side groups is polystyrene with —R_3Sn groups attached to the phenyl groups, as shown in Figure 6.22. Because these polymers contain no OSn bonds they are generally less hydrolyzable, and therefore generate less of a problem in the environment. It is thought that their hydrolyzability can be increased by the use of bulky R groups attached to the tin atoms. This raises the possibility of fine-tuning this important property.[48]

6.10 METAL COORDINATION POLYMERS

6.10.1 A Survey of Typical Structures

In this class of polymers, coordination occurs between an organic group in the polymer and a metal atom. This coordination can result in the metal atom being attached as a side group or part of a side group, or as an integral part of the chain backbone.[49-53]

A simple example of the first type of polymer is poly(vinyl pyridine), [—$CH_2CHC_5H_4N$—]$_x$, where the nitrogen atom in the —C_5H_4N side group has unpaired electrons available for coordination. Another example would be an ionic side chain, which could coordinate with a metal cation.

Much more research has been carried out with polymers in which the coordinated metal atom is part of the chain backbone. Typically, the metal atoms are copper, nickel, and cobalt. Oxygen atoms or carbon atoms adjacent to the metal atom provide the electrons required for the coordinate bond.[19] Polymers of this type are often rather intractable, for a variety of reasons. Specifically, insolubility can be a problem for species with moderate molecular weights. Also, coordination between chains can cause aggregation, and ligand exchange reactions with small molecules such as solvents can cause chain scission. However, in some favorable cases, the intramolecular coordination can be sufficiently strong for the polymer to be processed by the usual techniques such as spinning into fibers or extrusion into films.[19]

In other examples, compounds in which a metal atom is already coordinated in a molecule can be used as a comonomer in an addition polymerization. Two examples would be the vinyl ferrocene molecule shown in Figure 6.23, and the vinyl manganese complex shown in Figure 6.24.[19] An alternative approach involves condensation polymerization. For example, if the R group in the ferrocene unit shown in Figure 6.25 contains a hydroxyl group, it can be copolymerized with

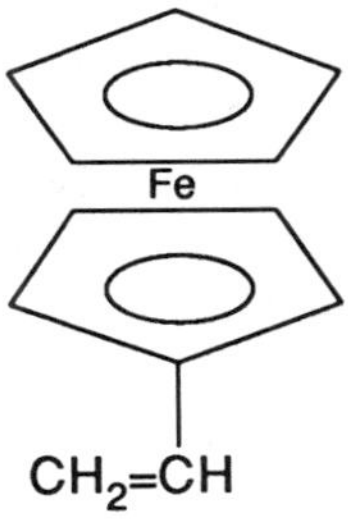

Figure 6.23 Vinyl ferrocene monomer.

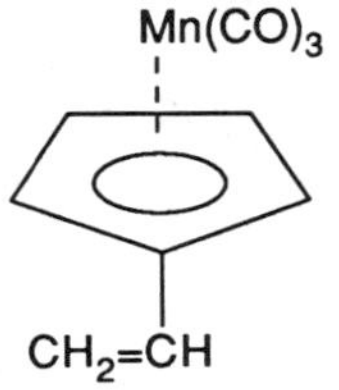

Figure 6.24 A vinyl manganese carbonyl monomer.

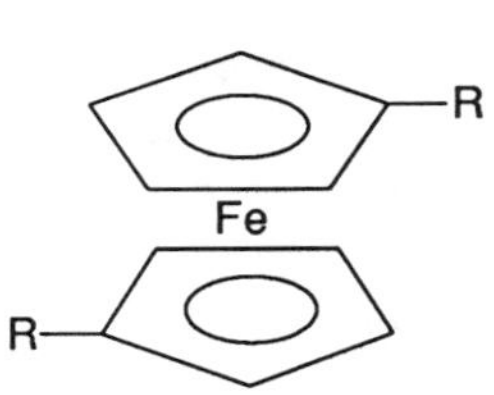

Figure 6.25 A ferrocene monomer having reactive R groups suitable for participation in a condensation polymerization.

Figure 6.26 A titanocene monomer having reactive chlorine atoms suitable for participation in a condensation polymerization.

a diacid chloride. If it is an acid chloride, it can be copolymerized with a diamine. This type of polymer is called a heteroannular chain. If only one of the rings in the repeat unit is in the backbone, the polymer is called homoannular.[7] Similarly, the titanium complex shown in Figure 6.26 can be copolymerized with diacids or diols.[19]

The preparation of a typical ferrocene condensation monomer is illustrated by the preparation of 1,1'-bis(beta-aminoethyl)ferrocene.[51] The starting material, ferrocene, is allowed to react with CH_3COCl in the presence of $AlCl_3$ as a catalyst, to link two methyl ketonic groups to the two ferrocene rings. These groups are then converted into carboxylic acid groups with the use of NaOCl, and then into methyl ester groups using CH_3OH. These groups can then be reduced to $-CH_2OH$ units with $LiAlH_4$, to yield the dialcohol already mentioned as a comonomer. They can be further converted to $-CH_2CN$ groups using KCN, which can then be reduced with $LiAlH_4$ to the desired $-CH_2CH_2NH_2$ groups. This disubstituted aminoethylferrocene can now be treated with various aromatic and aliphatic diacid chlorides or diisocyanates to yield ferrocene-containing polyamides and polyureas, respectively.[51]

Ferrocene also forms polymers with repeat units that consist simply of an iron atom bracketed by two cyclopentadienyl groups. These can be obtained by free radical polymerization in the vicinity of 200°C, but do not proceed beyond relatively low degrees of polymerization. They are stable up to 400°C, but their low molecular weights and limited solubility have prevented commercial applications.[7] However, this basic repeat unit also appears in a variety of copolymers, for example, with alkylene groups, carbonyl groups, amide groups, and urethane groups. Some of these materials have much higher molecular weights and may become of greater commercial significance.

Some metals form coordination polymers with sulfur that have a semiladder structure. One series is the bisdithiocarbamate polymers, which have the typical structure shown in Figure 6.27. The tetrahedral zinc polymer is called Zineb and the manganese one is called Maneb. Both are water-soluble fungicides. Other ladder-type structures occur in zirconium tetrasalicyclidene-1,2,4,5-tetraaminobenzene polymer and in bis(5,7-dichloroquinolinolato)quinoxaline-5,8-diolatotungsten(IV) polymer.[52]

Mercury, like zinc, appears in some chainlike structures, but apparently only in combination with other metals such as As, Sb, Nb, and Ta. Some of these have the characteristics of metallic conductors.[15] Rather high molecular weights have

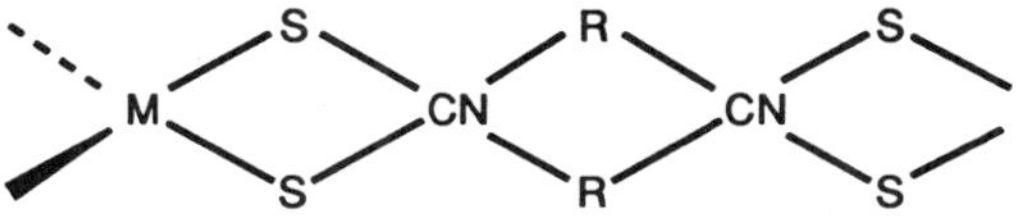

Figure 6.27   A bisdithiocarbamate polymer.

been obtained in the preparation of 6-coordinate cobalt (III) chelate polymers with acetylacetonato and leucinato ligands, 7-coordinate dioxouranium-(VI) dicarboxylate polymers, and 8-coordinate zirconium (IV) polymers with Schiff's base ligands.[52]

6.10.2 Metalloporphyrin Polymers

In the most important series of polymers of this type, the metallotetraphenylporphyrins, a metalloporphyrin ring bears four substituted phenylene groups X, as is shown in Figure 6.28. The metal atoms M in the structure are typically iron, cobalt, or nickel cations, and the substituents on the phenylene groups include $-NH_2$, $-NR_2$, and $-OH$. These polymers are generally insoluble. Some have been prepared by electro-oxidative polymerizations in the form of electroactive films on electrode surfaces.[54] The cobalt-metallated polymer is of particular interest since it is an electrocatalyst for the reduction of dioxygen. Films of poly(trisbipyridine)-metal complexes also have interesting electrochemical properties, in particular electrochromism and electrical conductivity.[53] The closely related polymer, poly(2-vinylpyridine), also forms metal complexes, for example, with copper(II) chloride.[55]

6.10.3 Cofacial Phthalocyanine Polymers

The phthalocyanine group is structurally related to the porphyrin group discussed in the preceding section. It consists of a series of interconnected rings containing nitrogen atoms, some of which face inward in such a way that they can complex very strongly with metal or metalloid atoms attracted to their unpaired electrons. This is illustrated in Figure 6.29, where typical M atoms are Si, Ge, or Sn. Dehydration of the "monomer" complexes induces condensation polymerization to yield a stacked "shish kebab" chain structure of the type shown in Figure 6.30. The backbone consists of the M atoms alternating with divalent atoms such as oxygen, as is shown in the sketch.[56-59] The degrees of polymerization can be a

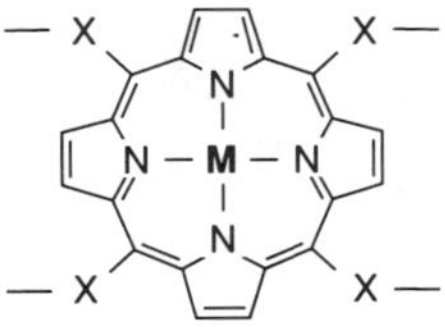

Figure 6.28 A substituted metalloporphyrin ring.

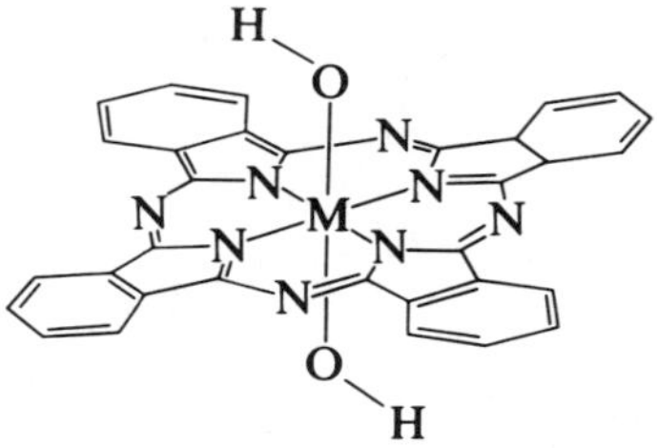

Figure 6.29 A dihydroxy metallophthalo-cyanine ring.

hundred or higher. Such polymers have excellent chemical and thermal stability. They can, for example, be dissolved in strong acids, frequently without degradation, possibly because protonation of some of the basic nitrogen atoms converts the chains into polyelectrolytes with sufficient coulombic repulsions to keep them dispersed long enough for processing. In this way it is possible to prepare fibers. As described shortly, these can become electrically conducting when treated with a dopant such as iodine.[19] They may also have interesting magnetic and electro-optical properties. With regard to electrical conductivity, these materials have the two major characteristics for producing a conducting arrangement of polymeric chains.[56-59] These are the presence of partially filled valence shells in the molecules, and the orientation of the rings into closely packed arrays, each ring with similar crystallographic and electronic environments. Thus the charge-carrying molecular subunits are covalently linked in a long chain, with cofacial orientations. The chains are relatively rigid and may be stiff enough to spontaneously form anisotropic liquid-crystalline phases similar to those used to spin aramid fibers such

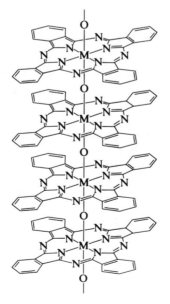

Figure   6.30 A   metallophthalocyanine chain.

as Kevlar®. It has been found that solutions of the siloxane-type cofacial polymer can be extruded alone or in combination with Kevlar® to form strong oriented fibers.[56-59] Doping of these fibers can be carried out either before or after the spinning process, to yield air-stable, electrically conducting fibers.

6.11 OTHER ORGANOMETALLIC SPECIES FOR SOL-GEL PROCESSES

The sol-gel process described in Chapter 4 often involves the use of organosilicates (alkoxysilanes), but a variety of other organometallic substances can be used for this purpose.[29,60-69] For example, titanates can be hydrolyzed, condensed, and polymerized in a series of steps that ultimately lead to titania. Aluminates yield alumina, borates give boria,[66] zirconates give zirconia, and so on. The possible variety of such reactions seems almost endless, and only a small fraction have been investigated to date. As mentioned in the discussion of silane-silica systems, much of this field lies on the fringes of inorganic polymer chemistry, and represents an interface with the area of high-performance ceramics.

6.12 REFERENCES

1. Stone, F. G. A.; Graham, W. A. G. *Inorganic Polymers;* Academic: New York, 1962.

2. Andrianov, K. A. *Metalorganic Polymers;* Interscience: New York, 1965.

3. Allcock, H. R. *Heteroatom Ring Systems and Polymers;* Academic: New York, 1967.

4. Borisov, S. N.; Voronkov, M. G.; Lukevits, E. Ya. *Organosilicon Heteropolymers and Heterocompounds;* Plenum: New York, 1970.

5. Van Dyke, C. H. In *Preparative Inorganic Reactions;* Jolly, W. L., Ed.; Interscience: New York, 1971.

6. Allcock, H. R. *Sci. Am.* **1974,** *230(3),* 66.

7. Zeldin, M. *Polym. News* **1976,** *3,* 65.

8. Elias, H.-G. *Macromolecules,* Part 2; Plenum: New York, 1977; Chapter 33.

9. Mark, J. E. *Macromolecules* **1978,** *11,* 627.

10. Carraher, C. E.; Sheats, J. E.; Pittman, C. U. *Organometallic Polymers;* Academic: New York, 1978.

11. MacGregor, E. A.; Greenwood, C. T. *Polymers in Nature;* John Wiley: New York, 1980; Chapter 8.

12. Peters, E. N. In *Encyclopedia of Chemical Technology;* Wiley-Interscience: New York, 1981; Vol. 13.

13. Carraher, C. E.; Sheats, J. E.; Pittman, C. U. *Advances in Organometallic and Inorganic Polymers;* Marcel Dekker, Inc.: New York, 1982.

14. Allcock, H. R. *Chem. Eng. News,* March 18, 1985; p 22.

15. Archer, R. D. In *Encyclopedia of Materials Science and Engineering;* Bever, M. B., Ed.; Pergamon: Oxford, 1986; Vol. 3.

16. Rheingold, A. In *Encyclopedia of Polymer Science and Engineering*, 2nd ed.; Wiley-Interscience: New York, 1987.

17. *Inorganic and Organometallic Polymers*, Zeldin, M.; Wynne, K. J.; Allcock, H. R., Eds.; ACS Symposium Series; American Chemical Society: Washington, DC, 1988; Vol. 360.

18. Pittman, C. U.; Sheats, J. E.; Carraher, C. E.; Zeldin, M.; Currell, B. *Preprints, ACS Division of Polymeric Materials Science and Engineering* **1989**, *61*, 91.

19. Allcock, H. R.; Lampe, F. W. *Contemporary Polymer Chemistry*, 2nd ed.; Prentice Hall: Englewood Cliffs, NJ, 1990; Chapter 9.

20. *Silicon-Based Polymer Science. A Comprehensive Resource;* Zeigler, J. M.; Fearon, F. W. G., Eds.; Advances in Chemistry Series; American Chemical Society: Washington, DC, 1990; Vol. 224.

21. Weil, E. D. In *Encyclopedia of Polymer Science and Engineering*, 2nd ed.; Wiley-Interscience: New York, 1987.

22. Dodge, J. A.; Manners, I.; Allcock, H. R.; Renner, G.; Nuyken, O. *J. Am. Chem. Soc.* **1990**, *112*, 1269.

23. Flory, P. J. *Statistical Mechanics of Chain Molecules;* Wiley-Interscience: New York, 1969.

24. Rheingold, A. L. *Homoatomic Rings, Chains, and Macromolecules of Main-Group Elements;* Elsevier: New York, 1977.

25. Laine, R. M.; Blum, Y. D.; Tse, D.; Glaser, R. In *Inorganic and Organometallic Polymers;* Zeldin, M.; Wynne, K. J.; Allcock, H. R., Eds.; ACS Symposium Series; American Chemical Society: Washington, DC, 1988; Vol. 360.

26. Seyferth, D.; Wiseman, G. H.; Schwark, J. M.; Yu, Y.-F.; Poutasse, C. A. In *Inorganic and Organometallic Polymers;* Zeldin, M.; Wynne, K. J.; Allcock, H. R., Eds.; ACS Symposium Series; American Chemical Society: Washington, DC, 1988; Vol. 360.

27. Lasocki, Z.; Dejak, B.; Kulpinski, J.; Lesnikak, E.; Piechucki, S.; Witekowa, M. In *Inorganic and Organometallic Polymers;* Zeldin, M.; Wynne, K. J.; Allcock, H. R., Eds.; ACS Symposium Series; American Chemical Society: Washington, DC, 1988; Vol. 360.

28. Lipowitz, J.; Rabe, J. A.; Carr, T. M. In *Inorganic and Organometallic Polymers;* Zeldin, M.; Wynne, K. J.; Allcock, H. R., Eds.; ACS Symposium Series; American Chemical Society: Washington, DC, 1988; Vol. 360.

29. Baney, R.; Chandra, G. In *Encyclopedia of Polymer Science and Engineering;* 2nd ed.; Wiley-Interscience: New York, 1987.

30. Seyferth, D. In *Inorganic and Organometallic Polymers;* Zeldin, M.; Wynne, K. J.; Allcock, H. R., Eds.; ACS Symposium Series; American Chemical Society: Washington, DC, 1988; Vol. 360.

31. Aitken, C.; Harrod, J. F.; Malek, A.; Samuel, E. *J. Organomet. Chem.* **1988**, *349*, 285.

32. Miller, R. D.; Michl, J. *Chem. Rev.* **1989**, *89*, 1359.

33. Miller, R. D., private communications.

34. Hallmark, V. M.; Zimba, C. G.; Sooriyakumaran, R.; Miller, R. D.; Rabolt, J. F. *Macromolecules* **1990**, *23*, 2346.

35. Welsh, W. J.; Johnson, W. D. *Macromolecules* **1990**, *23*, 1881.

36. Harrod, J. F. In *Inorganic and Organometallic Polymers;* Zeldin, M.; Wynne, K. J.; Allcock, H. R., Eds.; ACS Symposium Series; American Chemical Society: Washington, DC, 1988; Vol. 360.

37. Duda, A.; Penczek, S. In *Encyclopedia of Polymer Science and Engineering*, 2nd ed.; Wiley-Interscience: New York, 1987.

38. Labes, M. M.; Love, P.; Nichols, L. F. *Chem. Rev.* **1979**, *79*, 1.

39. MacDiarmid, A. G.; Heeger, A. J.; Garito, A. F. *McGraw-Hill Yearbook of Science and Technology: Polymer;* McGraw-Hill: New York, 1977.

40. Ellerstein, S. In *Encyclopedia of Polymer Science and Engineering*, 2nd ed.; Wiley-Interscience: New York, 1987.

41. *Polymer Handbook*, 3rd ed.; Brandrup, J.; Immergut, E. H., Eds.; Wiley-Interscience: New York, 1989.

42. Shaw, S. Y.; DuBois, D. A.; Nielson, R. H. In *Inorganic and Organometallic Polymers;* Zeldin, M.; Wynne, K. J.; Allcock, H. R., Eds.; ACS Symposium Series; American Chemical Society: Washington, DC, 1988; Vol. 360.

43. Paciorek, K. J. L.; Krone-Schmidt, W.; Harris, D. H.; Kratzer, R. H.; Wynne, K. J. In *Inorganic and Organometallic Polymers;* Zeldin, M.; Wynne, K. J.; Allcock, H. R., Eds.; ACS Symposium Series; American Chemical Society: Washington, DC, 1988; Vol. 360.

44. Paine, R. T.; Narula, C. K. *Chem. Rev.* **1990**, *90*, 73.

45. Nerula, C. K.; Paine, R. T.; Schaeffer, R. In *Inorganic and Organometallic Polymers;* Zeldin, M.; Wynne, K. J.; Allcock, H. R., Eds.; ACS Symposium Series; American Chemical Society: Washington, DC, 1988; Vol. 360.

46. Holmes, R. H.; Day, R. O.; Chandrasekhar, V.; Schmid, C. G.; Swamy, K. C. K.; Holmes, J. M. In *Inorganic and Organometallic Polymers;* Zeldin, M.; Wynne, K. J.; Allcock, H. R., Eds.; ACS Symposium Series; American Chemical Society: Washington, DC, 1988; Vol. 360.

47. Bellama, J. M.; Manders, W. F. In *Inorganic and Organometallic Polymers;* Zeldin, M.; Wynne, K. J.; Allcock, H. R., Eds.; ACS Symposium Series; American Chemical Society: Washington, DC, 1988; Vol. 360.

48. Al-Diab, S. S. S.; Suh, H.-K.; Mark, J. E.; Zimmer, H. *J. Polym. Sci., Polym. Chem. Ed.* **1990**, *28*, 299.

49. Foxman, B. M.; Gersten, S. W. In *Encyclopedia of Polymer Science and Engineering*, 2nd ed.; Wiley-Interscience: New York, 1987.

50. Pittman, Jr., C. U.; Carraher, Jr., C. E.; Reynolds, J. R. In *Encyclopedia of Polymer Science and Engineering*, 2nd ed.; Wiley-Interscience: New York, 1987.

51. Gonsalves, K. E.; Rausch, M. D. In *Inorganic and Organometallic Polymers;* Zeldin, M.; Wynne, K. J.; Allcock, H. R., Eds.; ACS Symposium Series; American Chemical Society: Washington, DC, 1988; Vol. 360.

52. Archer, R. D.; Wang, B.; Tramontano, V. J.; Lee, A. Y.; Ochaya, V. O. In *Inorganic and Organometallic Polymers;* Zeldin, M.; Wynne, K. J.; Allcock, H. R., Eds.; ACS Symposium Series; American Chemical Society: Washington, DC, 1988; Vol. 360.

53. Elliott, C. M.; Redepenning, J. G.; Schmittle, S. J.; Balk, E. M. In *Inorganic and Organometallic Polymers;* Zeldin, M.; Wynne, K. J.; Allcock, H. R., Eds.; ACS Symposium Series; American Chemical Society: Washington, DC, 1988; Vol. 360.

54. White, B. A.; Raybuck, S. A.; Bettelheim, A.; Pressprich, K.; Murray, R. W. In *Inorganic and Organometallic Polymers;* Zeldin, M.; Wynne, K. J.; Allcock, H. R., Eds.; ACS Symposium Series; American Chemical Society: Washington, DC, 1988; Vol. 360.

55. Lyons, A. M.; Pearce, E. M.; Vasile, M. J.; Mujsce, A. M.; Waszczak, J. V. In *Inorganic and Organometallic Polymers;* Zeldin, M.; Wynne, K. J.; Allcock, H. R., Eds.; ACS Symposium Series; American Chemical Society: Washington, DC, 1988; Vol. 360.

56. Marks, T. J.; Gaudiello, J. G.; Kellogg, G. E.; Tetrick, S. M. In *Inorganic and Organometallic Polymers;* Zeldin, M.; Wynne, K. J.; Allcock, H. R., Eds.; ACS Symposium Series; American Chemical Society: Washington, DC, 1988; Vol. 360.

57. Marks, T. J. *Science* **1985**, *227*, 881.

58. Marks, T. J. In *Frontiers in the Chemical Sciences;* Spindel, W.; Simon, R. M., Eds.; American Association for the Advancement of Science: Washington, DC, 1986.

59. Almeida, M.; Kanatzidis, M. G.; Tonge, L. M.; Marks, T. J.; Marcy, H. O.; McCarthy, W. J.; Kannewurf, C. R. *Solid State Commun.* **1987**, *63*, 457.

60. *Ultrastructure Processing of Ceramics, Glasses, and Composites;* Hench, L. L.; Ulrich, D. R., Eds.; Wiley-Interscience: New York, 1984.

61. *Better Ceramics Through Chemistry;* Materials Research Society Symposium Proceedings; Brinker, C. J.; Clark, D. E.; Ulrich, D. R., Eds.; North-Holland: New York, 1984; Vol. 32.

62. Klein, L. C. *Ann. Rev. Mater. Sci.* **1985**, *15*, 227.

63. *Science of Ceramic Chemical Processing;* Hench, L. L.; Ulrich, D. R., Eds.; Wiley-Interscience: New York, 1986.

64. Roy, R. *Science* **1987**, *238*, 1664.

65. Ulrich, D. R. *CHEMTECH* **1988**, *18*, 242.

66. *Ultrastructure Processing of Advanced Ceramics*, MacKenzie, J. D.; Ulrich, D. R., Eds.; John Wiley: New York, 1988.

67. Hench, L. L.; West, J. K. *Chem. Rev.* **1990**, *90*, 33.

68. Brinker, C. J.; Bunker, B. C.; Tallant, D. R.; Ward, K. J.; Kirkpatric, R. J. In *Inorganic and Organometallic Polymers;* Zeldin, M.; Wynne, K. J.; Allcock, H. R., Eds.; ACS Symposium Series; American Chemical Society: Washington, DC, 1988; Vol. 360.

69. Sakka, S.; Kamiya, K.; Yoko, Y. In *Inorganic and Organometallic Polymers;* Zeldin, M.; Wynne, K. J.; Allcock, H. R., Eds.; ACS Symposium Series; American Chemical Society: Washington, DC, 1988; Vol. 360.

Index

MEDICAL GRADE SILICONE SHEETING
IS BIOCOMPATIBLE, STRONG, AND EASY TO HANDLE

This sample of Silastic® Rx Medical Grade Sheeting from Dow Corning provides strength, ease of handling, and suitability for a variety of medical applications. The vulcanized silicone sheeting is subjected to the same stringent biosafety tests as other Dow Corning medical grade silicone materials. It meets or exceeds USP Class VI requirements and is subjected to 90 day implant studies.

Dow Corning will also offer Silastic® Rx Implant Grade Sheeting. Both the implant and medical grades will be available with or without Dacron® reinforcement in a variety of thicknesses.

Dacron is a registered trademark of E. I. du Pont de Nemours and Company.

Silastic® Rx Medical Grade Sheeting is individually packaged in 1' × 1' and 1' × 5' lengths. For information on custom sizing or formulating, please contact Dow Corning Corporation, Midland, MI 48686-0994.

Silastic® Rx Medical Grade and Implant Grade products are manufactured in a facility that is inspected by the United States Food and Drug Administration to the Good Manufacturer Practices. The facility is also registered to the International Organization for Standardization (ISO) 9002 Quality Standard.

DOW CORNING CORPORATION, MIDLAND, MICHIGAN   48686-0994